Multiphase Transport of Hydrocarbons in Pipes

Multiphase Transport of Hydrocarbons in Pipes

Juan J. Manzano-Ruiz
PetroConsulting and Associates
USA

Jose G. Carballo
USA

WILEY

Published by John Wiley & Sons, Inc., Hoboken, New Jersey.
Published simultaneously in Canada.

Library of Congress Cataloging-in-Publication Data:

Names: Manzano-Ruiz Juan, J., author. | Carballo, Jose G., author.
Title: Multiphase transport of hydrocarbons in pipes / Juan J. Manzano-Ruiz, Jose G. Carballo.
Description: Hoboken, New Jersey : Wiley, [2024] | Includes index.
Identifiers: LCCN 2023057850 (print) | LCCN 2023057851 (ebook) | ISBN 9781119888512 (cloth) | ISBN 9781119888529 (adobe pdf) | ISBN 9781119888536 (epub)
Subjects: LCSH: Multiphase flow. | Hydrocarbons.
Classification: LCC TA357.5.M84 M36 2024 (print) | LCC TA357.5.M84 (ebook) | DDC 665.5/44–dc23/eng/20240128
LC record available at https://lccn.loc.gov/2023057850
LC ebook record available at https://lccn.loc.gov/2023057851

Cover Design: Wiley
Cover Image: © Hirkophoto/Getty Images

Set in 9.5/12.5pt STIXTwoText by Straive, Chennai, India

To my Parents Juan Vicente and Carmen Teresa; my wife Lucy; my children Juan Carlos, Anabella, Andres Simon, and Raul Rodrigo; my beloved grandsons Juan Diego and Marcelo; my siblings Liliana and Juan Vicente; and my in-laws

August 2023 *Juan J. Manzano-Ruiz*

To my beloved wife Mariale; my cherished son Benjamin; and my parents Gilberto and Nicole

August 2023 *Jose G. Carballo*

Contents

Preface

Welcome to *Multiphase Flow in Pipelines*, an in-depth journey into the subject of the co-current flow of gas, liquids, and solids resulting from the production of hydrocarbons from underground reservoirs. This book is designed to be a comprehensive guide for both novices and experts seeking to gain a profound understanding of the core principles, best practices, and cutting-edge advancements in the realm of multiphase flow.

In today's fast-paced and ever-evolving technological landscape, staying ahead of the curve is vital for engineers, technicians, and students alike. The purpose of this book is to provide you with the practical knowledge and tools needed to understand field operations or to participate in projects to develop new fields.

We have chosen a variety of topics, blending theory, and practice, illustrated with examples taken from real projects. This book could be helpful to students embarking on an educational journey or to seasoned practitioners aiming to sharpen skills by understanding the first principles underlying the fluid flow and heat transfer models built into the commercial software available.

Key features of this book include the following:

Coverage: We explore all fundamental concepts and advanced topics related to multiphase flow in pipes that could add value to real projects or improve production performance. From basic models to the most intricate aspects, each chapter builds upon the previous one to foster a solid understanding.

Real-world applications: We believe in bridging the gap between theory and practice. As such, every concept discussed within these pages is accompanied by real-world applications, and examples to illustrate how the knowledge can be put to use effectively.

References: We include key references that provide in-depth knowledge supporting the material presented in each chapter.

Future horizons: A brief introductory chapter is included expressing the authors' views regarding potential technology evolution into the world of artificial intelligence, machine learning, data science, and artificial neural networks, with the humble hope that the reader may be initiated in the brilliant future developments in these amazing fields.

We encourage you to approach this book with curiosity, determination, and an open mind. Our hope is that the book triggers your interest in the transport of hydrocarbons through pipelines and empowers you to explore, innovate, and make significant contributions to the field. Should you have any feedback, questions, or suggestions, we look forward to your correspondence. Please refer to the companion website for color graphs included in this book.

2024, Katy
Texas

Juan J. Manzano-Ruiz
Jose G. Carballo

About the Authors

Juan J. Manzano-Ruiz, M.Sc., M.E., Ph.D., P.E., is a mechanical engineer by background, who gathers 40-plus years of experience in the energy industry, holding leadership and executive managerial positions with PdVSA-Intevep, BP-Amoco, Jacobs Engineering, Technip, and Bechtel O, G, and C. He earned three diplomas from the Massachusetts Institute of Technology and completed an MBA minor from the Sloan School of MIT. He taught undergraduate and graduate engineering courses for 16 years at universities in Caracas, Venezuela. He is currently a consultant and President of Petroconsulting and Associates LLC in Texas, USA (Email: petroconsultingllc@gmail.com).

Jose G. Carballo, M.Sc., P.E., is a flow assurance and hydraulics engineer, with 20 years of experience consulting for the energy industry. He is currently Vice President of FlowAssure Engineering in Texas, USA (Email: jose@flowassureng.com).

Acknowledgments

In the first place, I acknowledge the organizations that allowed me to learn and gain field experience in the subject of multiphase flow. These remarkable organizations include PdVSA-Intevep, BP-Amoco Venezuela, Jacobs Engineering, Technip, and Bechtel Oil, Gas, and Chemicals. The positions held at these organizations exposed me to practical situations comprising research and development, projects, field operations, and facilities construction. Thanks to such experiences I had the immense privilege of applying lessons learned from four giants in this field, namely Prof. Pete Griffith, the key advisor for my doctoral thesis at MIT; Dr. Jean-Marc Delhaye, a legend in the most rigorous analyses in this discipline; the late Prof. "Abe" Dukler; and Prof. Yehuda Taitel, who were the pioneers of the modern mechanistic methodology.

To my friend and co-author in this endeavor, **Jose G. Carballo**, M.Sc., P.E., who deserves the greatest acknowledgment for his work, thanks to his valuable 20 years of project experience, fine skills performing difficult simulations with commercial code, his ability to simplify field conditions, and for his great character and integrity. It has been a real pleasure working with Jose, who was my undergraduate student almost three decades ago.

I was extremely lucky to have a huge technical and editorial review collaboration in this book from **Maria A. Nass**, M.Sc., P.E., MBA, who happens to be Jose's wife. Maria has a long-standing and successful career in projects requiring advanced multiphase flow knowledge, while holding positions for Total, BP p.l.c., and currently for Subsea 7. Many thanks, Maria, for your outstanding contributions.

On a personal level, this work could not have been even attempted without the loving, permanent, and unconditional support from my family for the past two and a half years. Their time sacrifices do not go unrecognized because their backing was the true driver to bring this book to fruition. My endless love and heartfelt recognition to them all.

Juan J. Manzano-Ruiz,
Katy, Texas

I wish to convey my sincere appreciation to my dear friend, esteemed mentor, and lead author, Dr. Juan J Manzano-Ruiz, for graciously involving me in this project, and for his exceptional dedication in sharing his wealth of knowledge and expertise that has been instrumental in bringing this book to fruition.

Additionally, I express my sincere gratitude to my wife Mariale for her contribution to reviewing this text.

Jose G. Carballo
Katy, Texas

Nomenclature

List of Symbols

		SI units	FPS units
A	Cross-sectional area	m^2	ft^2
B	Formation volume factor	m^3/sm^3	bbl/stb
c	Speed of sound	m/s	ft/s
c_{solid}	Volumetric concentration of solids	% vol.	% vol.
C	Empirical constant of erosional velocity	$kg^{1/2}/s.m^{1/2}$	$lb^{1/2}/s.ft^{1/2}$
C_p	Specific heat coefficient at constant pressure	J/Kg K	Btu/lbmF
C_V	Valve's flow coefficient		$gpm/psi^{0.5}$
d	Pipe internal diameter	m	Inch
D	Pipe external diameter	m	inch
D_h	Hydraulic diameter	m	Inch
e	Fluid internal energy	J/kg	Btu/lbm
f	Friction factor	—	—
g	Gravitation acceleration	m/s^2	ft/s^2
g_c	Force–mass conversion factor	—	32.17 lbm ft/lbf s^2
G	Mass velocity	kg/m^2 s	lbm/ft^2 h
GOR	Gas–oil ratio at standard conditions	sm^3/sm^3	scf/stb
h	Fluid enthalpy	J/kg	Btu/lbm
h_f	Friction head	m	ft

(Continued)

		SI units	FPS units
h_i	Internal heat transfer coefficient	W/m^2 K	Btu/hr ft^2 F
h_L	Depth of liquid in stratified flow	m	ft
H	Enthalpy	J/kg	Btu/lbm
HFT	Hydrate formation temperature	°C or °K	°F or °R
H_L	Liquid holdup	—	—
ID	Internal diameter	mm	inch
j	Superficial velocity, volumetric flux	m/s	ft/s
j_{lg}	Slip drift flux	m/s	ft/s
k	Thermal conductivity	W/m K	Btu/h ft R
L	Pipeline length	m	ft
m	Mass flow rate	kg/s or kg/h	lbm/h
M	Amount of mass	kg	lbm
MW	Molecular weight	kg/kgmol	lbm/lbmol
n	Number of moles in ideal gas model	—	—
p	Pressure	Pa,abs or bara	psia
P_i	Phase-to-phase contact interface	m	inch
P_w	Wet (in contact with pipe wall) perimeter	m	inch
q''	Heat flux per unit area	W/m^2	Btu/h ft^2
Q	Flow rate	m^3/s	ft^3/s
Q	Heat flow rate	W	Btu/h
$\overline{R}$	Gas universal constant	J/kgmol K	lbf ft/lbmol R
R	Specific gas constant	J/kg K	lbf ft/lbm R
R_s	Solution gas ratio	sm^3/sm^3	scf/stb
s	Entropy	J/kg K	Btu/lb F
S	Slip factor	—	—
S	Specific gravity	—	—
S_g	Wet perimeter of gas phase	m	inch
S_{HT}	Heat transfer shape factor	m	ft
S_i	Perimeter of gas–liquid interface	m	inch
S_l	Wet perimeter of liquid phase	m	inch

		SI units	FPS units
S	Salinity	% w.	% w.
t	Time	s	s
T	Temperature	°C or °K	°F or °R
u	Actual fluid velocity	m/s	ft/s
V, v	Volume	m^3	ft^3
V_M	Volume of one molecular weight	m^3	ft^3
y_c	Critical pressure ratio (orifice)	—	—
y_i	Molar fraction of component "i" in gas phase	% m.	% m.
x	Mass quality	—	—
x	Direction of change for 1-D models	m	ft
x_i	Molar fraction of component "i" in liquid phase	% m.	% m.
z	Elevation (vertical) coordinate	m	ft
Z_c	Compressibility factor	—	—
z_i	Molar fraction of component "i" in the feed	% m.	% m.

Abbreviations

AERE	Atomic Energy Research Establishment, Harwell, Oxfordshire, UK
AGA	American Gas Association, Washington, D.C., USA
ANS	American Nuclear Society, Illinois, USA
API	American Petroleum Institute, Washington, D.C., USA
ASME	American Society of Mechanical Engineers, New York, N.Y., USA
bbl, bbls	barrels at local conditions
bpd,blpd	barrels of liquid per day
bopd	barrels of oil per day
CAPEX	capital expenditures
CD	cooldown transient
CS	control surface
CV	control volume
DC	drill center
DOC	dead-oil circulation

(Continued)

EOS	equation of state
EVR	erosion velocity ratio
FPSO	Floating Production Storage and Offloading Vessel
gpm	U.S. gallons per minute
GPSA	Gas Processors Suppliers Association, Tulsa, OK, USA
HC	Hydrocarbons
HOC	hot-oil circulation
IPR	inflow performance relationship
mscfd	thousands of standard cubic feet per day
mmscfd	millions of standard cubic feet per day
ODE	ordinary differential equation
OHTC	overall heat transfer coefficient
OPEX	operational expenditures
PDE	partial differential equation
ppg	density expressed in pounds per gallon
R&D	research and development
scf	cubic feet @ standard conditions
SPS	subsea production system
stb	barrels of liquid @ stock-tank conditions
TOTEX	combined CAPEX and OPEX expenditures
TPR	tubing performance relationship
TUFFP	Tulsa University Fluid Flow Projects, Tulsa, OK, USA
USC	upstream of choke condition
VIT	vacuum insulated tubing
VLP	vertical lift performance
WC	water cut

Dimensionless Groups

Fr	Froude number, $V/(gL)^{0.5}$
Gr	Grashof number, $L^3\rho^2\beta g\Delta T/\mu^2$
Nu	Nusselt number, hD/k
Pr	Prandtl number, $c_p\mu/k$
Ra	Rayleigh number, $g\beta\Delta Td^3/(\mu/\rho)^2$
Re	Reynolds number, $\rho DV/\mu$

Greek Letters

		SI units	FPS units
α	void fraction	—	—
β	volumetric quality	—	—
β_s	isentropic compressibility	Pa^{-1}	psi^{-1}
χ	Lockhart–Martinelli parameter	—	—
γ	specific gravity	—	—
Γ	volumetric mass transfer rate	kg/m^3 s	lbm/ft^2 hr
δ	layer thickness or property increment	mm	inch
δ	increment of a parameter	—	—
Δ	change in a property		
ε	wall roughness	mm	inch
η	efficiency	—	—
η	Joule–Thomson coefficient	K/Pa	F/psi
η	apparent viscosity of non-Newtonian fluid	s Pa	cP
θ	inclination with respect to horizontal direction	radian	radian
μ	dynamic viscosity	s Pa	cP
υ	specific volume	m^3/kg	ft^3/lbm
ξ	isentropic expansion coefficient	—	—
π	Pi number	3.1416	3.1416
ρ	density	kg/m^3	lbm/ft^3
σ	surface tension	N/m	dyne/cm
τ	shear stress	Pa	psi
τ	time	sec	sec
φ	two-phase multiplier	—	—
φ	fugacity coefficient	—	—
ϕ	subtended angle in stratified horizontal flow	radian	radian
ω	acentric factor	—	—

Subscripts

a	air phase
acc	acceleration
b	bubble point, bubble
$coat$	coating conductance or resistance
ch	choke
$crit, c$	critical condition
CS	control surface around a control volume
CV	control volume
$D\text{-}W$	Darcy–Weisbach friction factor
ext	external heat transfer to environment
$f, fric$	friction
f	fluid
$fric$	friction component
F	fanning friction factor
G, g, gas	gas phase
$geoth$	geothermal gradient
$grav$	gravitation
h	hydraulic diameter
H	homogeneous model
HC	hydrocarbon
HT	heat transfer
i	interface
int	internal heat transfer
ℓ	axial pipeline direction
L, l, liq	liquid phase
m, mix	mixture
o, oil	oil phase
p	pipe
pc	pseudo-critical
pr	pseudo-reduced
r	radial direction
r	reduced condition for pressure or temperature
$s, solid$	solid phase

(Continued)

s	slip, slip velocity
S	superficial flow (based on overall cross-section)
stag	stagnation (rest) condition
std	standard conditions: 15.56 °C and 1.01325 bara
total	total thermal conductance or resistance
w	water phase
wf	downhole well flowing
wh	wellhead
W	pipe's internal wall surface
x	axial pipeline direction
z	vertical (gravity) direction
1	inlet conditions
2	outlet conditions

Superscripts

″	double prime: property per unit flux area
¯	bar above variable: time average over a period

About the Companion Website

This book is accompanied by a companion websites.

www.wiley.com/go/Manzano-Ruiz/Multiphasetransport

The website includes:

- Coloured Figures

1

Introduction

1.1 What Is Multiphase Flow

Multiphase flow means the simultaneous transport of matter in the form of liquids, gases, solids, or any combination of these. The coexistence of at least two of these phases poses a complex time-dependent topology to describe even in the simplest cases. The spatial distribution of phases adopts the form of droplets, bubbles, solid particles, films, and chunks of liquids in an ever-changing process (time-dependent), determining the transfer phenomena of mass, species, thermal energy, and momentum. Figure 1.1 shows a sketch of typical flow patterns in horizontal flow.

Figure 1.1a corresponds to an evaporating liquid flowing from left to right. Figure 1.1b shows separated flows with distinctive continuous phases. Figure 1.1c shows dispersed flows of either bubbles in liquids or droplets in gas. Needless to say the current state of the art in multiphase pipeline flow modeling requires several approximations and simplifications.

An effortless way to tackle the topology conundrum is to identify flow regimes that show similar patterns and characteristics with a set of boundaries expressed in terms of key parameters or non-dimensional groups. Many flow pattern maps are available in the two-phase (gas–liquid) flow literature aimed to predict different flow orientations, that is, quasi-horizontal, vertical upward and downward, and inclined flow upward and downward. Chapter 4 of this book reviews the principal flow pattern maps proposed.

Multiphase flow in hydrocarbon production systems may include various phases, namely oil, gas, produced water, and solids (e.g. sand fines, asphaltene, and wax). In this book, the approach is to consider a single liquid phase made up of oil and produced water. Comprehensive coverage of solid–gas or

Multiphase Transport of Hydrocarbons in Pipes, First Edition.
Juan J. Manzano-Ruiz and Jose G. Carballo.

Companion Website: www.wiley.com/go/Manzano-Ruiz/Multiphasetransport

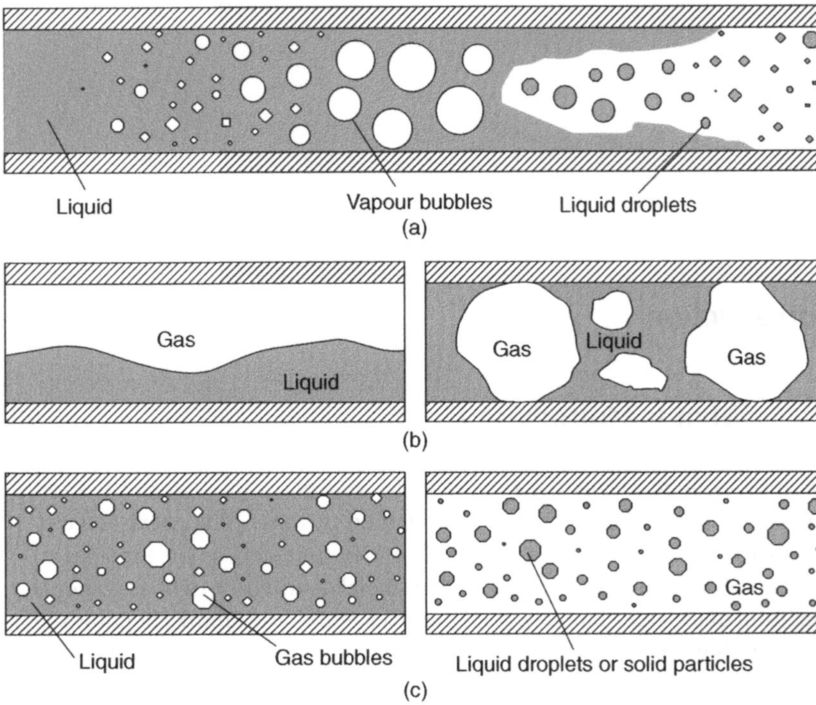

Figure 1.1 Two-phase flow in horizontal flowline.

solid–liquid transport in pipelines is beyond the scope of this book. However, a very brief introduction to the field of solid–fluid transport is included in Chapter 11. Thus, the main focus of this book is on *two-phase gas–liquid flow*.

Another important characteristic of multiphase flow is phase slippage, easily distinguished when an integral modeling approach is adopted. The differences in phases' velocities cause flow instabilities, flow regime transitions, and enhanced transport phenomena at the interfaces.

In this book, the *one-dimensional methodology* is followed. This means that conservation principles and closure equations will be expressed with equations aligned to the pipeline axis. This simplification stems from the cross-sectional integral application to obtain representative one-dimensional (1-D) properties and parameters for each phase. Such a methodology, on the one hand, simplifies the governing set of equations but, on the other hand, introduces the need for additional closure equations to account for the phases' interactions.

1.2 Single- and Multicomponent Fluids

Multiphase flow is present in the steam–water phases of the boiling water reactors (BWR, Figure 1.2) of the nuclear industry, for which intense R&D work has been done during the past 70 years. Introduction to the field of nuclear power generation utilizing the technology of BWR can be found in the texts by Lahey and Moody (1993) and Tong and Tang (1997). The boiling water within the BWR experiences a variety of successive changes in flow regime, ranging from bubble, plug, slug, churn, annular, and misty flow, inside vertical tubes. There is inclusive a possibility to have reverse liquid flow within the tubes, a phenomenon known as flooding.

Due to advances in the nuclear power generation industry, the study of single-component two-phase flow has enjoyed impressive R&D financing, producing a wealth of valuable experimental data and proposed models to match the data.

Vapor–liquid phases are also present in evaporators and condensers of a mechanical refrigeration system. Flows of steam and hot water are also observed in oilfield distribution systems to stimulate heavy-oil production (enhanced oil recovery technique). All these examples have in common that working fluid is a simple substance with well-known properties, described by simple thermodynamics relationships.

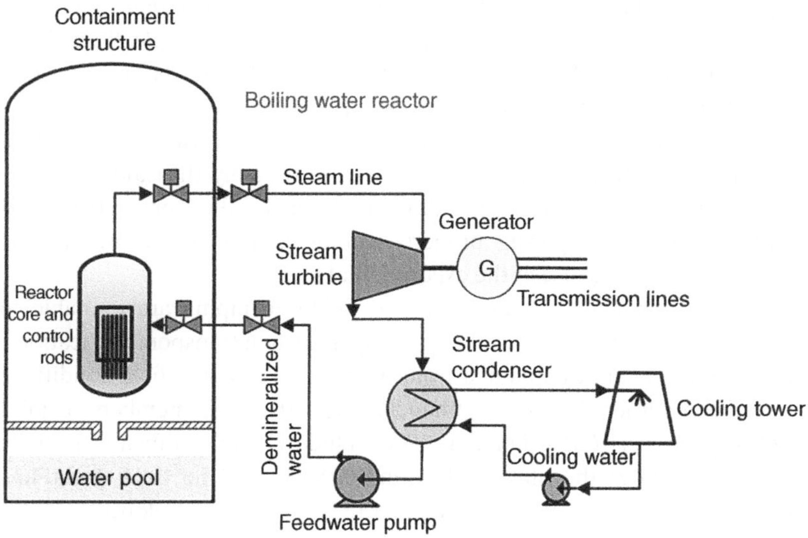

Figure 1.2 Boiling water reactor with steam–water flow.

Multicomponent mixtures are completely different fluids from the standpoint of the thermodynamics description and mathematical handling. The common hydrocarbon produced from a reservoir comprises thousands of single components, subject to attraction and repulsion intermolecular forces. The prediction of fluid thermodynamic and transport properties is in general a challenge in comparison to a single-component system. Chapter 3 addresses the challenges and issues of *compositional models* required to describe the flow performance of hydrocarbons in pipeline systems.

1.3 Challenges to Model Multiphase Flow

This section summarizes the majority of the challenges experienced in attempting to predict the two-phase flow performance of oil and gas through a production system. As pointed out before, the flow topology and the complexity of the phases' interaction hinder rigorous analysis, for which available simplified models provide approximate answers in many situations. A brief comment on the primary challenges follows.

1.3.1 Experiments and Scale-up

The need to add closure equations to complement the conservation principles still requires laboratory testing to generate reliable data. The issue with lab testing is threefold: *test fluid(s)*, *pressure*, and *scale*. Traditional modeling fluids at the test labs have been air and water because of cost, availability, and safety. Most flow pattern maps were developed using air–water mixtures at different mass ratios to cover a wide range of fluid velocities in vertical and horizontal flow. Unfortunately, the fluid properties of air–water mixtures are quite different from the fluid properties of oil and natural gas, requiring property-scaling parameters or mechanistic approaches to make use of the experimental data.

Very few data at high pressure and field-size piping are available to validate the two-phase flow models for hydrocarbon transport applications. Proprietary databases obtained from hydrocarbon flow at field conditions have been gathered with the financial support of the members of joint industry projects (JIP). Renown JIPs include OLGA's software, initially developed under the Statoil, IFE, and SINTEF JIP in 1984; LedaFlow Improvements to Flow Technology (LIFT), initially developed in 2013 under the Chevron, ConocoPhillips, ExxonMobil, Shell, Statoil, TotalEnergies, and Woodside JIP; and the University of Tulsa Fluid Flow Projects (TUFFP) since 1973. Thus, the most reliable technical information about multiphase flow in pipes is reserved for the JIP members. Additional data obtained from research in the academic world make their way into journal papers although to a limited extent. For these reasons, the most prestigious commercial codes in the market today tap experimental data through the JIP to fill the gap left between conservation principles and the number of unknowns, that is, valuable closure equations.

1.3.2 Multidimensional Fluid Flows

The standard practice to deal with the intractable three-dimensional field of multiphase flow has led most practitioners to focus on 1-D models. This strategy simplifies the governing equations for system's sections with large L/D ratio and makes the set amenable to mathematical solution. However, there is a price for this simplification in terms of additional closure equations to compensate for the loss of multidimensional information. In general, the 1-D models rely on the cross-sectional integral methodology.

Given that most production systems are exceedingly long compared to the piping ID, the 1-D axis-line approach has gained overwhelming acceptance for multiphase flow simulations. More recently, since 2016, two-dimensional (2-D) and quasi-three-dimensional (Q3D) analyses are becoming commercially available, obviously adding noticeable computation time if the full production system is simulated. For the simulation of specific pipeline accessories (e.g. a choke) with sudden changes in flow properties and geometry, the preferred option is computational fluid dynamics (CFD) code, enabling either Eulerian or Lagrangian formulations of the gas–liquid–solid flow field over a huge amount of nodes/cells.

The *hold-up* property (or its converse, the void fraction) captures part of the essence of the topology challenge. The 1-D hold-up is defined as the ratio of the cross-sectional pipe area occupied by the liquid(s) phase to the total cross-sectional area. This parameter embodies the relative weight of liquid(s) and gas phases to estimate important mixture properties such as density, viscosity, and thermal conductivity.

The other challenge of the flow topology is to identify distinctive flow regimes with common similarities in phase distribution. The objective is to have a limited number of flow patterns with common characteristics to apply specific models that capture the core of each regime. The best option to focus on flow topology is to identify either *dispersed flows* or *separated flows* (Brennen, 2005). A dispersed flow comprises *bubbles* in a continuous liquid phase or *droplets* in a continuous vapor phase. The dispersed flow regime is more difficult to represent due to the numerous particles interacting with the continuous phase and among themselves. The separated flow consists of two continuous phases exchanging transport phenomena (mass, momentum, and energy).

1.3.3 Time Fluctuations

The standard multiphase flow modeling approach rests on the assumption that the time scale of interest is orders of magnitude larger than the turbulence frequency of the smallest eddies. Nevertheless, the fluctuating nature of intermittent flow and the phases' topology pose daunting challenges.

Rigorous modeling of the instantaneous multiphase flow properties and parameters is unfeasible. A time-filtering methodology, similar to the traditional velocity fluctuations smoothing technique for turbulent flows, is necessary to obtain a well-approximated set of equations with reasonable computation time. We can speculate that commercial codes work utilizing time-filtering techniques for multiphase flow parameters, but without access to the code program, it is not possible to confirm such conjecture.

1.3.4 Slow Transients

One important simplification of the dynamic performance of multiphase flow is the distinction between *fast* and *slow* transients. Transient operations are planned and executed to change the production flow rate from state A to state B.

An example of a transient operation is a well startup, originating from a shut-in state and aiming to achieve a target production rate. This example may be regarded as a slow transient because the rate of change in fluid properties is gradual, and there are no appreciable compressibility phenomena. Similar examples are quite common in the industry, such as well shutdown, system planned shutdown, ramp-up, and system pigging, among others. In general, most transient field operations could be classified as slow transients and are characterized by their time and velocity scales.

On the other hand, a fast transient is triggered by an abrupt change in operation, involving substantial elastic phenomena (i.e. compressibility) in the bulk of the fluid. In fast transients, fluid compressibility dominates over the friction and gravity factors, and wave perturbations travel at very high velocities labeled the *speed of sound*. Such waves are usually modeled assuming that the fluid compressibility is an elastic phenomenon transmitted via isentropic processes. Fast transients have very small characteristic time and very large characteristic velocity (i.e. speed of sound), quite the opposite of slow transients. One example of a fast transient is when bulk fluid is subject to exceedingly high acceleration due to the rupture (loss of containment) of a pipeline operating at high pressure, followed by the system inventory blowdown. In this example, at the rupture location, gas could attain the speed of sound, and the flow is considered "choked," which means that the phenomenon is not controlled by the downstream (surroundings) conditions but rather by the upstream conditions.

It is important to point out that localized choked flow may occur at pipeline accessories, having a small flow cross area, despite the pipeline being operated under a steady state. This scenario happens when gas flows through a choke valve partly open, and the production flow rate remains constant because the fluid(s) cannot flow faster than the speed of sound.

A simple criterion to identify a slow transient from a fast transient is the criterion expressed by equation (1.1), if it holds across the entire production system.

$$\left(\frac{u}{c}\right)_{mix} \ll 1 \tag{1.1}$$

1.3.5 Compositional Simulation

An exact representation of the fluid properties is never possible with state-of-the-art thermodynamic modeling, although impressive approximations are nowadays feasible with cubic equations of state (*EOS*) and newer PC-SAFT EOS. Exact representation means, in this regard, the capability to precisely match all laboratory results of the PVT routine assays. For this reason, before multiphase simulations of multicomponent pipeline flow are attempted, it is necessary to *characterize the fluid* composition with a number of pseudo-components to enable the equation to state to reproduce the lab assays performed. This characterization process is painstaking and does not always guarantee that all lab test results can be replicated with the EOS chosen.

The customary way to generate look-up fluid properties tables has been to perform pressure-temperature flashing calculations. This is the preferred approach to save time during thermo-hydraulics modeling, instead of calculating thermodynamic equilibrium at each point and for every time step. However, there are several instances in which the compositional simulation is needed at the excessive cost of computational time; one example is when commingling vastly different streams at a node. A newer trend that adds accuracy to the properties estimations for single-component systems is based on pressure–enthalpy relationships.

1.4 Hydrocarbon Flow

Multicomponent mixtures of hydrocarbons show a variety of pipe flow characteristics not found in single-phase flow. At one end, we have *live oil* carrying dissolved light hydrocarbons that are released from solution by pressure and temperature changes. This process generates a two-phase flow with marked characteristics of oil properties. Once the dissolved gas and any possible water are removed in a process plant, the remaining oil phase is known as *dead oil* and shows fluid properties of single-phase flow, just as the gas and water will do as well.

On the other end, we have gas-condensate fluids with prevalent fluid properties of the gas phase, although along the production system will drop liquid hydrocarbon condensate. Between these two limits, any combination of fluid properties can arise adding to the complexity of the simulations.

1.5 Modeling Approaches

The challenges aforementioned define a strategy to model hydrocarbon flows based on a set of abridging assumptions. These assumptions have proved to yield reasonable modeling results in agreement with field measurements while making the set of equations tractable for calculations. The principal assumptions are as follows:

- The representation of the thermodynamic fluid properties using cubic EOS,
- The interpolation of fluid properties from look-up P–T tables to save computing simulation time,
- The use of a point-to-point compositional model to predict fluid properties for unusual cases such as fluid blending, gas-condensate flows, and processes with high acceleration effects (e.g. blowdown),
- One-dimensional fluid flow to simulate long spans of pipelines,
- Slow-transient operations without wave phenomena,
- Division of production system domain into cells small enough to consider uniform properties within the control volume,
- Explicit integration schemes with small time-integration steps to satisfy the Courant–Friedrich–Lewy (CFL) condition (Peyret, 1983),
- Implicit integration schemes to warrant computational stability with methodologies such as Crank–Nicholson (Peyret, 1983), but at the cost of handling large sparse matrices,
- Closing the set of equations by means of complementing closure equations and jump conditions (e.g. hold-up correlations),
- Using flow-regime-specific correlations to close a system of equations (mechanistic approach),
- Heat transfer to the environment is accounted for through an energy equation, in order to predict temperature changes along the flowline which introduce major changes in phases' composition, and

- Neglecting heat-transfer transient effects in pipeline steel and thermal-insulation coatings (to be addressed later).

In the next sections, use of the assumptions listed is made to present two-phase flow models applicable to quasi-horizontal, inclined, and vertical flow.

References

Brennen, C. (2005). *Fundamentals of Multiphase Flow*. Cambridge, UK: Cambridge University Press.

Lahey, R. and Moody, F. (1993). *The Thermal-Hydraulics of a Boiling Water Nuclear Reactor*, 2e. La Grange Park, Illinois, USA: American Nuclear Society.

Peyret, R. (1983). *Computational Methods for Fluid Flow*. New York, NY, USA: Springer-Verlag.

Tong, L. and Tang, Y. (1997). *Boiling Heat Transfer and Two-Phase Flow*, 2e. Washington, D.C., USA: Taylor and Francis Publishers.

2

Fundamentals of Multiphase Flow

The following sections address the basis for having a solid foundation to simulate multiphase flow in pipelines that unfolds in subsequent chapters. The topics covered herein are as follows:

- Examples of multiphase flow in the oil and gas industry
- Nomenclature of multiphase flow
- Strategy to model multiphase flow
- Measurement techniques for two-phase flow

2.1 Multiphase Flow in the Production of Oil and Gas

In the oil and gas industry, the production activity from the reservoir to the fluid processing facilities might encompass several phases, namely oil, natural gas, produced water, sand fines, and potentially other solid particles resulting from the precipitation of heavy ends.

Multiphase flow may be present in various places along the production path of oil and gas. First, in the reservoir, there may exist a gas cap when the trapped fluid is below the bubble point, leading to the production of a two-phase flow from the reservoir. Multiphase flow in porous media is beyond the scope of this book and will not be addressed. Second, in the tubing string of a production well, gas may progressively separate from the bulk of the hydrocarbon liquid due to the pressure loss in the ascending flow. In this case, the wellhead flowing pressure would be below the bubble point of the mixture. The flow may adopt different patterns (Figure 2.1) depending on the relative amount of gas and liquid, as well as on the average velocities

Multiphase Transport of Hydrocarbons in Pipes, First Edition.
Juan J. Manzano-Ruiz and Jose G. Carballo.

Companion Website: www.wiley.com/go/Manzano-Ruiz/Multiphasetransport

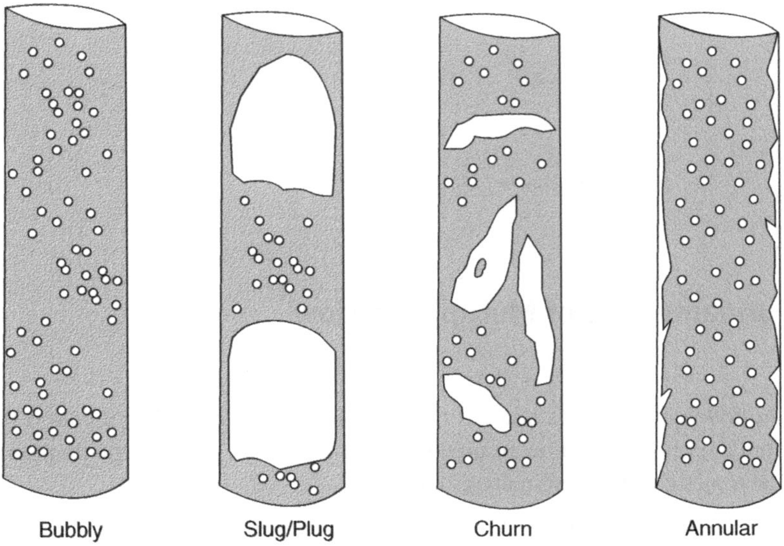

Figure 2.1 Flow regimes in vertical flow.

of each phase. Third, the production system is quite likely to have flowing mixtures of multiphase flow, in particular at the production riser of offshore facilities.

The regimes in vertical flow are strongly influenced by gravity and tend to form axisymmetric patterns. The presence of phases in the well string determines local mixture density, hold-up, and total pressure drop. Therefore, the well production depends on the driving pressure gradient imposed by both the single- and the two-phase flow sections. The latter is always larger than the single-phase flow gradient, and the point within the well at which the bubble point is reached is key to knowing which regime is dominating the performance.

Downstream of the well choke, there is a production system comprising flowlines (and a riser for offshore wells). In offshore fields, subsea wells tie well trees to a subsea manifold via subsea flowlines. For onshore fields with multiple wells, a gathering system collects production from the cluster via flowlines that deliver the hydrocarbons to a battery station (Figure 2.2) where gas and liquids are separated.

The flowlines convey production in multiphase form and again the higher pressure gradients determine the amount of production delivered per well and per cluster. Some authors like to visualize the total pressure gradients, which need to be overcome with the reservoir energy, by a series of *flow resistances*, as an analog to an electrical circuit with resistors in series

Figure 2.2 Partial view of onshore gathering system. Source: Max Phillips (Jeremy Buckingham MLC)/Flickr/CC BY 2.0.

and parallel. The production flow going through such flow resistances is equivalent to the electrical current. Such flow resistances are highly nonlinear and variable, in comparison to common electrical resistors, and are a function of several flow variables that will be addressed in later chapters. The complete set of flow resistances (or "*chokes*") from the reservoir to the processing facility is also known as the *choke model* and serves well to quickly envision ways to enhance production flow through the reduction of the "chokes."

Subsea field developments are more complex compared to onshore fields, from the standpoint of the subsea equipment, controls, and installation in a harsh environment. The subsea production system (SPS) can adopt four typical arrays, namely a cluster of wells, satellite well, daisy-chain, and template of wells. These subsea gathering system configurations are shown in Figure 2.3.

The selection of the development architecture depends on the drilling location of the wells' tophole. If the well's locations are very distant from each other, a daisy-chain development would suit best because the wells will not have to be of extended reach (high deviation wells). If the reservoir production targets can be reached from a single drill center (DC), the wells' template will save CAPEX for flowlines, jumpers, and umbilical; however, the wells' drilling and completion costs will be higher because of the longer measured depth (MD). If the wells' cost is extremely sensitive to drilling

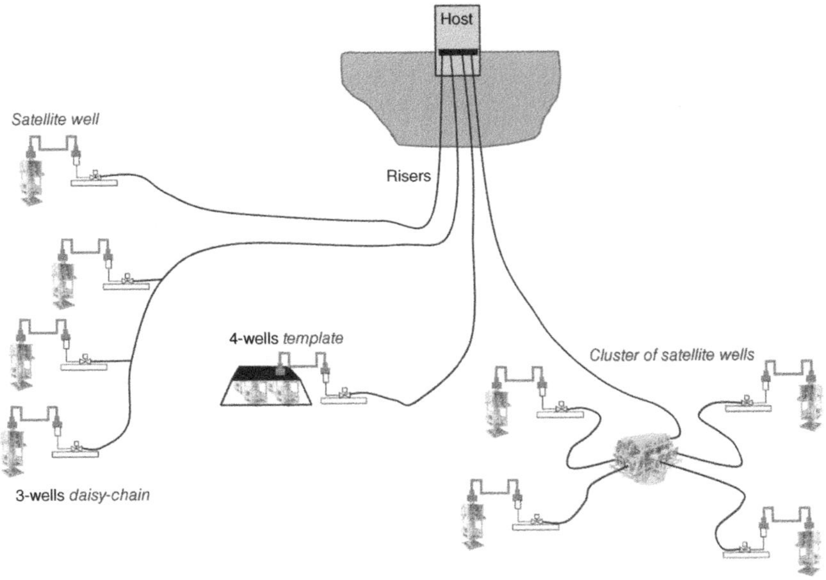

Figure 2.3 Subsea gathering system options.

deviation and the drill sites are not too far apart, then a cluster satellite development could minimize overall TOTEX. The latter term refers to the total cost of the facility over the life cycle, that is, CAPEX and OPEX. Finally, isolated wells can be produced as satellite tiebacks. In summary, the subsea development design is a compromise between wells' TOTEX (the highest cost of all) and the SPS TOTEX, which are driven by the location of the drilling sites.

Regardless of the subsea development design, all fields will produce multiphase flow through the flowlines and risers. Long flowlines in deepwater, exposed to low seawater temperatures and large pressure drops, are most likely to endure the consequences of multiphase transport. However, there are some exceptional examples of long subsea flowlines built in shallow waters, posing minimum hydrate formation risk despite the presence of produced gas and water due to the low internal pressure and the warm seawater environment.

From an operations perspective, onshore gathering systems are easier to produce than their subsea counterparts. Onshore gathering systems can be installed in trenches for enhanced safety and soil thermal insulation. The cost of trenches is extremely low compared to the installation equipment cost in offshore applications. Should any repair or system intervention of the onshore gathering system be necessary, its cost will be a small fraction of the cost of a similar intervention in a subsea gathering system.

2.2 Multiphase Flow Concepts

Multiphase flow in pipelines requires a specific nomenclature for the dynamic magnitudes involved. Two-phase flow is a subset of multiphase flow, and in this section, we focus on the nomenclature applicable to gas–liquid flows. The nomenclature presented herein can be extrapolated to other analogous two-phase flows such as solid–liquid (slurries), solid–gas, and oil–water. Moreover, the same definitions introduced below can be extrapolated to more cumbersome three- and four-phase flow applications. For this reason, we depart from full multiphase flow at this place and focus on two-phase gas–liquid flow, with opportune comments to extend models to multiple phases.

The following sub-sections address the most important parameters for gas–liquid pipe flow following the practical and customary *one-dimensional (1-D) model* approach (Wallis, 2020), based on the concept of cross-area averaging. Also, the conventional use of the statistical stationary process by Delhaye (Bergles et al., 1981) is utilized everywhere to simplify the time variation of magnitudes in the models.

2.2.1 Void Fraction and Hold-up

Consider the following phase-density function (Bergles et al., 1981) denoting the presence (P_k) of either phase at a specific point of the flow field at a time "t":

$$P_k(\vec{x}, t) = \begin{bmatrix} 1 \text{ if at point } \vec{x} \text{ phase "}k\text{" is present at time "}t\text{"} \\ 0 \text{ if at point } \vec{x} \text{ phase "}k\text{" is absent at time "}t\text{"} \\ k = l \text{ (liquid), or } g \text{ (gas)} \end{bmatrix} \tag{2.1}$$

The fraction of time (over a period of time "τ") the gas phase occupies the control volume at the location $\vec{x}$ is the *local instantaneous void fraction* α (Lahey and Moody, 1993) defined as

$$\alpha(\vec{x}, t) \equiv \frac{1}{\tau} \int_{t}^{t+\tau} P_k(\vec{x}, t')dt' \tag{2.2}$$

This concept can be applied to a finite control volume within the pipe, as it is sometimes used to measure the *instantaneous volumetric void fraction* by section isolation, to yield

$$\alpha(t) \equiv \frac{\iiint\limits_{V_g} dv}{\iiint\limits_{V_g+V_l} dv} = \frac{V_g(t)}{V_g(t) + V_l(t)} \tag{2.3}$$

The 1-D models make extensive use of the *practical* void fraction concept at cross-sections of the flowlines. The angled parentheses ⟨⟩ are used to signify the area average of a multiphase property, but often such notation is dropped for simplification but with the same meaning. For statistically stationary processes (i.e. steady-state analysis approximation), time integrated over a prolonged period of time "τ," the *cross-sectional stationary void fraction* (Lahey and Moody, 1993) becomes

$$\alpha = \langle\alpha\rangle \equiv \frac{1}{\tau}\int_t^{t+\tau} \left(\frac{\Delta x \iint\limits_{A_g} dA_{CS}}{\Delta x \iint\limits_{A_g+A_l} dA_{CS}} \right) dt' = \frac{A_g}{A_g + A_l} = \frac{A_g}{A} \tag{2.4}$$

The concept of the stationary void fraction at a cross-section leads to the equivalent *hold-up* concept collectively used in the oil and gas industry (Brill and Mukherjee, 1999):

$$H_L = \langle H_L \rangle \equiv 1 - \langle\alpha\rangle \tag{2.5}$$

For hydrocarbon mixture flows through flowlines, it is customary to adopt the cross-sectional stationary void fraction approximation in the hydrodynamic equations, even for slow transient processes, by implicitly assuming that the integration period of time τ is far smaller than the time constant for the transient. In other words, the cross-sectional void fraction "adjusts" quickly to local perturbations in comparison to the longer time constant of the transient phenomenon, and any time variation of the local section void fraction follows the slow-transient trend.

2.2.2 Phase Velocity, Slip Factor, and Volumetric Flow

The instantaneous local phase velocity vectors for vapor and liquids are, respectively, represented in this book by

$$\vec{u_g}(\vec{x}, t) \text{ and } \vec{u_l}(\vec{x}, t) \tag{2.6}$$

For the 1-D models presented in this book, the following area-averaged and statistically stationary approximations are utilized for the *phase velocity*:

$$u_g = \langle u_g \rangle \equiv \frac{1}{A_g \tau}\int_t^{t+\tau} \left(\iint\limits_{A_g} \vec{u_g}(\vec{x}, t) \cdot \overrightarrow{dA_{CS}} \right) dt' \tag{2.7}$$

$$u_l = \langle u_l \rangle \equiv \frac{1}{A_l \tau} \int_t^{t+\tau} \left(\iint_{A_l} \vec{u_l}(\vec{x}, t) \cdot \overrightarrow{dA_{CS}} \right) dt' \tag{2.8}$$

One fact about gas–liquid flow is that in general the gas phase velocity is larger than the liquid phase velocity, and it is accounted for with the slip velocity concept

$$u_s = \langle u_s \rangle \equiv \langle u_g \rangle - \langle u_l \rangle \tag{2.9}$$

This is true for any orientation except for incipient bubbly flow in horizontal flowlines and for certain applications with downward flow. The ratio of gas-to-liquid phase velocities is known as slip factor S,

$$S \equiv \frac{\langle u_g \rangle}{\langle u_l \rangle} = \frac{u_g}{u_l} \tag{2.10}$$

The 1-D statistically stationary volumetric flows of gas and liquids are defined based on the phase velocity concepts expressed by equations (2.7) and (2.8), yielding

$$Q_g = \langle Q_g \rangle \equiv \frac{1}{\tau} \int_t^{t+\tau} \left(\iint_{A_g} \vec{u_g}(\vec{x}, t) \cdot \overrightarrow{dA_{CS}} \right) dt' = u_g A_g \tag{2.11}$$

$$Q_l = \langle Q_l \rangle \equiv \frac{1}{\tau} \int_t^{t+\tau} \left(\iint_{A_l} \vec{u_l}(\vec{x}, t) \cdot \overrightarrow{dA_{CS}} \right) dt' = u_l A_l \tag{2.12}$$

$$Q \equiv Q_g + Q_l \tag{2.13}$$

These time- and area-averaged concepts are ordinarily used for long flowlines and risers modeling with gradual transient operations. Thus, the previous rigorous presentation of the parameters' definition will be left aside in the following chapters to concentrate on the practical application of these principles while understanding the implicit assumptions.

2.2.3 Volumetric Flux and Superficial Velocities

The essence of the 1-D model allows us to define the *volumetric flux* (Collier and Thome, 1994), best known in the multiphase flow literature as *superficial velocity*, as the ratio of volumetric flow and overall cross-sectional area. As such, the volumetric flux provides a lower bound estimate of phase velocity because it is based on a constant flow area (pipe). The definitions for vapor and liquids are

$$j_g = \langle j_g \rangle \equiv \frac{\langle Q_g \rangle}{A} = \frac{4Q_g}{\pi d^2} = \alpha\, u_g \tag{2.14}$$

$$j_l = \langle j_l \rangle \equiv \frac{\langle Q_l \rangle}{A} = \frac{4Q_l}{\pi d^2} = (1-\alpha)u_l \tag{2.15}$$

The mixture or total volumetric flux is defined as

$$j = \langle j \rangle \equiv \langle j_g + j_l \rangle = j_g + j_l \tag{2.16}$$

The use of the superficial velocity concept is amply utilized for flow pattern maps, the drift–flux model, and the separate flow models, as will be covered in a later chapter.

2.2.4 Density

The density property can be defined with the 1-D model, for each phase and for the mixture. The area-averaged and statistically stationary density definitions are

$$\rho_g = \langle \rho_g \rangle \equiv \frac{1}{\tau A_g} \int_t^{t+\tau} \left(\iint_{A_g} \rho_g dA_{CS} \right) dt' \tag{2.17}$$

$$\rho_g = \langle \rho_g \rangle \equiv \frac{1}{\tau A_l} \int_t^{t+\tau} \left(\iint_{A_l} \rho_l dA_{CS} \right) dt' \tag{2.18}$$

$$\rho_m = \langle \rho_m \rangle \equiv \frac{1}{\tau A} \int_t^{t+\tau} \left(\iint_{A_g} [\rho_g dA_{CS}] + \iint_{A_l} [\rho_l dA_{CS}] \right) dt'$$

$$= \rho_g \alpha + \rho_l (1-\alpha) \tag{2.19}$$

2.2.5 Mass Flows

The area-averaged and statistically stationary mass flow rate definitions, according to the 1-D model, are

$$m_g = \langle m_g \rangle \equiv \iint_{A_g} \rho_g \vec{u_g}(\vec{x}, t) \cdot \overrightarrow{dA_{CS}} = \rho_g u_g \alpha A \tag{2.20}$$

$$m_l = \langle m_l \rangle \equiv \iint_{A_l} \rho_l \vec{u_l}(\vec{x}, t) \cdot \overrightarrow{dA_{CS}} = \rho_l u_l (1-\alpha) A \tag{2.21}$$

$$m_m = A[\rho_g u_g \alpha + \rho_l u_l (1-\alpha)] \tag{2.22}$$

Dividing equation (2.20) by equation (2.21) yields a useful mass conservation principle for area-averaged and statistically stationary flows to solve for the void fraction parameter when there is a slip-factor correlation available:

$$\frac{m_g}{m_l} = \left(\frac{\rho_g}{\rho_l}\right)\left(\frac{\alpha}{1-\alpha}\right)\frac{u_g}{u_l} = \left(\frac{\rho_g}{\rho_l}\right)\left(\frac{\alpha}{1-\alpha}\right)S \tag{2.23}$$

2.2.6 Mass Velocity

The concept of mass velocity relates mass flow rate to constant flow area. This parameter enables us to express principles in terms of superficial velocity. The definitions are

$$G_g \equiv \frac{m_g}{A} = \rho_g u_g \alpha = \rho_g j_g \tag{2.24}$$

$$G_l \equiv \frac{m_l}{A} = \rho_l u_l(1-\alpha) = \rho_l j_l \tag{2.25}$$

$$G \equiv \frac{m}{A} = G_g + G_l = \rho_g j_g + \rho_l j_l \tag{2.26}$$

2.2.7 Mass Quality and Volumetric Quality

The mixture quality in terms of mass and volume is defined by the following expressions:

$$x_m \equiv \frac{m_g}{m_l + m_g} = \frac{m_g}{m} \tag{2.27}$$

$$\beta_m \equiv \frac{Q_g}{Q} \tag{2.28}$$

2.2.8 Drift Velocities and Drift Fluxes

The gas and liquid *drift velocity* represents the relative velocity of either phase with respect to the volumetric flux (mixture superficial velocity), defined by

$$u_{gj} = \langle u_{gj}\rangle \equiv \langle u_g - j\rangle \cong u_g - j \tag{2.29}$$

$$u_{lj} = \langle u_{lj}\rangle \equiv \langle u_l - j\rangle \cong u_l - j \tag{2.30}$$

The *drift-flux* of one phase is the volumetric flux relative to a surface moving at the mixture superficial velocity, defined by the following expressions:

$$j_{gl} = \langle j_{gl}\rangle \equiv \langle \alpha(u_g - j)\rangle = \langle j_g(1-\alpha) - j_l\alpha\rangle \cong j_g(1-\alpha) - j_l\alpha \tag{2.31}$$

$$j_{lg} = \langle j_{lg} \rangle \equiv \langle (1 - \alpha)(u_l - j) \rangle = \langle j_l \alpha - j_g (1 - \alpha) \rangle \cong j_l \alpha - j_g (1 - \alpha) \tag{2.32}$$

$$j_{gl} = \langle j_{gl} \rangle \equiv \langle u_s \alpha (1 - \alpha) \rangle \cong u_s \alpha (1 - \alpha) \tag{2.33}$$

The equations above are useful to introduce the drift–flux model in a later chapter.

2.3 Modeling Strategy

The solution of the set of rigorous two-phase flow equations is absolutely impractical, and the application of simplified models is necessary. This section summarizes a brief description of the major approximations that have produced good simulation results.

2.3.1 Area Averaging and One-dimensional Flow

For flowlines and risers, the ratio of flow length to pipe's internal diameter is dominant. Therefore, the expected changes (fluid and flow properties) in the axial direction are greater than the changes in radial or circumferential directions. However, multiphase flow through accessories such as elbows, tees, and valves experiences rapid acceleration and property changes in all directions.

The practice of dealing with area-averaged properties, that is, the 1-D model, is standard in the simulation world of multiphase flow. Zuber and Findlay's (1965) model, which will be reviewed in Chapter 5, proposes the concept of correcting the 1-D modeling approach by incorporating factors that account for cross-sectional variation of the parameters. Pipeline singularities (e.g. pipeline accessories such as valves, elbows, and expansions) may or may not influence the performance simulation. Long flowlines are most commonly insensitive to changes introduced by accessories. Short flowlines, on the contrary, can be affected by accessories and require special simulation tools such as CFD (computational fluid dynamics).

2.3.2 Pseudo-steady Flow

The modeling strategy for *normal production* of multiphase flow is to assume the actual flow as *pseudo-steady*, filtering out the inevitable fluctuations in parameters that field production has. In other words, the pseudo-steady-state hypothesis asserts that fluid properties and other parameters are time-invariant. Parameters change only along the flow path in the production system.

2.3.3 Time Averaging of Turbulent Fluctuations

All multiphase flow models include the time averaging of turbulent fluctuations in the governing equations. The time averaging is performed over a period of time larger than the turbulence frequency of the smallest eddies. Thus, in transient simulations of multiphase flow, the unsteadiness of the flow is captured through the time-derivative term without turbulent fluctuations.

2.3.4 Mechanistic Approach

The Lagrange modeling of multiphase flow (Brennen, 2005) is sometimes attempted for *dispersed type* of flows. This methodology seeks to represent the flow field of small particles immersed in a continuous phase (droplets in gas phase, or bubbles in liquid phase), including the interaction among particles. Obviously, this task is quite challenging and requires lots of computing power. The modern approach is to use commercial CFD codes with adapting mesh and dynamic updating of boundary conditions.

The other approach is to express the governing equations following the Euler methodology jointly with closure equations. To facilitate the burden of the set of equations, a mechanistic approach is taken, whereby the equations are tailored to conform to the most probable flow regime present. Thus, after identification of the prevailing flow pattern, specific models applicable to such regime are used to close the system of equations. In this regard, the flow regimes considered can vary from a very concise set of dispersed versus separate patterns, to more elaborate sets comprising 4–5 different regimes. The downside of this approach is the necessary prediction capacity (uncertainty) of the flow pattern maps for any flow orientation and the lack of established knowledge concerning multiphase flow development length (after changes in direction or passing through system components).

2.3.5 Boundary and Initial Conditions

Typically, the integrated modeling of a hydrocarbon production system, framed by the reservoir conditions at the start point and by the host at the delivery point, defines the achievable production flow rate. Such conditions become the boundary conditions (BC) to solve the conservation equations.

At the reservoir end, at the well's perforations, the fluid temperature is fairly close to the constant reservoir temperature. The flowing bottomhole pressure (bhp) of the well is, however, changing in time due to reservoir's mass depletion (i.e. reservoir performance decline). The bottomhole pressure is lower than the reservoir pressure due to the pressure drops through the porous media and the completion perforations, fracking, and sand screen. Such pressure loss is flow-dependent, and it is known as the *well's drawdown*. Thus, pressure and temperature are referred to as the *upstream boundary* conditions.

At the host facilities, the arrival constant pressure set at the first-stage separator becomes the *downstream boundary condition* for the production system. At this end, temperature is not fixed, and it floats depending on the thermal-hydraulic performance of the overall system. The arrival temperature though is an important parameter in flow assurance studies to identify potential flowing issues related to paraffins, hydrates, and emulsions.

For transient process simulations, a system's initial condition must be defined. The initial condition may be fluid(s) at rest as in the case of system startup after shutdown. In this case, the initial condition specification shall include system's profiles of densities, hold-up, temperature, and pressure.

Another common initial condition is steady-state flow followed by transient flow ramp-up, flow ramp-down, or system shutdown. In this case, the simulation tool is switched from the steady-state flow, with all the property distribution information (i.e. the initial conditions), into transient simulator mode. Transients, in general, are process shifts between two stable operating conditions.

2.3.6 Boundary-value and Marching-scheme Strategies

The boundary conditions described in Section 2.3.5 cast the simulation as a boundary-value problem, that is, the system performance is driven by the two-end conditions. This type of problem poses challenges for the nonlinear set of governing equations.

One common solving strategy is to adopt a *marching scheme* starting from upstream BC, assuming a mass flow rate and solving cell by cell the discretized domain, to estimate the arrival fluid properties at the delivery point. This numerical approach is referred to as a "*shooting method*" (Gerald, 1980) and seeks to estimate the right mass flow rate by several trials trying to match the arrival pressure at the host.

The alternative to the marching scheme is to state conservation in terms of finite difference equations for each cell of the discretized domain and solve the resulting set of linearized equations by numerical methods. Usually, the resulting set of equations creates sparse matrices requiring optimized techniques to minimize storage and accelerate convergence by Gauss–Jordan or *LU* techniques (Gerald, 1980).

2.3.7 Closure Equations

Conservation equations have more unknowns than the number of equations, and the need for model closure is a classic problem of multiphase flow modeling. Such closure equations are a blend of empirical data and mechanistic models, with value limited by the degree of similitude between the experimental information and the problem attempted to solve. Nevertheless, the lack of similar field data (fluids, actual conditions) that have been used in the majority of the available correlations has not dissuaded workers to apply them in the simulations despite the differences because of the lack of better information.

The typical closure equations added to the conservation principles include correlations for void fraction of bubbles in the liquid phase, shear stress at the pipe wall (for each phase), interfacial shear stress between the phases, void fraction of droplets in gas phase, and mixture enthalpy.

2.4 Two-Phase Flow Measurements

Uncertainty in the results obtained from simulators is not only associated with the uncertainty in the simplifications made of the conservation principles (e.g. one-dimensional model, thermal equilibrium) but also related to the uncertainty in the data utilized to generate the closure correlations. The data collected also have measurement uncertainty that feeds into the correlations. Moreover, comparing field data gathered with instruments (also subject to uncertainty readings) has to be assessed with judgment while comparing to simulation results.

This section addresses the role of experimental measurements and their accuracy in providing useful information to the operators.

2.4.1 Well Testing and Production Allocation

Wells have to be independently and periodically tested for reservoir surveillance, reservoir management, and to comply with local legal regulations. During the execution of the test, it is extremely important to have the well in stable flowing condition. The total field production and the isolated well test help estimate the production allocation factors. The allocation factors are necessary to allow the individual contribution of every single well within the overall production gathered.

The task of a well test is to measure the flow rate of gas, oil, and water at a particular set of choke beans, separator pressure, and fluid temperature at the separator. For onshore wells, it is common to have a test separator at the wellsite of a cluster of wells. For subsea wells, there is usually a separate testing flowline that conveys only the production from the well tested to a separator installed at the topside facilities. Sampling ports are provided at the test site to collect fluid samples for composition analyses (i.e. surface–fluid recombination). Well-calibrated flow meters, pressure transmitters, and temperature transmitters provide local and remote data for recording and calculations. Alternatively, local *multiphase flow meters* can be installed near a cluster of wells, either onshore or subsea, to waive the use of a test separator and remotely measure the desired parameters in real time, as described in the section below.

The information gathered serves for

- Logging cumulative production per well,
- Verifying the well is producing the scheduled plan,
- Estimating allocation factors,
- Estimating reserves left,
- Updating well's decline curve,
- Calculating local gas–oil ratio (GOR),
- Calculating oil–water ratio (OWR),
- Sampling and identifying changes in hydrocarbon composition,
- Monitoring increments in produced water or gas flow rates,
- Noticing presence of foam, asphaltene, and emulsions, and
- Identifying impending need for well's workover job.

A well to be tested is isolated from the rest and is aligned to the test flowline ending at a test separator downstream. The well testing process requires a test header, isolation valves, and a dedicated test flowline to reach a test separator downstream. A sketch of a three-phase test separator is shown in Figure 2.4.

2.4.2 Multiphase Flow Meter (MPFM) and Field Application

The in-line multiphase flow meter technology underwent advanced development in the 1990s aiming to fill gaps in field applications such as

- Remote location of wells and clusters, far from processing centers,
- Multiphase flow meter (MPFM) built into the well's Christmas tree,
- Platform topsides with limited space for test separator,

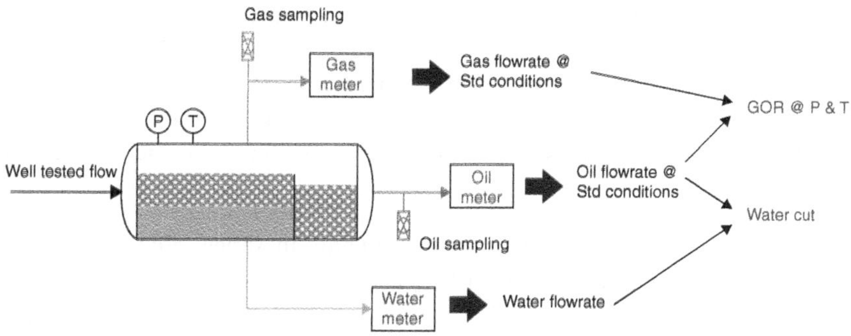

Figure 2.4 Test separator.

- Optimize the choke setting for each producing well with instantaneous fluids' flow rate measurement, and
- Quickly monitor water cut (WC) with minimal risk of emulsions.

The MPFM can be used independently in a field application or integrated with a conventional test separator for periodic calibrations. Figure 2.5 shows a well-test scheme with separator and MPFM; this scheme allows aligning the well to be tested and selecting either the separator or the MPFM to record the measurement.

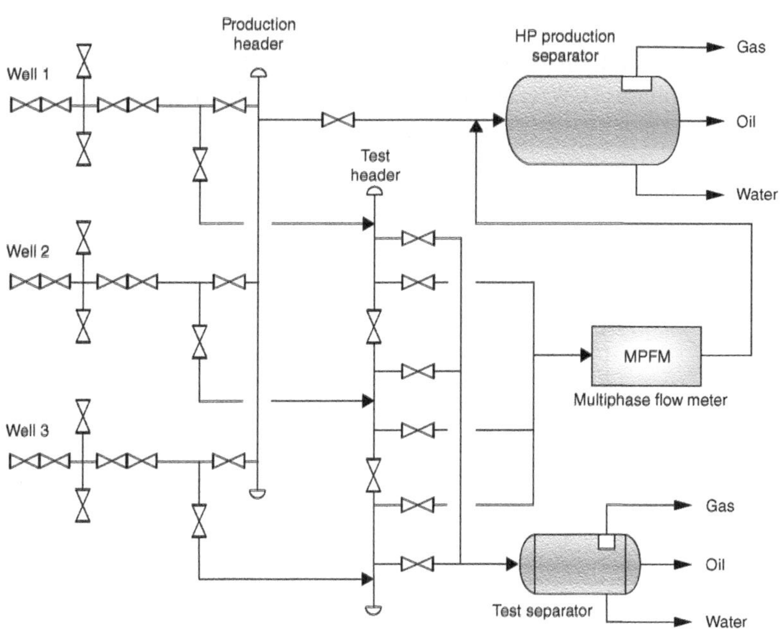

Figure 2.5 Integrated MPFM and test separator.

References such as Falcone et al. (2010) and Folgero et al. (2013) describe the fundamentals of the various components comprising the MPFM and the level of uncertainty to expect from the measurements taken.

In general, there are four types of MPFM, depending on the type of devices utilized to generate measurement signals. These types (Falcone et al., 2010) are as follows:

- Homogenizer followed by two instruments in series, plus a sampling instrument. The homogenizer helps create a pseudo-homogeneous flow pattern of droplets and bubbles, with minimal slip among the phases. Thus, a homogeneous equilibrium model (HEM) can be applied to provide mass and momentum conservation equations. The instruments in series can be a venturi tube and a gamma-densitometer unit to estimate average mixture velocity and density. The third instrument operating on a fluid sample provides an estimate of the oil–water ratio (i.e. WC).
- The next option is similar to the previous one of a homogenizer but with three instruments in series and no sampling. The third instrument should be able to estimate the WC, for instance, a neutron attenuation device.
- The third option does not have a homogenizer and requires up to five instruments if no periodic calibrations are made. The instrumentation is more sophisticated and appropriate to laboratory experimentation rather than fieldwork.
- The fourth alternative is symbolic because it comprises the separation of the three phases O/W/G and becomes the test separator option addressed before.

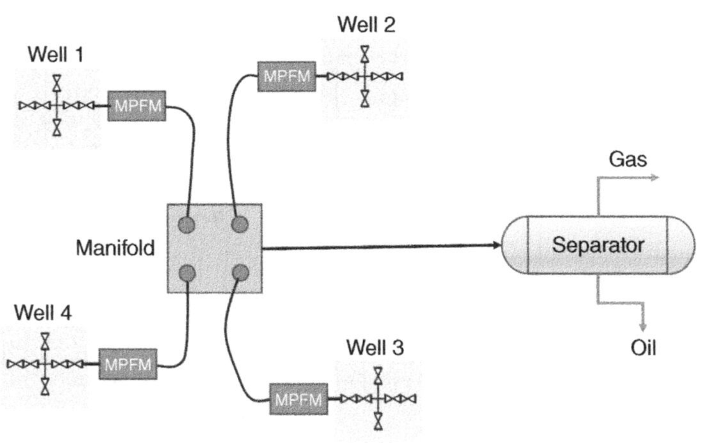

Figure 2.6 Field array of MPFMs.

The value of the MPFM becomes evident for remote or subsea fields where the test separator option requires pad space or dedicated test lines. Figure 2.6 shows a sketch of an array of MPFM feeding remote data to the control room of downstream processing facilities.

For detailed information about the application of MPFM at a full scale, please refer to "Handbook of Multiphase Flow Metering," The Norwegian Society for Oil and Gas Measurement, Norway (2005).

2.4.3 Density Meters

There are several measurement techniques applicable for laboratory experiments and field tests. All measurement devices require specific correlations with respect to a reference standard to improve the accuracy of the measurement. For instance, reference conditions are pipe full of the liquid transported and pipe full of the gas transported, both at stagnant conditions. These two limits set the measurement bounds required for a good density estimate under flowing conditions.

By far the foremost methodology applicable at field level is gamma densitometry (usually referred to as *gamma-d*). The gamma-densitometer (Falcone et al., 2010) could have a single-collimated high-energy beam, but in general it is a multi-beam (3–7 beams) device to sweep the pipe's cross-section. A source of Cesium-137 radiates the beam, which is attenuated by the flowing multiphase mixture, and the radiation is captured by detector photo-multiplier tubes. The attenuated energy from the radiation beam is detected with a device diametrically positioned on the piping (or coating) surface. An algorithm (previously calibrated for the pipe size) applied to the weighted averaged beams' measurements produces an estimate for the area-averaged mixture density. The attenuation phenomenon involves three basic physical processes, namely photoelectric, Compton effect, and pair production.

At the laboratory level, several techniques have been implemented such as

- Weighted section of pipe utilizing strain gauges,
- Vibrating section of pipe by means of electromagnetic forces to generate a response in the frequency domain,
- Acoustic attenuation using an emitter and a receiver, and
- Electrical impedance to measure the capacitance of the multiphase medium with electrodes. This technique has also been applied for liquid–liquid flows (oil–water) in MFM.

2.4.4 Void Fraction Meters

For field applications, the hold-up, or its converse void fraction, can be inferred utilizing a gamma-densitometer device (Hewitt, 1978), usually with five beams crossing the flow area. The gamma-d is normally part of a multiphase flow meter, discussed in Section 2.4.2, to estimate the local void fraction.

At the laboratory scale, a set of consecutive quick-closing valves (Hewitt, 1978; Dukler and Taitel, 1982) may trap a sample of a flowing multiphase mixture. The trapped sample is evaluated to measure the volume of each phase and therefore derive the hold-up and the WC of the mixture. The valves can be fitted in the horizontal or the vertical direction but should have an extremely fast closing time.

The laboratory techniques discussed in the previous sections have also been utilized with varying degrees of success to estimate local void fraction at the lab to generate empirical correlations.

2.4.5 Momentum Flux Meters

The venturi-tube meter is the favorite choice for field and lab applications because of the wealth of data for single- and two-phase flow. In combination with an upstream homogenizer, it is a good device to estimate the momentum flux ρU_m^2, jointly with the HEM (see Chapter 3).

A venturi tube is also a common instrument found in the MFM. Other momentum flux devices (or pressure drop devices), such as the orifice plate and the flow nozzle, may yield analogous results with proper calibration and correlation, but the venturi-tube device offers greater erosion resistance (e.g. against sand fines) and better momentum recovery due to its shape.

2.4.6 Flow Pattern Identification

Hewitt (1978) and Dukler and Taitel (1982) reviewed the techniques to identify the boundaries of flow regimes in pipe flow. The research in this area is not free of controversy because of the potential subjectivity of the regimes identified by different workers. It is fair to say that the relative position of the flow regimes within a 2-D flow map is universal among the number of maps proposed; the potential controversies stem from the regime transitions observed and the uncertainty associated with such transitions. Most flow maps have been derived from lab data gathered in small-diameter piping with air–water mixtures.

The methodologies applied to determine pattern transitions are as follows:

- Still photography, high-speed photography, and videos. This technique requires transparent piping (which usually require low-pressure flows) and clean, transparent fluids.
- X-ray absorption fluctuations. Probability density functions of the output signal can be interpreted to match some flow regimes. This technique can be applied to non-transparent piping and fluid flows at high pressure.
- Analysis of pressure fluctuations. A quartz pressure transducer can monitor and follow rapid pressure fluctuations. This method is capable of differentiating between dispersed and separated flow regimes.
- Wall shear stress measurement with the electrochemical method. This is a laboratory technique monitoring the fluctuations in shear stress and establishing corresponding correlations for each regime.

In practice, field hydrocarbon production operations do not pursue identifying flow regimes with actual measurements. Instead, multiphase flow simulators have programmed flow pattern maps to implement mechanistic models.

2.4.7 Uncertainty in Multiphase Flow Measurement

The uncertainty in the multiphase flow measurements has to be viewed from the standpoint of laboratory data gathering and from the standpoint of oil and gas field applications. Obviously, multiphase data gathering must have minimal uncertainty to be of any value. However, the spatial distribution of the phases (i.e. flow regime) poses quite a challenge to obtain accurate measurements.

The paper by Folgero et al. (2013) discusses at length the sources of error present while utilizing an MFM and the uncertainty level in the measured mass flow rate of hydrocarbons. It is extremely important to work with measurements on a mass basis because it is conserved, whereas volumetric measurements depend upon varying local conditions of pressure and temperature. The error propagation in the mass flow rate of hydrocarbons can be estimated using equation (2.34), if all contributing measurements follow normal distributions (i.e. follow truly random performances),

$$\frac{\delta m_{HC}}{m_{HC}} = \begin{bmatrix} \left(\frac{m_o}{m_{HC}}\right)^2\left(\frac{\delta\rho_o}{\rho_o}\right)^2 + \left(\frac{m_g}{m_{HC}}\right)^2\left(\frac{\delta\rho_g}{\rho_g}\right)^2 + \left(\frac{m_o}{m_{HC}}\right)^2\left(\frac{\delta Q_l}{Q_l}\right)^2 + \\ + \left(\frac{m_g}{m_{HC}}\right)^2\left(\frac{\delta Q_g}{Q_g}\right)^2 + \left(\frac{\rho_o m_w}{\rho_w m_{HC}}\right)^2\left(\frac{\delta WC}{WC}\right)^2 \end{bmatrix}^{0.5} \quad (2.34)$$

based on the following relationships:

$$m_{HC} = m_o + m_g \tag{2.35}$$

$$m = m_{HC} + m_w \tag{2.36}$$

$$WC \equiv \frac{m_w}{m_o + m_w} \tag{2.37}$$

Thus, the uncertainty in the hydrocarbon's total mass depends on the error propagation caused by the following five uncertainty variables:

- Oil density uncertainty $\delta\rho_o$,
- Gas density uncertainty $\delta\rho_g$,
- Liquid volumetric flow rate δQ_l,
- Gas volumetric flow rate δQ_g, and
- Water cut uncertainty δWC.

Calibration of MFM at field conditions is either too difficult or impossible, depending on the environment of the system. The industry recommendation is to provide an MFM per well and a set of two additional MFM at the topsides or central processing facility. These two other units being within a process facility may be connected to the first-stage separator and to a test header connected to a test separator. Such measurement system serves the purpose of frequent calibration of the MFM units within the battery limit of the process facility. In turn, the total mass flow rate of hydrocarbons measured by the calibrated battery limit MFMs is used to calibrate the sum of all field MFM total mass flow rates of hydrocarbons. This methodology allows to periodically generate mass correction factors to estimate the allocation factors per well.

In summary, to minimize the uncertainty in production mass flow rate of hydrocarbons and produced water, it is necessary to constantly update the PVT data of the fluids (e.g. taking samples at the test separator) and the thermodynamic estimation of fluids' density, as well to minimize measurement uncertainty in the volumetric flow rate of all fluids produced.

2.5 Future Measuring Trends

What should be expected to improve the accuracy of multiphase flow field measurements, in particular, flow rate and density? A glimpse at the future trend is presented in the excellent monograph by Falcone et al. (2010). A combination of advanced artificial neural network (ANN), tuned and calibrated with field data, and a learning algorithm to minimize performance deviations could produce far better predictions than the output from real-time operations dynamic flow simulators, or from other empirically determined correlations. The power of ANN methodologies enables them to use the available data and minimize the field uncertainty in a way equivalent to how the human brain thinks, but at dazzling speed and to the extent of evaluating all feasible options. Field data are usually split to "train" the ANN algorithm, usually with the help of fuzzy logic algorithms, and the other part of the database serves to test the accuracy of the trained algorithm.

References

Bergles, A.E., Collier, J.G., Delhaye, J.M. et al. (1981). *Two-Phase Flow and Heat Transfer in the Power and Process Industries*. New York, NY, USA: McGraw Hill Book Company.

Brennen, C. (2005). *Fundamentals of Multiphase Flow*. Cambridge, UK: Cambridge University Press.

Brill, J. and Mukherjee, H. (1999). *Multiphase Flow in Wells*. Richardson, TX, USA: Society of Petroleum Engineers Inc.

Collier, J. and Thome, J. (1994). *Convective Boiling and Condensation*, 3e. Oxford, UK: Clarendon Press.

Dukler, A. and Taitel, Y. (1982). *A Short Course on Principles of Modelling Gas-Liquid Flow and on Modern Measuring Methods*," Chapter IX. University of Houston, Houston, TX, USA.

Falcone, G., Hewitt, G.F., and Alimonti, C. (2010). *Multiphase Flow Metering*, 1e. Amsterdam, The Netherlands: Elsevier Publishing.

Folgero, K., Lyng, E., Kocbach, J., Kjell-Eivind, F., Kleppe, K., and Abro, E., Uncertainty Analysis of Multiphase Flow Meters Used for Allocation Measurements. Paper Presented at the 31st International North Sea Flow Measurement Workshop, Tønsberg, Norway, (October 22–25) 2013.

Gerald, C. (1980). *Applied Numerical Analysis*, 2e. Reading, MA, USA: Addison-Wesley Publishing Company.

The Norwegian Society for Oil and Gas Measurement (2005). *Handbook of Multiphase Flow Metering*. Oslo, Norway: The Norwegian Society for Oil and Gas Measurement.

Hewitt, G.F. (1978). *Measurement of Two-Phase Flow Parameters*. London, UK: Academic Press.

Lahey, R. and Moody, F. (1993). *The Thermal-Hydraulics of a Boiling Water Nuclear Reactor*, 2e. La Grange Park, IL, USA: American Nuclear Society.

Wallis, G. (2020). *One-Dimensional Two-Phase Flow*. Garden City, NY, USA: Dover Publications Inc.

Zuber, N. and Findlay, J. (1965). Average Volumetric Concentration in Two-Phase Flow Systems. *ASME J. Heat Transfer* 87: 453–468.

3

Hydrocarbon Fluid Properties and Thermodynamics

3.1 Phase Behavior

A fluid's phase behavior refers to the relationship between pressure, temperature, and volume. The fluid may be a pure compound, a mixture of two compounds (a binary system), or a more complex mixture of many compounds – a multicomponent fluid.

To capture the essence of the *P–V–T* relationship, it is often presented in graphical form with scales of pressure versus temperature or pressure versus volume. However, the most useful form of the relationship is an EOS validated with empirical data.

3.1.1 Empirical Data

Experiments conducted in high-pressure cells allow the generation of empirical *P–V–*T data under thermodynamic equilibrium conditions. The *PVT* cells let to change the pressure by varying it with a liquid mercury pump. Figure 3.1 illustrates a constant pressure experiment by removing the mass of mercury from the cell while heating it to evaporate the liquid and expand the volume.

To perform these laboratory tests, it is necessary to have a representative sample of the production fluid. Although sampling in general is a routine operation, either at surface level or downhole a well, it is difficult to ensure the sample is truly representative of the production stream. Refer to the monograms of Whitson and Brule (2000) and Pedersen et al. (2015) for ample description of field sampling methodologies.

Figure 3.2 shows another typical laboratory test, a constant temperature experiment, by removing the mass of mercury from the cell while heating to evaporate the liquid and expand the vapor volume.

Multiphase Transport of Hydrocarbons in Pipes, First Edition.
Juan J. Manzano-Ruiz and Jose G. Carballo.

Companion Website: www.wiley.com/go/Manzano-Ruiz/Multiphasetransport

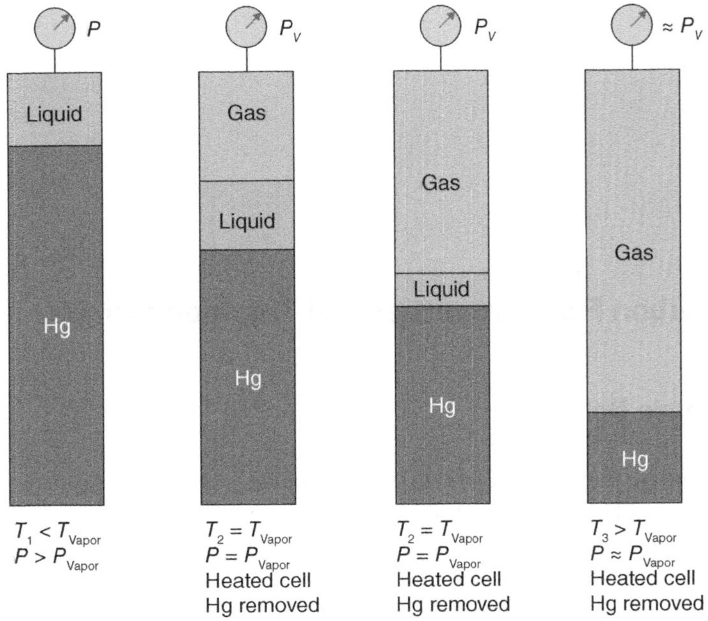

Figure 3.1 Constant vapor pressure test.

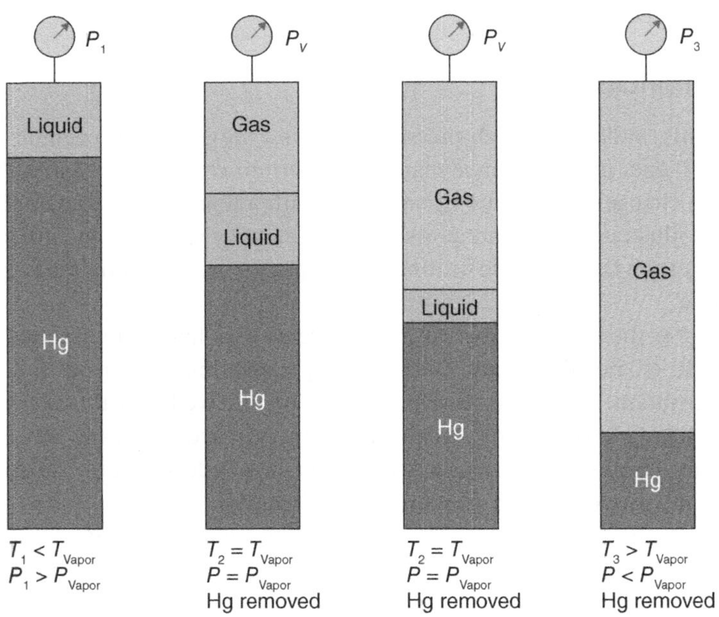

Figure 3.2 Constant temperature test.

3.1.2 Pure Compound Phase Behavior

PVT cell experiments as described in the previous section allow to generate vapor–liquid data to gain understanding of the phase transitions. Figure 3.3 shows isobaric transitions for a pure substance, below and above the critical point. The vapor curve, represented by pressure versus temperature scales, shows how a liquid state can transform into a vapor phase by increasing temperature at constant pressure or vice versa.

Changes in temperature at pressures above the critical point take place without noticeable variations at the vapor–liquid interface. Below the critical pressure, the phase change presents bubble formation at the interface when the liquid temperature is increased.

Similarly, a liquid state can transform into a vapor state at constant temperature by reducing the pressure, as illustrated in Figure 3.4. Again, pressure changes above the critical temperature take place without distinctive vapor–liquid interface variations.

Vapor pressure data are valuable for single-component fluid flow in pipelines. However, hydrocarbon fluids are a much more complex blend of thousands of components, posing a challenge to modeling the thermodynamic behavior in the pressure–temperature domain.

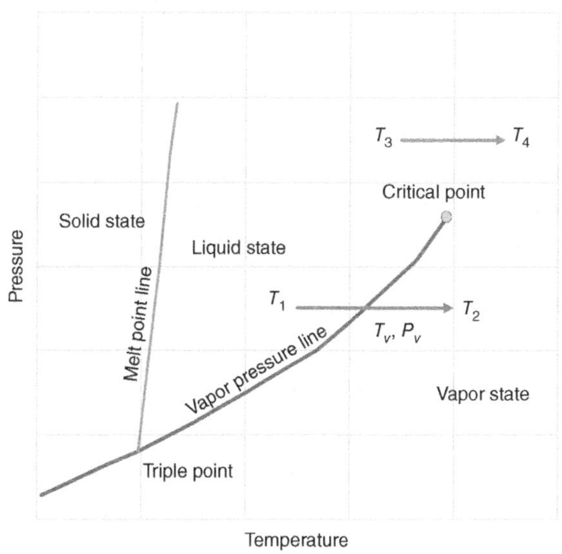

Figure 3.3 Simple-component phase change at constant pressure.

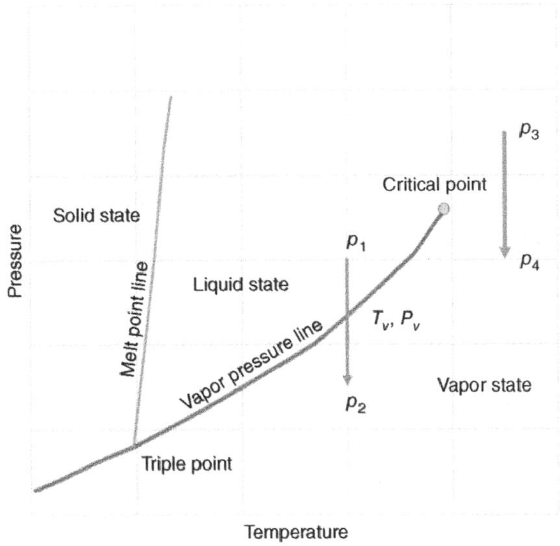

Figure 3.4 Simple-component phase change at constant temperature.

3.1.3 Binary Systems

Blending two hydrocarbons at a time and determining the phase envelope provide insight into the contribution of different boiling points toward the thermodynamic performance of more complex mixtures. Thermodynamic states enclosed within a certain envelope are coexisting states of vapor and liquid in equilibrium. Sage and Lacey (1939) performed pioneering research work with alkane hydrocarbon binary systems and also with nitrogen, carbon dioxide, and hydrogen sulfide, determining the critical loci for such systems. Thus, a multicomponent mixture of hydrocarbons and other components can be considered as the superposition of individual binary mixtures with distinctive phase envelopes.

3.1.4 Multicomponent Fluid Phase Behavior

Multicomponent hydrocarbon fluids in general are a rich fusion of hundreds or thousands of components. The simulation of multicomponent mixture properties requires focusing on the modeling of the dominating components and adjusting based on *PVT* empirical data obtained from lab tests, to derive a representative thermodynamic EOS.

Notice that the critical point of the mixture does not coincide with the cricondenbar point. The bubble point and the dew point lines delimit the condensate, the gas–liquid, and the gas regions in the P–T map. Figure 3.5 shows a handy tool to visualize the different phase areas crossed by a

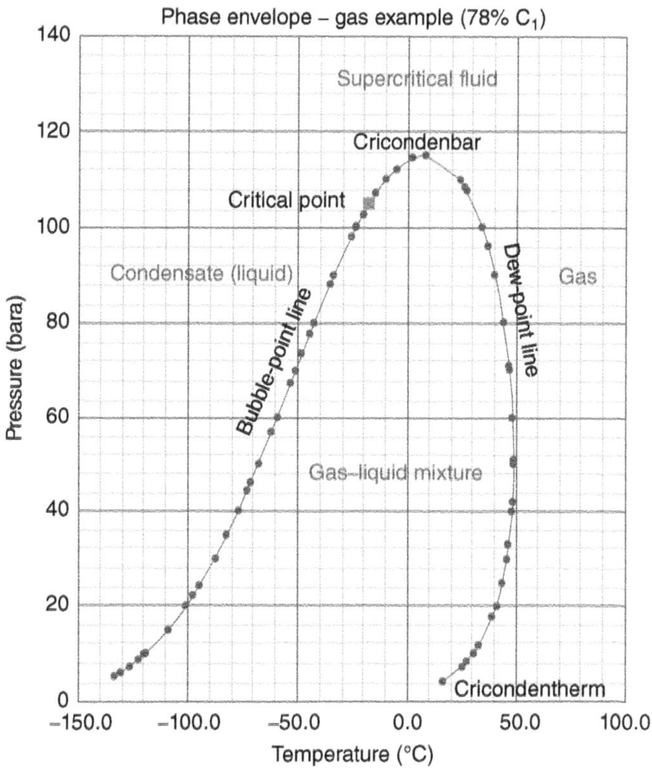

Figure 3.5 Example of gas mixture phase envelope.

production fluid flowing through a pipeline. Notice the region above the phase envelope denominated *supercritical* fluid. Fluids under supercritical conditions (also referred to as *dense fluid*) exhibit blended properties between the liquid and the vapor states, without clear transitions. The assessment of supercritical fluid properties is beyond the scope of this book. Consult Stephan and Lucas (1979) for a comprehensive review of dense fluid properties. For a revision of the impact of supercritical fluid properties in the simulation of a gas pipeline, please refer to Manzano-Ruiz et al. (2014).

The hydrocarbon phase envelope in the pressure–temperature graph could be understood as the mix of the individual binary envelopes addressed in the previous section. Figure 3.5 shows an example of a natural gas phase envelope, with a cricondenbar of 115 bar absolute and cricondentherm of 50 °C. The gas composition is presented in Table 3.1.

The interesting phenomenon of retrograde condensation in gas-condensate fluids can be illustrated on a phase envelope graph. Figure 3.6 shows the isothermal path 1–2–3–4–5, whereby lowering gas pressure

Table 3.1 Example of a natural gas composition.

Component	Mole %
Methane	78.33
Ethane	9.65
Propane	6.63
i-Butane	0.98
n-Butane	2.70
i-Pentane	0.63
n-Pentane	0.64
n-Hexane	0.37
Heptane	0.07
	100.00

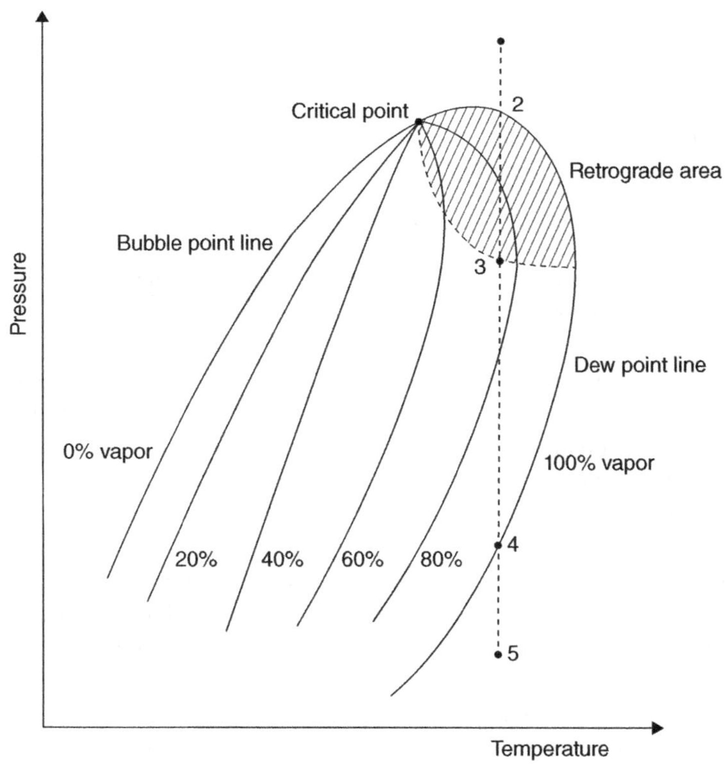

Figure 3.6 Retrograde condensation phenomenon.

from point "1" to point "2" induces the first drops of condensate to appear. Further pressure reductions below point "2" produce more condensate, until reaching a maximum drop-out at point "3." Further isothermal pressure reductions on the mixture reverse the condensation process, progressively starving the condensate phase, until reaching point "4" where the mixture becomes fully vapor again. At temperatures lower than the critical point, no retrograde condensation occurs.

3.2 Physical Properties

This section describes the principal thermodynamic and transport properties required to complement the conservation principles used to model multiphase flow in pipelines. Some of these properties have a direct effect on momentum or energy conservation equations, but there are other properties, herein addressed, that have an indirect influence on multiphase flow modeling via changes in the flow pattern. The properties needed apply to all phases, namely oil, gas, produced water, and emulsions as well.

The emphasis in this section is on approximate correlations that have an empirical basis, usually referred to as "*black oil*" correlations. Such an approach, albeit with accuracy limitations, provides a fast way to generate fluid properties in simulation work. The following section focuses on models based on thermodynamic fundamental relationships labeled EOS. Despite the more theoretical basis of the EOS, there is a minimum of experimental input to incorporate the molecular force effects.

3.2.1 Gas Compressibility Factor

The gas compressibility factor "Z" accounts for the real gas effects associated with molecules' attractions and repulsions while using the modified ideal gas model, as follows:

$$pV = ZnRT \tag{3.1}$$

An isothermal example of the compressibility factor behavior is shown in Figure 3.7, displaying the equilibrium shift among repulsive and attractive molecular forces (i.e. the compressibility factor shows a minimum in the plot).

Based on the concept of the corresponding states, Standing and Katz (1944) plotted the gas compressibility factor as a function of the reduced temperature (nondimensional absolute temperature, referenced to the critical temperature) and the reduced pressure (nondimensional pressure,

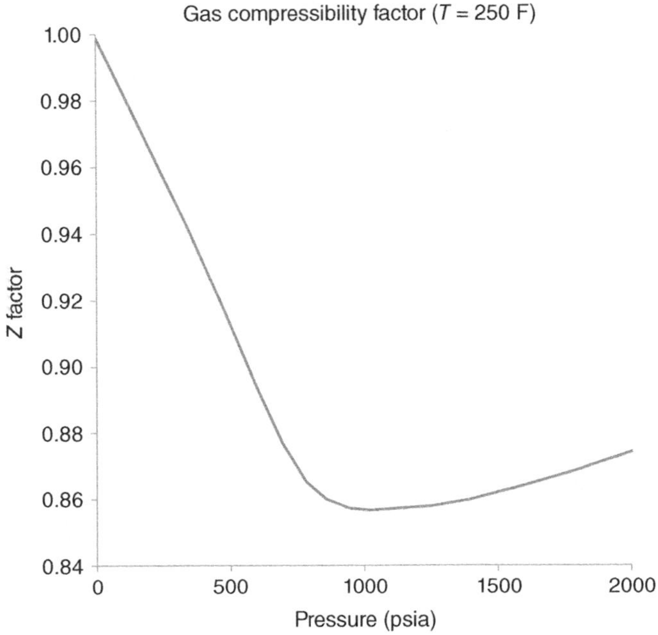

Figure 3.7 Isothermal performance of gas compressibility factor.

referenced to the critical pressure). An alternative to the Standing and Katz (1944) and GPSA (2012) compressibility factor graph is the correlation published by Dranchuk and Abou-Kassem (1975), herein replicated in terms of the reduced temperature, pressure, and density. The parameters used in this correlation are summarized in Table 3.2.

$$Z = 1 + c_1\rho_r + c_2\rho_r^2 - c_3\rho_r^5 + c_4 \tag{3.2}$$

$$\rho_r \equiv \frac{0.27p_r}{ZT_r} \tag{3.3}$$

$$c_1 = A_1 + \frac{A_2}{T_r} + \frac{A_3}{T_r^3} + \frac{A_4}{T_r^4} + \frac{A_5}{T_r^5} \tag{3.4}$$

$$c_2 = A_6 + \frac{A_7}{T_r} + \frac{A_8}{T_r^2} \tag{3.5}$$

$$c_3 = A_9\left(\frac{A_7}{T_r} + \frac{A_8}{T_r^2}\right) \tag{3.6}$$

$$c_4 = A_{10}\left(1 + A_{11}\rho_r^2\right)\left(\frac{\rho_r^2}{T_r^3}\right)\exp\left(-A_{11}\rho_r^2\right) \tag{3.7}$$

Table 3.2 Dranchuk and Abou-Kassem correlation parameters.

A_1	0.3265
A_2	−1.0700
A_3	−0.5339
A_4	0.0157
A_5	−0.0517
A_6	0.5475
A_7	−0.7361
A_8	0.1844
A_9	0.1056
A_{10}	0.6134
A_{11}	0.7210

The nondimensional reduced properties T_r and P_r are calculated with the critical constants T_{crit} (absolute temperature) and P_{crit}, for pure substances, or with pseudo-critical properties (P_{pc} and T_{pc}) for gas mixtures using the following equations:

$$T_r \equiv \frac{T}{T_{crit}} \tag{3.8}$$

$$P_r \equiv \frac{P}{P_{crit}} \tag{3.9}$$

Pseudo-critical properties for mixtures can be estimated with Sutton (1985) correlations using the gas-specific gravity as follows:

$$P_{pc} = 756.8 - 131\gamma_g - 3.6\gamma_g^2 \tag{3.10}$$

$$T_{pc} = 169.2 - 349.5\gamma_g - 74\gamma_g^2 \tag{3.11}$$

3.2.2 Gas and Oil Density

The gas density can be estimated from equation (3.12) with the modified ideal gas equation, the compressibility factor, and the gas molecular weight MW_g

$$\rho_g = \frac{ZRT}{p} \tag{3.12}$$

The gas standard conditions are 101.325 kPa (14.7 psia) and 15.6 °C (60 °F). The nondimensional specific gravity concept refers the gas density to the air density both at standard conditions:

$$\gamma_g \equiv \left(\frac{\rho_g}{\rho_a}\right)_{std} = \frac{MW_g}{28.97} \tag{3.13}$$

The oil density being produced from the reservoir initially experiences a reduction in value when the pressure is reduced (because of volume expansion) until reaching the bubble point at the corresponding temperature. Further pressure reductions below the bubble point produce an increment in oil density because of gas-in-solution liberation. This effect is shown in Figure 3.8.

An old but accurate correlation to estimate the oil density, as a function of pressure and temperature and when the fluid composition is available, was published by Standing and Katz (1942a,1942b), based on the standard conditions pseudo-liquid density and pressure and temperature corrections. Whitson and Brule (2000) provide an explicit correlation that closely fits the Standing–Katz graph, and it is easy to program.

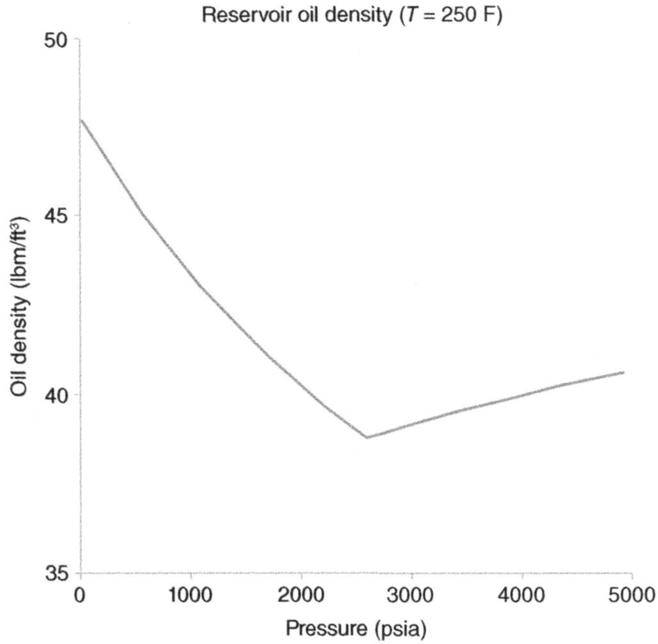

Figure 3.8 Oil density changes around the bubble point.

3.2.3 Gas, Oil, and Water Viscosity

One of the most used correlations to estimate gas viscosity is Lee et al. (1966) herein presented by equations (3.14) and (3.15) – with μ_g in centipoise, T in degree Rankine, and ρ_g in g/cm^3. Other standard correlations can be found in Whitson and Brule (2000).

$$\mu_g = \frac{(9.379 + 0.01607M_g)T^{1.5}}{10^4(209.2 + 19.26M_g + T)} \exp\left[\phi \rho_g^{(2.447-0.2224\phi)}\right] \tag{3.14}$$

$$\phi = 3.448 + \frac{986.4}{T} + 0.01009M_g \tag{3.15}$$

A graph of liberated gas viscosity from a reservoir fluid is shown in Figure 3.9 at a constant temperature.

Oil viscosity estimation depends on whether the fluid is undersaturated or saturated with associated gas. Saturated oils are modeled based on the viscosity of dead oil (gas-free oil) and solution gas/oil ratio. Undersaturated oils are correlated with bubble point viscosity estimates. Figure 3.10 shows a plot of oil viscosity at reservoir conditions.

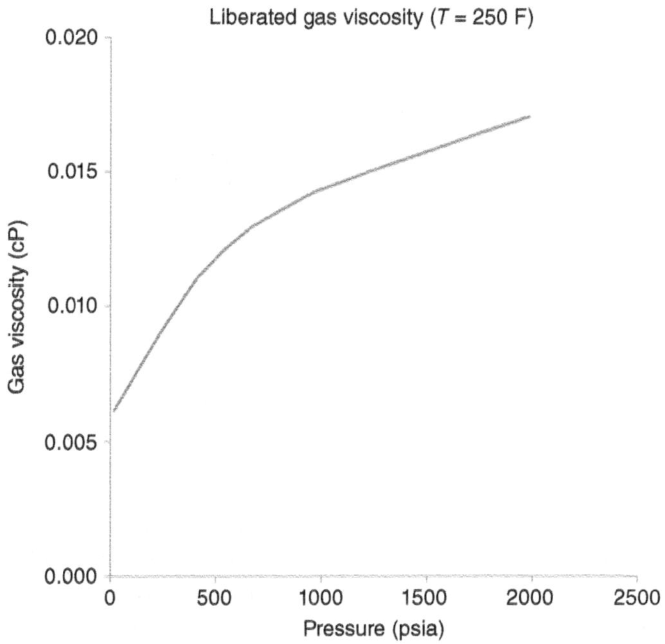

Figure 3.9 Pressure effect upon gas viscosity at constant temperature.

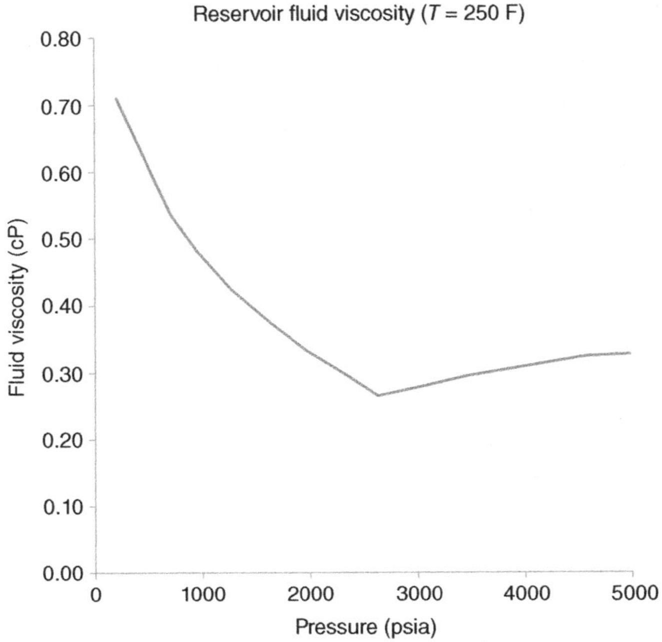

Figure 3.10 Oil viscosity at constant temperature.

McCain (1990) correlated produced water density and viscosity with the salinity "S" (weight % solids) by the following equations (density in kg/m^3, viscosity in centipoise, and temperature in degree centigrade):

$$\rho_w = 999 + 7.026S + 0.02564S^2 \tag{3.16}$$

$$\mu_w = \left(109.6 - 8.406S + 0.3133S^2 + \frac{8.722}{10^3}S^3\right)(1.8T + 32)^{\phi} \tag{3.17}$$

$$\phi = -1.122 + \frac{2.640}{10^2}S - \frac{6.795}{10^4}S^2 - \frac{5.471}{10^5}S^3 + \frac{1.556}{10^6}S^4 \tag{3.18}$$

3.2.4 Solution Gas Ratio (R_s)

The produced oil phase always carries gas in solution, labeled associated gas, and the specific property to characterize the amount of gas in solution at a specific temperature and pressure is the solution gas-oil ratio R_s. Figure 3.11 shows how the solution gas-oil ratio changes with pressure at reservoir temperature.

The rigorous method (Ahmed, 2007) to evaluate the solution gas-oil ratio is based on the oil stock-tank specific gravity at standard condition, the oil

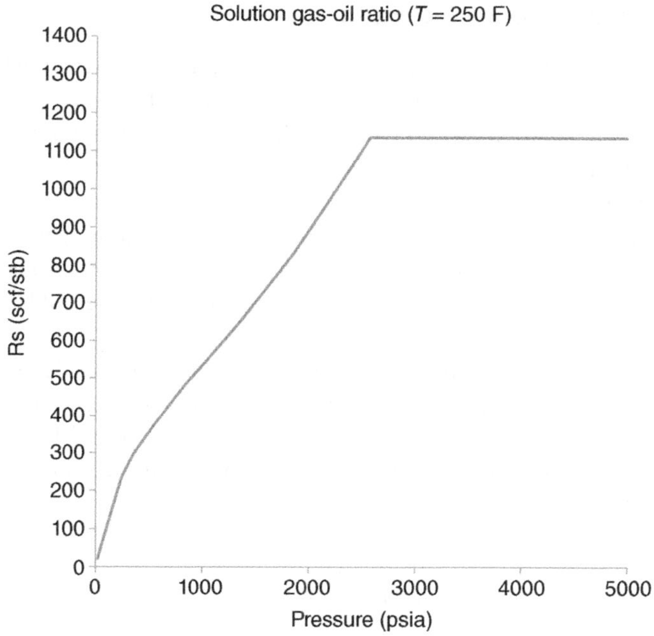

Figure 3.11 Solution gas ratio at a constant temperature.

formation volume factor (see Section 3.2.6), the oil density (kg/m^3) at the pressure and temperature conditions, and the gas-specific gravity averaged between the separator and the stock-tank conditions. The equation is

$$R_s = \frac{B_o \rho_o - 1000\gamma_o}{1.2\gamma_g} \tag{3.19}$$

3.2.5 Bubble Point

The bubble pressure correlations are based on the solution gas-oil ratio, the gas-specific gravity, the oil gravity expressed in API, and the temperature. McCain (1990) recommends using the gas-specific gravity from a sample taken at a separator. Ahmed (2007) listed the best-known bubble point correlations in full detail. These correlations are Standing, Lasater, Whitson and Brule, Vasquez-Begg, Glaso, Marhoun, and Petrosky-Farshad. Here, we present as an example the *Standing correlation* for the bubble point pressure (R_s in sm^{3}gas/sm^{3}oil, pressure in bar abs., temperature in degree centigrade):

$$p_b = 1.255\left[10^{(0.00162T-0.0125API+0.0288)}\left(\frac{5.615R_s}{\gamma_g}\right)^{0.83} - 1.4\right] \tag{3.20}$$

3.2.6 Oil Formation Volume Factor (B_o)

The oil formation volume factor is a nondimensional parameter, determined during the *PVT* laboratory tests, which measures the ratio of one barrel of oil (including gas in solution) at reservoir conditions to the volume of the dead oil at standard conditions.

The oil in reservoir conditions, at the start of production, usually is *undersaturated*, meaning that the liquid phase cannot hold more gas in solution. This can be appreciated in Figure 3.11 where the solution gas ratio remains constant despite increments in pressure above the bubble point.

During production, the oil volume at reservoir conditions expands due to pressure reduction, which is reflected in Figure 3.12 by an increment in B_o until reaching the bubble point. At pressures below the bubble point, the oil phase volume contracts because gas begins to come out from the solution; consequently, the value of B_o diminishes.

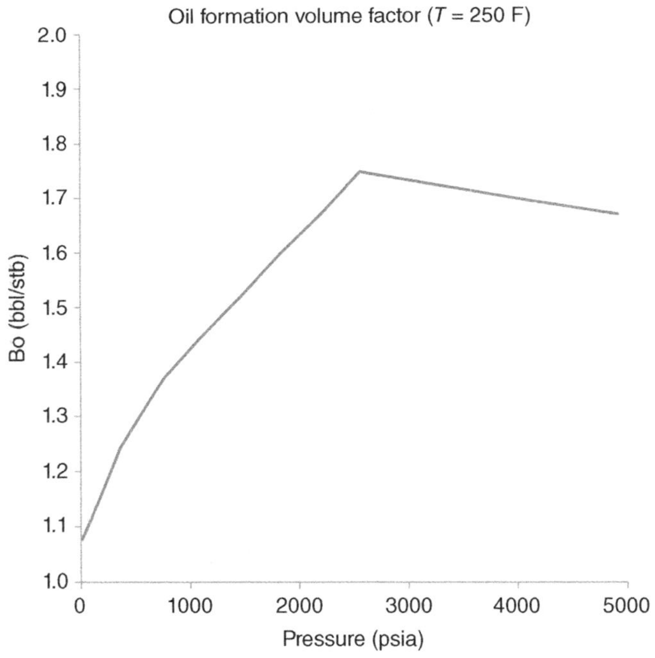

Figure 3.12 Oil formation value at a constant temperature at a constant temperature.

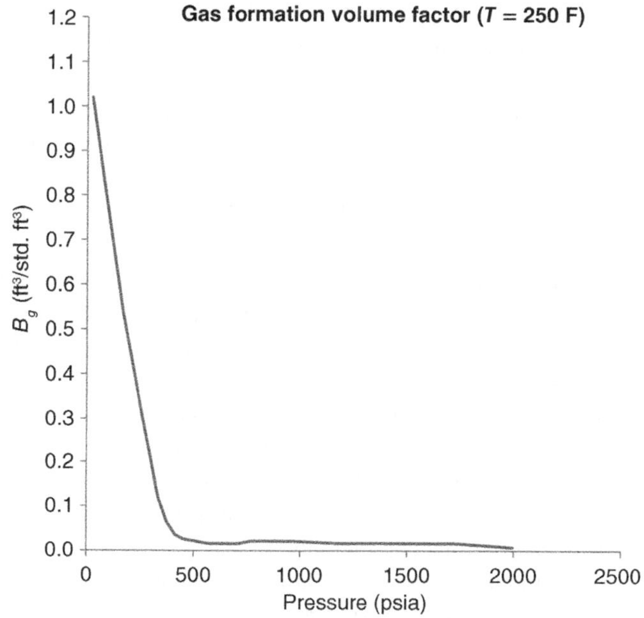

Figure 3.13 Gas formation volume factor at constant temperature.

A correlation for oil formation volume factor reported by Ahmed (2007) follows, where the temperature is in degree Kelvin and R_s is in sm^3 gas/sm^3 oil.

$$B_o = 0.9759 + 0.000951\left[2.25T + R_s\sqrt{\frac{S_g}{S_o}}\right]^{1.2} \tag{3.21}$$

3.2.7 Gas Formation Volume Factor (B_g)

The gas behavior in terms of volume changes is contrary to the oil volume changes. Figure 3.13 reveals that B_g remains constant above the bubble point but grows rapidly below the bubble point pressure. This behavior can be explained by the enhanced compressibility of the gas phase.

3.3 Equation of State (EOS)

Without any doubt, the modern trend to generate the necessary fluid properties rests on thermodynamic relationships based on equations of state. There are numerous EOS proposed in the literature since the last quarter of the nineteenth century, with varying degrees of success and complexity to attempt to reproduce the data obtained at laboratories. This section covers the mainstream equations of states that have found a solid place of application for multiphase flow in pipelines. Many technical references are available addressing this important topic of EOS, for instance, Ahmed (2007), McCain (1990), Firoozabadi (1999), and Whitson and Brule (2000).

In general, an EOS is a relationship between pressure, temperature, and volume (i.e. density) with several coefficients that are tuned to match *PVT* data. One example of such adjusting parameters, the binary-interaction parameters (BIP), is covered in Section 3.3.8.

The EOS can be processed using thermodynamic relationships (James Clerk Maxwell relationships) to derive various key thermal properties such as enthalpy, specific heat, Joule–Thomson (JT) parameter, and speed of sound. Furthermore, EOS is also used to obtain the important transport properties of viscosity and thermal conductivity.

Mixtures of pure components pose a simulation challenge to the model chosen to derive properties based on mixing rules (Danesh, 1998).

Despite the remarkable success of using EOS in multiphase flow simulation work, it is important to point out two main downsides of this practice. First, EOS does require broad tuning work to match *PVT* data before relying on its predictive capability. Second, the computing time of multiphase flow code employing EOS for point-to-point calculations (compositional modules) is orders of magnitude larger than simulation codes that rely on pre-estimated tables of fluid properties or partly on black oil correlations. These two downsides demand careful judgment to define the required accuracy in the fluid properties and the nature of the simulation, either steady-state or transient simulations.

3.3.1 Ideal Gas

The simplest EOS is without doubt the ideal gas model proposed by Clasius–Clapeyron in the nineteenth century, combining previous work done by Boyle, Charles, Avogadro, and Gay-Lussac. The ideal gas behavior is devoid of intermolecular forces and has a compressibility factor equal to 1. In real applications, the ideal gas model finds a place at extremely low pressures and elevated temperatures, and conditions necessary to neglect the intermolecular forces of attraction and repulsion. The model equation for ideal gas is equation (3.1) skipping the compressibility factor Z.

3.3.2 Van der Waals

The Dutch scientist Johannes Diderik van der Waals published his famous EOS in his Doctoral Dissertation in 1873. This remarkable research contribution was not only a thermodynamic accomplishment in van der Waals's career, but he also proposed the model of the corresponding state based on critical properties.

The value of the Van der Waals EOS is the surprisingly uncomplicated way to account for the non-ideality of real gases influenced by intermolecular forces. Presently, the Van der Waals EOS is not directly used in multiphase flow simulations. This model is referred to as a two-parameter cubic EOS because it has three roots for the gas volume within the two-phase domain. This pioneering approach inspired the way to develop the cubic EOS with enhanced accuracy, the most popular models currently in ample usage, such as the Redlich–Kwong and Peng–Robinson EOS.

The impact of this EOS was to consider a term (co-volume b) to account for molecule volume and the corrective term a/V^2 for the attractive intermolecular forces. The EOS proposed is

$$p = \frac{RT}{V - b} - \frac{a}{V^2} \tag{3.22}$$

Applying equation (3.22) at the critical point and comparing the equation form with equation (3.1), it can be demonstrated that the compressibility factor is a constant (Z_c) equal to 0.375 for any gas according to this EOS. Real gases though have lower Z_c because of the complexity of the intermolecular forces which are not well characterized by the simple model expressed by equation (3.22). At low pressure, the gas volume V is large compared to the co-volume, and the attractive forces are weak. Thus, the gas behavior follows the ideal gas law, equation (3.1), with the compressibility factor equal to 1.

The model of the corresponding states proposed by Van der Waals opened the way to generalize gas behavior simulation scaled with the critical properties via reduced nondimensional parameters, see equations (3.8) and (3.9). This scaling modeling allowed us to capture, in a single plot such as Figure 3.9, the intricate behavior of the compressibility factor.

3.3.3 Benedict–Webb–Rubin

The BWR EOS is an old (1931) but accurate model with a number of empirical constants that are derived from laboratory tests, although many simple hydrocarbon constants are already tabulated (McCain, 1990). Despite the accuracy that can be attained with the BWR EOS to estimate thermodynamic parameters, its use is limited because of the intractable mixture's rules. The EOS is

$$p = \frac{RT}{V_M} + \frac{B_0RT - A_0 - C_0/T^2}{V_M^2} + \frac{bRT - a}{V_M^3} + \frac{a\alpha}{V_M^6} + \frac{c}{T^2V_M^3}\left(1 + \frac{\gamma}{V_M^2}\right)\exp\left(-\frac{\gamma}{V_M^2}\right) \quad (3.23)$$

Please refer to McCain (1990) for the tabulated constants for alkanes to C_7, some non-hydrocarbons, and some olefins.

3.3.4 Redlich–Kwong

Redlich and Kwong (1949) proposed an innovative cubic EOS inspired by the work of Van der Waals, well received because of its simplicity and remarkable success in modeling thermodynamic properties. Two decades later, Soave (1972) improved the constants of the Redlich–Kwong model making it a favorite for practitioners of multiphase flow and process engineering, becoming the Soave–Redlich–Kwong (SRK) EOS. The SRK model for a simple substance is

$$\left[p + \frac{a_T}{V_M(V_M + b)}\right](V_M + b) = RT \quad (3.24)$$

The constants a_T and b are determined from the following equations and the *Pitzer acentric factor* ω (McCain, 1990):

$$a_T = 0.42747\frac{(RT_c)^2}{p_c}[1 + (0.480 + 1.574\omega - 0.176\omega^2)(1 - \sqrt{T_r})]^2 \quad (3.25)$$

$$b = 0.08664\frac{RT_c}{p_c} \quad (3.26)$$

$$\omega = -\left(1 + \log_{10}\frac{p_{vap}}{p_c}\right) \quad (3.27)$$

3.3.5 Peng–Robinson

Peng and Robinson (1976) published a slightly different EOS to the SRK EOS. The volume term was modified in order to improve the predictions of liquid density. PR EOS for a simple substance is

$$\left[p + \frac{a_T}{V_M(V_M + b) + b(V_M + b)}\right](V_M + b) = RT \quad (3.28)$$

$$a_T = 0.45724\frac{(RT_c)^2}{p_c}[1 + (0.480 + 1.574\omega - 0.176\omega^2)(1 - \sqrt{T_r})]^2 \quad (3.29)$$

$$b = 0.07780\frac{RT_c}{p_c} \tag{3.30}$$

3.3.6 Other Equations of State

The use of cubic EOS for multiphase flow outweighs the use of any other. However, accuracy requirements have motivated us to develop more advanced EOS for specific applications. Two of the best-known equations of state are *GERG* (Kunz and Wagner, 2012) developed for natural gas applications and *PC-SAFT* (Gross and Sadowski, 2001) based on statistical mechanics but inspired by the *virial equation* of state (McCain, 1990). *GERG-2008* EOS is often used for natural gas custody transfer applications and also for estimating LNG properties. *PC-SAFT* is an EOS that is gaining more attention in flow assurance applications at substantial risk of wax and asphaltene deposition.

Another important EOS is Lee and Kesler (1975), based on the BWR EOS and focused on properties deviation from reference substances and scaling following the law of corresponding states. It is recognized that the LK EOS provides the best estimates of enthalpy.

3.3.7 Equilibrium and Fugacity Coefficient

Thermodynamics is a science based on equilibrium states, and for the case of two coexisting phases, it establishes the principle of balanced chemical potentials. What this means is that there is an equal flux of vapor molecules and liquid molecules changing phase at any time. The function that measures the potential is the fugacity "f" function, and its ratio with the system pressure is known as the fugacity coefficient "ϕ." The importance of the fugacity concept is that under liquid–vapor equilibrium, for a pure substance or for a mixture of pure substances, the liquid fugacity function should match the vapor fugacity function.

It is easy to demonstrate (McCain, 1990) that cubic EOS such as SRK and PR can be transformed into a cubic equation in terms of compressibility factor, namely

$$Z^3 - (1 - B)Z^2 + (A - 2B - 3B^2)Z - (AB - B^2 - B^3) = 0 \tag{3.31}$$

where the constants A and B are given by

$$A = \frac{a_T p}{(RT)^2} \tag{3.32}$$

$$B = \frac{bp}{RT} \tag{3.33}$$

and the parameters a_T and b are obtained from equations (3.25) and (3.26), or equations (3.29) and (3.30), depending on the EOS chosen.

Equation (3.31) is cubic, and the three roots may be either three real roots or one real root and two conjugate complex roots. For liquid and vapor in equilibrium, three real roots can be solved from equation (3.31), applied to both phases. The smallest root is the liquid compressibility factor Z_L, and the largest root is the gas compressibility factor Z_G.

When the pressure and temperature are known for a vapor–liquid equilibrium state of a pure substance, the compressibility factors allow to calculate the densities of either phase from the following equations:

$$\rho_L = \frac{pM_L}{Z_L RT} \tag{3.34}$$

$$\rho_G = \frac{pM_G}{Z_G RT} \tag{3.35}$$

If either temperature or pressure of the vapor–liquid equilibrium is unknown, a trial-and-error solution is required to match the fugacity coefficients of the phases, by assuming the unknown parameter. The liquid and vapor fugacity coefficients are

$$\phi_L = \exp\left[(Z_L - 1) - Ln(Z_L - B) - \frac{0.3536A}{B} Ln\left(\frac{Z_L + 3.828B}{Z_L - 1.828B}\right)\right] \tag{3.36}$$

$$\phi_G = \exp\left[(Z_G - 1) - Ln(Z_G - B) - \frac{0.3536A}{B} Ln\left(\frac{Z_G + 3.828B}{Z_G - 1.828B}\right)\right] \tag{3.37}$$

3.3.8 Mixing Rules

Mixing rules are necessary to obtain weighted-average properties for real production fluids. The fluids from oil and gas wells have thousands of components that pose a challenge to characterize their phase behavior. Judgment is required to focus on the dominating components and their interactions through mixing rules applied to the EOS. There have been several mixing rules proposed to estimate multicomponent mixture properties, but by far the most popular is Kay's rule (Kay, 1936) of mixing for its simplicity. This rule defines gas pseudo-critical properties based on the molar fractions and the components' critical properties as follows:

$$p_{pc} = \sum_{i=1}^{i=n} y_i p_{c,i} \tag{3.38}$$

$$T_{pc} = \sum_{i=1}^{i=n} y_i T_{c,i} \tag{3.39}$$

$$V_{pc} = \sum_{i=1}^{i=n} y_i V_{c,i} \tag{3.40}$$

Kay's mixing rule applied to the *liquid phase* reads as follows:

$$a_{T,L} = \sum_{i=1}^{i=n} \sum_{j=1}^{j=n} [x_i x_j \sqrt{a_{T,i} a_{T,j}} (1 - \delta_{i,j})] \tag{3.41}$$

$$A_L = \frac{a_{T,L} p}{(RT)^2} \tag{3.42}$$

$$b_L = \sum_{i=1}^{i=n} x_i b_i \tag{3.43}$$

$$B_L = \frac{b_L p}{RT} \tag{3.44}$$

enabling to estimate the liquid compressibility factor and the liquid fugacity coefficient from the following equations:

$$Z_L^3 - (1 - B_L) Z_L^2 + \left(A_L - 2B_L - 3B_L^2\right) Z_L - \left(A_L B_L - B_L^2 - B_L^3\right) = 0 \tag{3.45}$$

$$\phi_{L,j} = \exp\left[(Z_L - 1)\varsigma_{L,j} - Ln(Z_L - B_L) - \frac{0.3536 A_L}{B_L}(\psi_{L,j} - \varsigma_{L,j}) \times Ln\left(\frac{Z_L + 3.828 B_L}{Z_L - 1.828 B_L}\right)\right] \tag{3.46}$$

$$\psi_{L,j} \equiv \frac{(2\sqrt{a_{T,j}}) \sum_{i=1}^{i=n} x_i (1 - \delta_{i,j}) \sqrt{a_{T,i}}}{a_{T,L}} \tag{3.47}$$

$$\varsigma_{L,j} \equiv \frac{b_j}{b_L} \tag{3.48}$$

The parameter $\delta_{i,j}$ is known as the BIP which accounts for the interactions among different molecules of the multicomponent mixture. In general, in regular simulations, it is assumed that $\delta_{i,j} = 0$; however, these parameters serve to adjust (i.e. tune) the EOS model to improve the match with laboratory *PVT* data. Lists of typical $\delta_{i,j}$ parameters are offered by Whitson and Brule (2000) and Firoozabadi (1999). The Kay's mixing rule applied to the gas phase is equivalent to equations (3.41) through (3.48) making the corresponding changes in fluid properties.

3.4 C_7^+ Characterization

The objective of the characterization task is to expand the heavy-end fraction of the fluid mixture seeking to get a better match with *PVT* data with the selected EOS. A customary way to report production fluid composition is a list of alkanes from C_1 through n-C_6, N_2, CO_2, H_2S, and C_7^+. The heavy ends comprising the higher molecular weight substances are usually lumped into the C_7^+ fraction and reporting the equivalent molecular weight and density of such fraction.

Properties simulation based on this type of composition normally yields poor results because lumping thousands of components into the single C_7^+ fraction simply is too ambitious. Therefore, a characterization of the C_7^+ fraction is routinely performed to generate a number of pseudo-components, with their own molecular weight and density data, which will provide flexibility to match laboratory tests.

A source of generalized properties of petroleum groups is Whitson and Brule (2000), providing average boiling temperature, specific gravity, acentric factor, critical properties, and molecular weight for carbon numbers from C_7 to C_{45}^+.

There are a few methods to characterize the C_7^+ fraction, for instance, Whitson and Brule (2000), Ahmed (2007), and Pedersen et al. (2015). These methods divide the plus fraction into several carbon number fractions (pseudo-components) and define their properties, guaranteeing that the molar-averaged molecular weight and density match the C_7^+ fraction respective property data.

3.5 Thermal and Transport Properties

The prediction of thermal properties is currently dominated by thermodynamic relationships derived from EOS. This is covered in the next four subsections.

Transport properties depend on laboratory data, as described in Section 3.2.3 for the already addressed dynamic viscosity property.

3.5.1 Enthalpy

The enthalpy property, as a function of pressure and temperature, can be derived (Riazi, 2005) from the first law of thermodynamics and the definition of heat capacity (see next section) to obtain the change in enthalpy, equation (3.49), between two states:

$$dH = C_p dT + \left[V - T\left(\frac{\partial V}{\partial T}\right)_p\right] dp \tag{3.49}$$

The *residual enthalpy function* is evaluated at constant temperature and measures the departure of said property from the reference state of ideal gas as follows:

$$[H - H^{(i)}]_T = \int_0^p \left[V - T\left(\frac{\partial V}{\partial T}\right)_p\right] dp \tag{3.50}$$

The nondimensional residual enthalpy function can be obtained by evaluating the bracket term under the integral sign with an appropriate EOS. This nondimensional ratio becomes

$$\frac{[H - H^{(i)}]_T}{RT} = -T\int_0^p \left(\frac{\partial Z}{\partial T}\right)_p \frac{dp}{p} \tag{3.51}$$

It has become customary to introduce the scaling parameters of the corresponding states principle into equation (3.51) and use the Lee and Kesler (1975) EOS, as shown in the next equation. The latter provides the best estimates of enthalpy function as compared to using cubic equations of state.

$$\left[\frac{H - H^{(i)}}{RT_c}\right]_{T_r} = -T_r^2\int_0^p \left(\frac{\partial Z}{\partial T_r}\right)_{p_r} \frac{dp_r}{p_r} \tag{3.52}$$

The enthalpy property can be obtained from equation (3.53) with the help of tabulated results for the residual enthalpy, Lee and Kesler (1975).

$$H = H^{(i)} + RT_c\left[\frac{H - H^{(i)}}{RT_c}\right]_{T_r} \tag{3.53}$$

3.5.2 Heat Capacity

The heat capacity property is defined by equation (3.54):

$$C_p \equiv \left(\frac{\partial H}{\partial T}\right)_p \tag{3.54}$$

Using Maxwell's relationships, it can be shown (Riazi, 2005) the following equation ready for evaluation with a proper EOS:

$$C_p = -T\left(\frac{\partial^2 V}{\partial T^2}\right)_p \tag{3.55}$$

Similarly, it is convenient to evaluate the residual heat capacity through an isothermal path for easier calculation:

$$C_p - C_p^{(i)} = -\int_0^p \left[T\left(\frac{\partial^2 V}{\partial T^2}\right)_p\right]_T dp \tag{3.56}$$

3.5.3 Joule–Thomson Coefficient

The Joule–Thomson (J–T) coefficient is defined by equation (3.57) as the change in temperature due to a pressure change at constant enthalpy. Gases typically chill during a pressure reduction and oils usually warm during the pressure expansion. This behavior evolves because the J–T coefficient changes sign at a point called the inversion temperature. Most gases have a high inversion point as compared to the local temperature and therefore when they expand the temperature drops. On the contrary, oils have a low inversion point compared to the local temperature, and thus they warm on expansion. The J–T coefficient property is particularly useful to characterize throttling processes such as passage through a choke valve. The partial derivative is determined with the help of a proper EOS.

$$\eta \equiv \left(\frac{\partial T}{\partial p}\right)_H \tag{3.57}$$

3.5.4 Speed of Sound

The speed of sound is a property associated with the fluid compressibility under the assumption of an isentropic process. Phenomena that take place at high velocity may experience drastic changes in fluid elasticity. A common way to measure the fluid elastic is using the isentropic compressibility factor defined by equation (3.58):

$$\beta_s \equiv \frac{1}{\rho}\left(\frac{\partial \rho}{\partial p}\right)_s \tag{3.58}$$

The isentropic compressibility is closely related to the definition of *speed of sound*, equation (3.59). The speed of sound is an important physical parameter measuring how fast perturbations in the flow field can propagate. The application of equation (3.59) to single-phase flow is straightforward but not so for multiphase flow (please refer to Section 7.3 addressing critical flow):

$$c = \sqrt{\left(\frac{\partial p}{\partial \rho}\right)_S} = \frac{1}{\sqrt{\rho\beta_s}} \tag{3.59}$$

When the fluid velocity reaches the local speed of sound, the phenomenon of choking sets in. One typical scenario where large velocity might develop is in small flow areas or constrictions. Any pressure reduction downstream of the choked point becomes insensitive to the upstream fluid flow because perturbations created downstream cannot migrate upstream because the fluid travels at the speed of sound.

Very rarely, the flow chokes in multiphase flow through pipelines, except in choke valves with large pressure drops across it, or at the location of a pipe rupture with loss of containment.

3.5.5 Thermal Conductivity

Prediction methodologies for thermal conductivity are not well developed as compared to other properties. Pedersen et al. (2015) have condensed the state of the art of estimating thermal conductivity based on the law of corresponding states and some complex rules applicable to hydrocarbon mixtures. An old but effective correlation for oils in general was published by Cragoe (1929) herein reproduced (temperature in degree centigrade and thermal conductivity in Wm-K),

$$k \simeq \frac{1.406}{S_o}(1 - 0.00054T) \tag{3.60}$$

3.5.6 Surface Tension

Oil surface tension can be estimated with the following correlation (Riazi, 2005), where K_w is the Watson parameter, T_r is the reduced temperature, and the surface tension is in dyne-cm:

$$\sigma = \frac{673.7(1 - T_r)^{1.232}}{K_W} \tag{3.61}$$

3.6 Types of Hydrocarbon Fluids

It is customary to classify hydrocarbons into five generic groups with common characteristics and corresponding modeling approaches. These five groups are black oils, volatile oils, retrograde gas condensate, wet gas, and dry gas. Not surprisingly, there are no sharp boundaries among these five types of hydrocarbons and the following subsections depict generic phase behaviors.

3.6.1 Black Oils

Black oil fluid refers to a multicomponent hydrocarbon mixture with a critical temperature relatively higher than the reservoir temperature of the fluid. In other words, the locus of dew points is shifted far to the right of the isotherm crossing the reservoir conditions. This means that the production process never crosses the dew point line when entering the two-phase

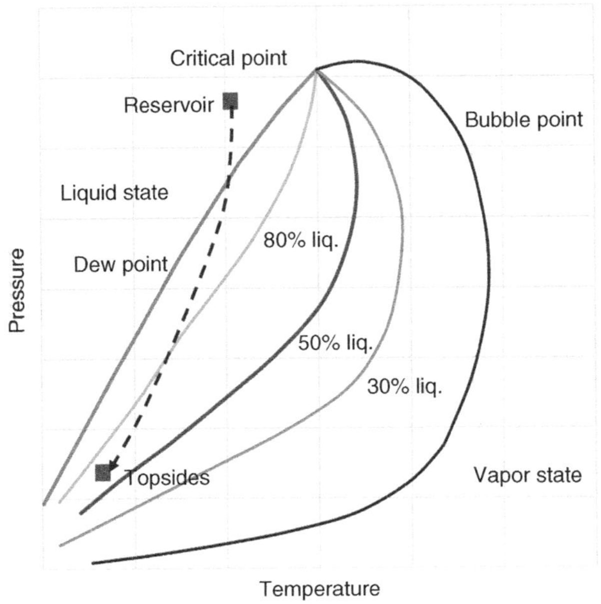

Figure 3.14 Black oil typical phase envelope.

envelope. The hydrocarbon expansion process goes from liquid state to two-phase state passing the bubble point line.

In general, the black oil fluid could be modeled well with black oil correlations previously addressed, sparing somehow the need to employ more time-consuming EOS estimations. Figure 3.14 shows a sketch of how a typical black oil fluid envelope would look like.

3.6.2 Volatile Oils

Volatile oil fluid is a lighter hydrocarbon mixture with a critical temperature closer to the reservoir temperature of the fluid but still larger than the reservoir temperature. The locus of dew points is also shifted to the right of the isotherm crossing the reservoir conditions, signaling that the production process never crosses the dew point line. The hydrocarbon expansion process also goes from liquid state to two-phase state passing the bubble point line.

In general, the volatile oil fluid needs to employ the EOS approach because the neighborhood of the critical point requires greater accuracy for good modeling. Figure 3.15 illustrates a typical volatile oil fluid envelope.

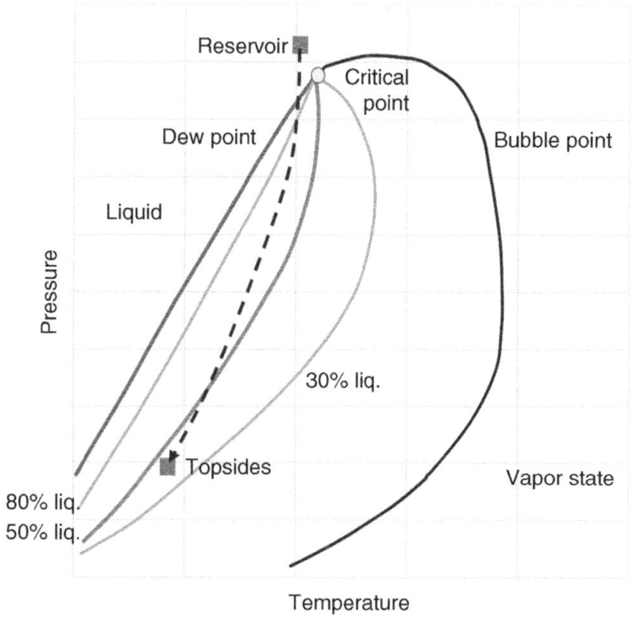

Figure 3.15 Volatile oil phase envelope.

3.6.3 Retrograde Gas Condensate

The retrograde gas fluid mixture has a reservoir temperature larger than the critical temperature but lower than the cricondentherm. The locus of bubble points is shifted to the left of the isotherm crossing the reservoir conditions, signaling that the production process will cross the dew point line into the two-phase state.

The retrograde gas fluid not only will need to employ an accurate EOS approach, but the flowline calculations will require compositional modeling.

Figure 3.16 shows a sketch of a typical retrograde gas fluid envelope.

3.6.4 Wet Gas

The wet gas fluid mixture has a reservoir temperature lower than the cricondentherm. Retrograde behavior may or may not be observed. The two-phase state is approached by crossing the dew point.

The wet gas fluid can be modeled with the EOS approach, or with black oil models for gases (if there is no retrograde behavior). Figure 3.17 shows a sketch of a typical wet gas fluid envelope.

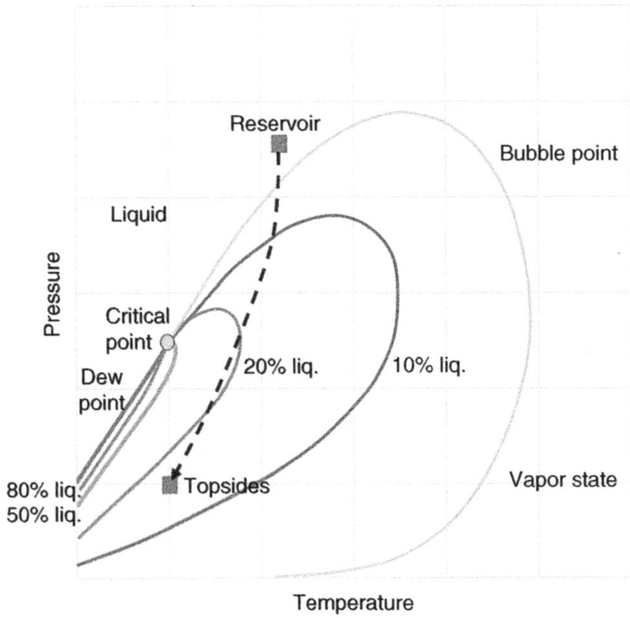

Figure 3.16 Retrograde gas phase envelope.

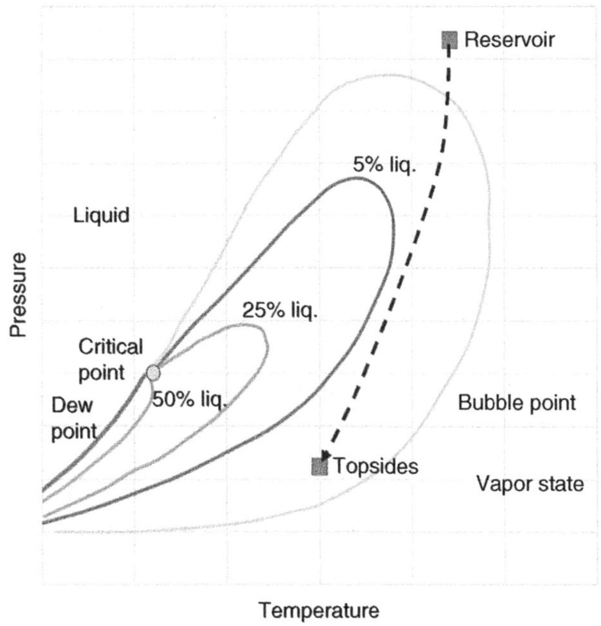

Figure 3.17 Wet gas phase behavior.

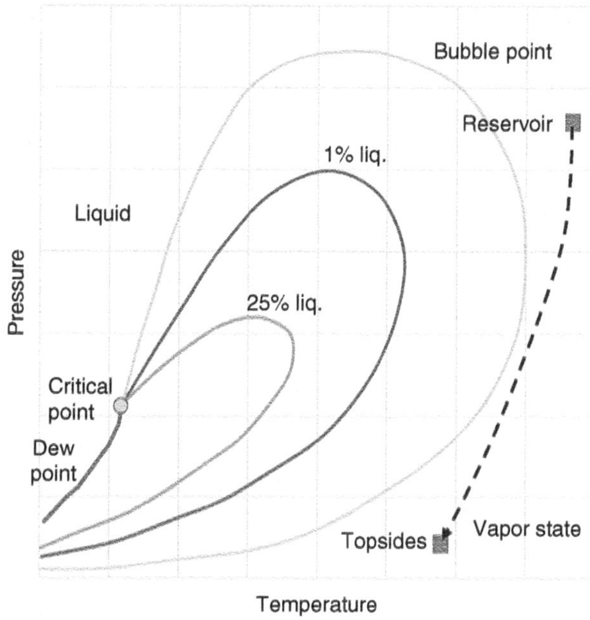

Figure 3.18 Dry gas phase behavior.

3.6.5 Dry Gas

The dry gas fluid mixture has a reservoir temperature higher than the cricondentherm. Two-phase states might or might not be observed in the field application because of the large fraction of light alkanes in the fluid mixture.

The dry gas fluid can be well modeled with the simpler EOS (3.1) and the compressibility factor correlation or graph. Figure 3.18 illustrates a typical dry gas fluid envelope.

3.7 PVT Analyses

This section reviews the typical PVT lab tests performed to gain an understanding of the production fluid. Although the majority of the tests are focused on the fluid performance at reservoir conditions, before production starts and during depletion, useful data are gathered to tune and characterize the EOS of state chosen to model the fluid. The following subsections briefly address the typical tests captured in the PVT report.

Obviously, a paramount requisite to perform the PVT work is to have available a good-quality and representative sample, in enough quantity to enable several tests. This implies not only the sampling procedure in the field but also the sample preservation during the transporting to the lab. Samples in general can be gathered downhole the well, or at the wellhead, or at a nearby separator equipment on the topsides or at the well pad. During the sampling process, it is necessary to have a stable (quasi-steady-state) production flow. Flowrate(s), pressure, and temperature at the sampling port shall be recorded with the greatest accuracy possible, in order to recreate the downhole conditions as much as possible.

3.7.1 Constant Composition Expansion (CCE)

CCE is a standard PVT test that allows to determine the saturation point ($\boldsymbol{P_b}$) at isothermal reservoir conditions (usually the *CCE* reports only the reservoir temperature) by gradually reducing the PVT cell pressure. A volumetric relationship as a function of pressure is determined during the test as illustrated in Figure 3.19.

The PVT cell equipment utilized to conduct a CCE test is sketched in Figure 3.20.

The test results from a CCE are plotted in Figure 3.21, from which the saturation point can be determined.

3.7.2 Constant Volume Depletion (*CVD*)

In general, this kind of test is performed in the lab for volatile oil and retrograde gas fluids. This test simulates a depletion process, for a constant volume reservoir producing rich gas, while in thermodynamic equilibrium with liquid fraction present. It differs from the differential liberation test

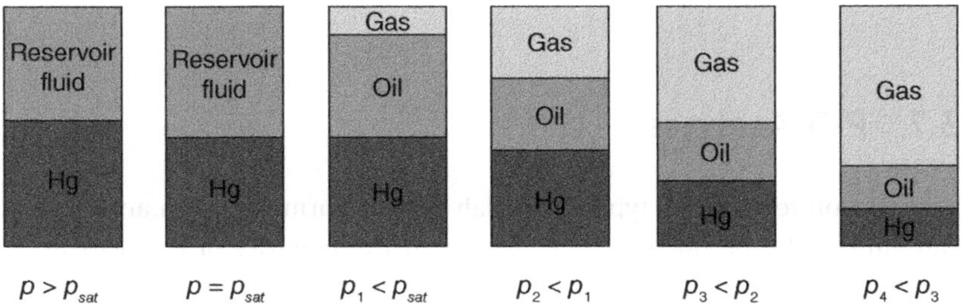

Figure 3.19 Constant composition expansion at constant temperature.

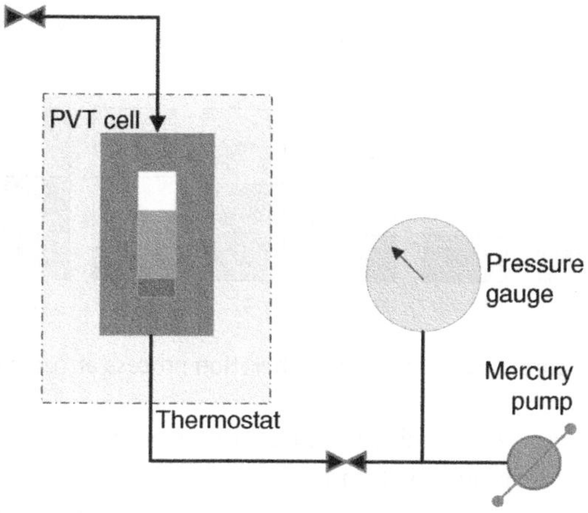

Figure 3.20 PVT cell equipment for CCE tests.

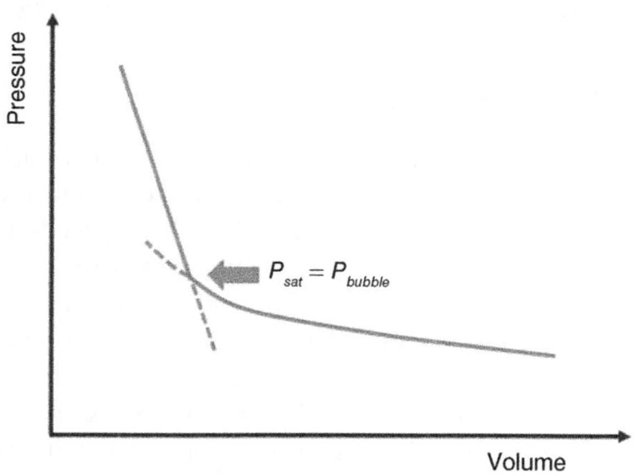

Figure 3.21 Saturation point determination for a black oil from CCE test results.

(see next subsection) in that no gas is withdrawn. This test is performed for fluids where large volumes of gas are released from solution during pressure expansion.

3.7.3 Differential Liberation (*DL*)

DL reproduces the reservoir depletion process, and the test involves isothermal withdrawal of mass from the *PVT* cell. Thus, this test gives the reservoir engineers a hint about how the fluid will change during the production life cycle. Volumetric and compositional data are obtained from the *DL* test. Typical parameters derived from the test are the relative volumes of gas and oil for each pressure level, the gas in solution R_s, and the gas compressibility

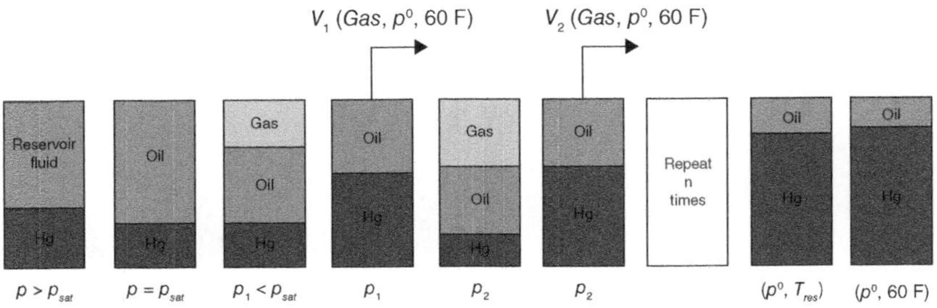

Figure 3.22 Differential liberation process at constant temperature.

factor Z at each pressure. In summary, the *DL* could provide a wealth of information to attempt to match with a tuned EOS. A tuned EOS is critical to feed accurate properties to the mass, momentum, and energy balances of the simulation task. A sketch of the *DL* process is shown in Figure 3.22.

3.7.4 Separator Test

The *PVT* study may include a separator test, at a single or at different pressure levels. The objective of this work is to mimic in the lab the separation stages of a future field operation. Samples of liquids and gas are taken from each stage of separation to determine gas and oil composition. Gas–oil ratio (*GOR*) is calculated at each stage, and in particular at stock-tank conditions, and adjusted for standard temperature. This information is useful to minimize oil tank boil-off during storage.

Separator lab tests can be used to check the equilibrium constants obtained with EOS flash simulations and build trust in the property modeling for steady-state and dynamic simulations.

3.7.5 Viscosities

The two more common laboratory tests for oil viscosity are the rolling ball and the capillary tube viscometer. The first method (*RBV*) comprises a sphere inside a tube with minimal clearance, which rolls down a specific length (between two detectors) under a tube inclination; the travel time of the rolling ball is recorded and used in a correlation for this meter.

The second method (*CTV*) measures the pressure drop across the capillary length, for a given and controlled flowrate of the fluid (using a constant flowrate pump). The oil viscosity is determined by assuming laminar flow and using Poiseuille equation for pipe flow. Figure 3.23 illustrates the equipment of a capillary tube viscometer.

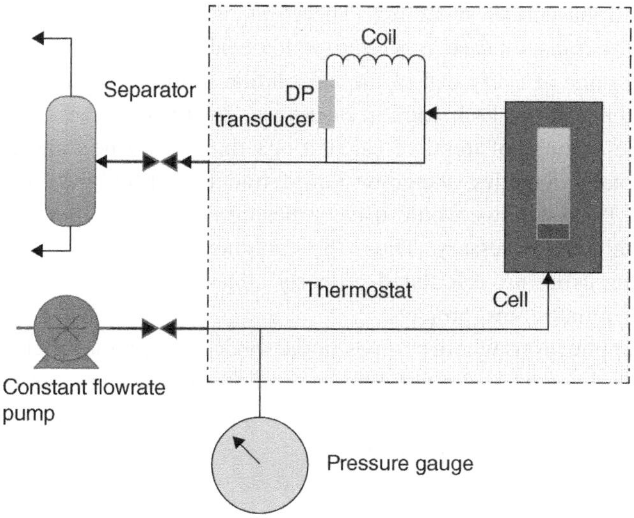

Figure 3.23 Capillary tube viscometer.

3.8 Strategy for Modeling Fluid Properties

The importance of estimating the thermodynamic and transport phenomena properties has been emphasized throughout this chapter. However, there are practical questions to address, for example,

- What is better, a compositional tracking module or property tables to interpolate?
- Does the black oil model have any place in modeling, or the use of EOS is necessary?
- How should we handle a case where different streams, with different compositions, meet/blend at a point in the system (e.g. manifolds, daisy-chain tie-in)?
- Are there shortcuts for reducing the computing time of compositional modeling (e.g. curve fitting)?

The questions above concern not only the novice to the field of multiphase flow modeling in pipelines but also the skilled practitioner who has access to commercial tools. The answers provided here are generic because each case should be evaluated independently.

3.8.1 Compositional versus Tabular

Most modern commercial codes come with an integrated properties' package (from other suppliers) that seamlessly interfaces with the calculation engine for multiphase flow modeling. Therefore, while setting up the steady-state or the transient case, it is required to decide if a compositional analysis will be done at every cell of the calculation domain (the overall production system). If we look back at Sections 3.3.4 through 3.5.2, we realize the huge amount of iterative calculations that would be done at each cell to obtain densities, vapor or liquid fraction, enthalpies, and heat capacities. Therefore, we would use compositional tracking modules only when absolutely necessary. Thus, the economy in computational time achieved by using pre-calculated property tables and point-to-point interpolation dominates most projects.

Examples of applications where compositional modeling accuracy makes a difference include commingling of different composition streams, retrograde gas applications, operation near the critical point of the production fluid, and deposition of wax or asphaltene.

3.8.2 EOS or Black Oil Models

EOS is more accurate (although more complex) and has a wider range of applications than black oil models that originated from a database with limited validity. Nevertheless, there are certain cases where black oil models do a decent job and are orders of magnitude faster to calculate as compared to EOS. Examples of potential application of black oil models are dry gas fluids, black oil fluids, and sometimes wet gases as well.

3.8.3 Blending of Different Streams

Modeling a tie-in point of two or more different composition streams (e.g. well clusters/manifolds) poses quite a challenge. Although pre-calculated interpolation tables of properties may be available for the blending streams, the issue is to know in advance the commingling ratio of the streams, that is, the mass flowrates of each stream are unknown and depend on the thermo-hydraulic analysis of the overall production system – it becomes an iterative trial-and-error problem. In these cases, the use of a compositional tracking module is indicated at the excessive cost of computational time.

A more rudimentary approach to avoid compositional tracking and solve the commingling issue of two streams would be creating a set of interpolation of tables for the blended ("child") stream as a function of the commingling ratio of the "parent" streams.

- Evaluate fluid properties for each parent stream using interpolation tables upstream of the tie-in point,

- Create tables of properties for a family of blending ratios say, by steps of 10% change,
- Carry the integration of the system of differential equations using the corresponding boundary conditions but assuming some mass flowrate,
- Interpolate mass flowrate once two assumed ratios bracket the solution,
- With the interpolated mass flowrate, do the interpolations of properties for the child stream based on the properties of the parent streams.

3.9 Commercial Software

There are many commercial codes capable of performing accurate thermodynamic calculations with a multiplicity of equations of state to choose from. Given the difficulty of performing such calculations, it is quite common to integrate properties estimating software into multiphase flow simulators with seamless performance. This integration enables compositional analyses point-to-point but at a high computational (i.e. time) cost.

To save computational time most codes offer the option of creating pre-calculated tables of properties for interpolation during the execution of the dynamic multiphase flow code.

Please refer to the Appendix of this book for a brief summary of commercial codes.

References

Ahmed, T. (2007). *Equations of State and PVT Analysis*. Houston, Texas, USA: Gulf Publishing Company.

Cragoe, C. (1929). *Thermal Properties of Petroleum Products*. Washington: United States Government Printing Office.

Danesh, A. (1998). *PVT and Phase Behavior of Petroleum Reservoir Fluids*. Amsterdam, The Netherlands: Elsevier.

Dranchuk, P. and Abou-Kassem, J. (1975). Calculation of Z Factors for Natural Gases using Equations of State. *J. Pet. Technol.* 36–36.

Firoozabadi, A. (1999). *Thermodynamics of Hydrocarbon Reservoirs*. New York, New York, USA: McGraw-Hill.

GPSA (2012). *Engineering Data Book*, 13e, vol. FPS Vol. I and II. Tulsa, Oklahoma, USA: GPSA.

Gross, J. and Sadowski, G. (2001). Perturbated-Chain SAFT: And Equation of State Based on Perturbation Theory for Chain Molecules. *Ind. Eng. Chem. Res.* 40: 1244–1260.

Kunz, O. and Wagner, W. (2012). The GERG-2008 Wide Range Equation of State for Natural Gases and Other Mixtures: An Expansion of GERG-2004. *J. Chem. Eng. Data* 57: 3032–3091.

Lee, A., Gonzalez, M., and Eakin, B. (1966). The Viscosity of Natural Gases. *J. Pet. Technol.* 237: 997–1000.

Lee, B. and Kesler, M. (1975). A Generalized Thermodynamic Correlation Based on Three-parameter Corresponding States. *AICHE J.* 21: 510–527.

Kay, W. (1936). Density of Hydrocarbon Gases and Vapors at High Pressure and Temperature. *Ind. Eng. Chem.* 28: 1014–1019.

Manzano-Ruiz, J., Lu, Y., Janardhanan, K., and Liang, C., "Natural-Gas Supercritical Fluid Properties Characterization for Pipeline Applications," GPA Convention Mobilizing Midstream: A Global Connection, Dallas, Texas, USA (2014).

McCain, W. (1990). *Petroleum Fluids*, 2e. Tulsa, Oklahoma, USA: PennWell Books.

Pedersen, K., Christensen, P., and Shaikh, J. (2015). *Phase Behavior of Petroleum Reservoir Fluids*, 2e. Boca Raton, Florida, USA: CRC Press.

Peng, D. and Robinson, D. (1976). A New Two-constant Equation of State. *Ind. Eng. Chem. Fundam.* 15 (1): 59–64.

Redlich, O. and Kwong, J. (1949). On the Thermodynamics of Solutions – An Equation of State. *Chem. Rev.* 44: 233–244.

Riazi, M. (2005). *Characterization and Properties of Petroleum Fractions.* West Conshohocken, PA, USA: ASTM International.

Sage, B. and Lacey, W. (1939). *Volumetric and Phase Behavior of Hydrocarbons.* Stanford, California, USA: Standford University Press.

Soave, G. (1972). Equilibrium Constants From a Modified Redlich-Kwong Equation of State. *Chem. Eng. Sci.* 27 (6): 1197–1203.

Standing, M. and Katz, D. (1942a). Density of Natural Gas. *Trans. AIME* 146: 140–149.

Standing, M. and Katz, D. (1942b). Density of Crude Oils Saturated with Natural Gas. *Trans. AIME* 146: 159–165.

Standing, M. and Katz, D. (1944). Vapor-Liquid Equilibria of Natural Gas-Crude Oil Systems. *Trans. AIME* 155: 232–245.

Stephan, K. and Lucas, K. (1979). *Viscosity of Dense Fluids.* New York, New York, USA: Plenum Press.

Sutton, R. (1985). *Compressibility Factors for High-Molecular Weight Reservoir Gases.* Richardson, TX, USA: SPE Paper 14265.

Whitson, C. and Brule, M. (2000). *Phase Behavior.* Richardson, TX, USA: Society of Petroleum Engineers Inc.

4

Multiphase Flow Patterns

4.1 Gas–Liquid Flow

The processes a hydrocarbon mixture is exposed while flowing through a production system causes pressure and temperature drops. One can visualize the thermodynamic path of the mixture in a pressure–temperature plot to verify what saturation line (either bubble or dew point) will be crossed while entering the gas–liquid region. Crossing the phase envelope into a two-phase region produces a few small bubbles (bubble point) or liquid droplets (dew point) which start to increase in number while agglomerating into larger volume particles.

Figure 4.1 shows a picture of a diabatic flow (with heat transfer at the boundary) transitioning from a few isolated bubbles in the continuous liquid phase (bubbly flow), followed by formation of a large centerline bubble (plug flow), followed by rupture of the large bubble and gas channeling at the centerline (churn flow), and finally liquid swept radially out with a clear core of centered gas phase and a few small droplets in it (annular-mist flow).

It is obvious from Figure 4.1 that it is impossible to computationally track every fluid particle in the flow domain at every instant of time. Two-phase flow modeling requires simplifications and assumptions that are focused on each generic flow regime, to make tractable the solution of the system of equations. These simplifications center on properties time-variance, liquid–gas interface shear and mass transfer, and cross-sectional averaging of fluid and flow properties.

Multiphase Transport of Hydrocarbons in Pipes, First Edition.
Juan J. Manzano-Ruiz and Jose G. Carballo.

Companion Website: www.wiley.com/go/Manzano-Ruiz/Multiphasetransport

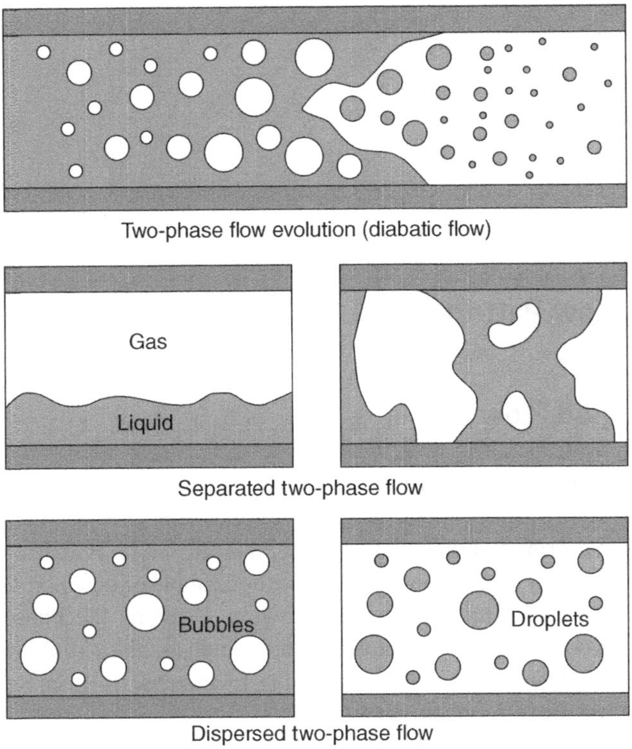

Figure 4.1 Evolution of gas phase in two-phase horizontal flow.

4.2 Dispersed versus Separated Flow Regimes

One striking and high-level characteristic of two-phase flow in pipelines arising from flow observations is that regimes could be broadly classified in either dispersed or separated flows. *Dispersed* refers to a pattern comprising many small particles (bubbles or droplets) surrounded by a connected continuous phase, as depicted in Figure 4.1 at the bottom. *Separated* two-phase flow implies distinctive streams of continuous phase, with entrained particles, interacting between them, as shown in the middle of Figure 4.1. These two-phase flow regimes could preserve their phase distribution characteristics when the heat transfer at the boundary is limited or negligible (i.e. adiabatic flow).

When heat transfer is considerable at the piping boundary (i.e. diabatic flow such as evaporation or condensation), the flow characteristic evolves along the flow path. The top picture in Figure 4.1 shows an evolving regime of two-phase flow.

Dispersed and separated flow regimes are different mechanisms driving the balance of forces at play. Dispersed flows are dominated by surface tension and drag forces acting on the particles entrained in a continuous medium. Separated flows are driven by fluids' density difference or by the fluid motion in turbulent flow (Brennen, 2005).

4.3 Vertical and Inclined Flow Maps

Multiphase upward flow in pipelines, either vertical or inclined, shows a trend toward axisymmetric patterns. Two extreme regimes that exhibit this clear trend of symmetry are the bubbly and the annular upward flow. Some slight departure from symmetry is shown by intermediate regimes such as plug-slug and churn flows. For slow liquid velocity in vertical flow, a remarkable symmetrical Taylor bubble (Taylor, 1950) rises while liquid slides on the sides of the bubble. Plug flow resembles the flow of a Taylor bubble but often shows a "skirt" end with a trail of small bubbles. Refer to Figure 2.1 for the types of vertical two-phase upflow.

Patterns such as slug and churn are intermittent in nature, meaning that there is some approximation to cycling flow where a pattern is repeated periodically. However, the prediction of the liquid slug frequency and the length of the slug is quite challenging using principles, and empirical rules are often used.

Upward flow in inclined pipelines departs from the symmetry observed in true vertical flow. Bubbles tend to move driven by buoyancy toward the upper zones of the pipe. However, it is customary to use the vertical flow pattern maps to identify regimes when the inclination is less than $\pm 10°$ with respect to the vertical direction. Figure 4.2 sketches the common patterns in inclined flow.

One of the first flow maps developed for vertical upward flow is Hewitt and Roberts (1969). Although this map was set up from small diameter tubing and air/water fluids, it is still in use to qualitatively estimate flow regime.

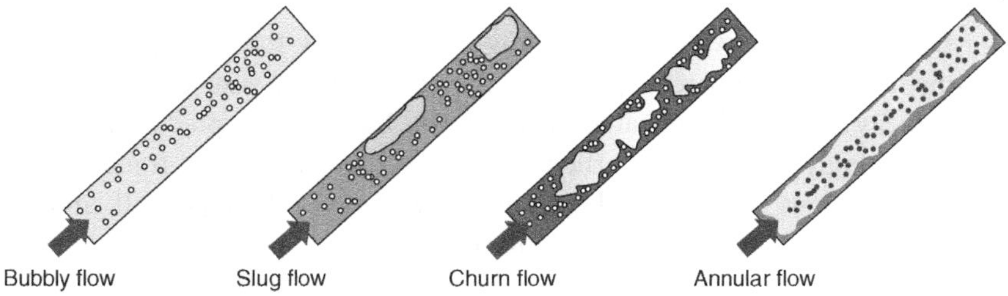

Figure 4.2 Typical flow patterns of upward-inclined flow.

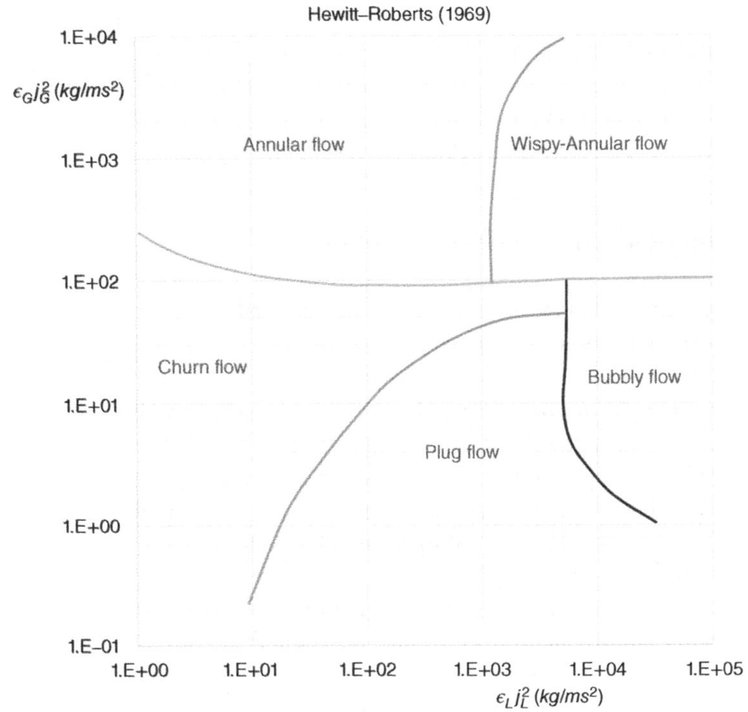

Figure 4.3 Vertical flow map. Source: Hewitt and Roberts (1969)/U.S. Department of Energy/Public domain.

The coordinates of this map represent the superficial momentum of liquid and gas. Figure 4.3 shows this flow map.

One more recent flow pattern map, with a mechanistic background, was published by Taitel et al. (1980). Figure 4.4 shows this mechanistic flow map.

4.4 Quasi-Horizontal Flow Maps

Horizontal flow regimes are greatly influenced by the gravity force. Two distinctive limits are experimentally observed. At low gas superficial velocity and high liquid superficial velocity bubbly flow predominates. At exceedingly high gas superficial velocities there is a clear trend toward annular flow, characterized by liquid flow close to the pipe wall, and gas moving faster around the pipe centerline.

For two-phase pipe flow subject to significant heat transfer rates (i.e. diabatic flow) at the boundary wall, for example, liquid evaporation in a

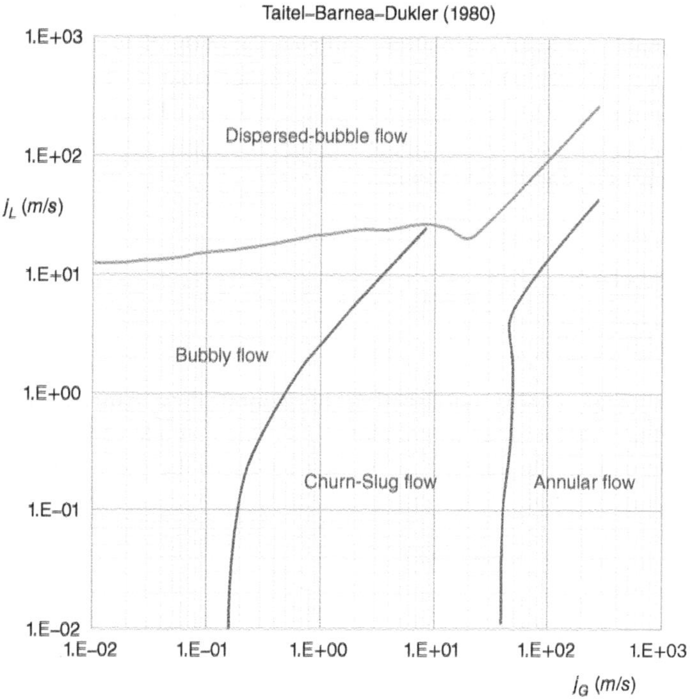

Figure 4.4 Vertical flow map. Source: Adapted from Taitel et al. (1980).

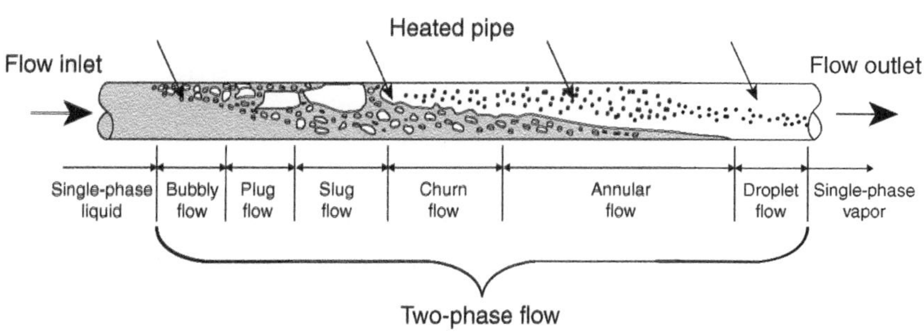

Figure 4.5 Diabatic gas–liquid horizontal flow in regime transition. Licensed under CC BY.

heat exchanger, the flow may experience several pattern transitions in short spans, as depicted in Figure 4.5.

Given the complexity of the flow dynamics when more than one phase is present, the interest of researchers in classifying the flow regimes began in the 1940s. Baker (1954) published the first flow pattern map based on field-scale oil and gas pipeline observations. Although this map is in disuse, it is still amenable for hand-calculation confirmation of the flow regime in small-size piping (≈ 50 mm). The coordinates for Baker's map are the nondimensional parameters λ and ψ based on fluid properties (see Figure 4.6). The parameters are calculated with the following equations and with the water and air properties at standard conditions:

$$\lambda \equiv \sqrt{\frac{\rho_g}{\rho_a}\frac{\rho_l}{\rho_w}} \quad (4.1)$$

$$\psi \equiv \left(\frac{\sigma_w}{\sigma_l}\right)\left[\frac{\mu_l}{\mu_w}\left(\frac{\rho_w}{\rho_l}\right)^2\right]^{1/3} \quad (4.2)$$

Mandhane et al. (1974) published a horizontal flow pattern map (Figure 4.7) that has found great popularity due to its simple coordinates (superficial velocity). The authors based their study on the most comprehensive database available at the time, the AGA-API Two-Phase Flow Data

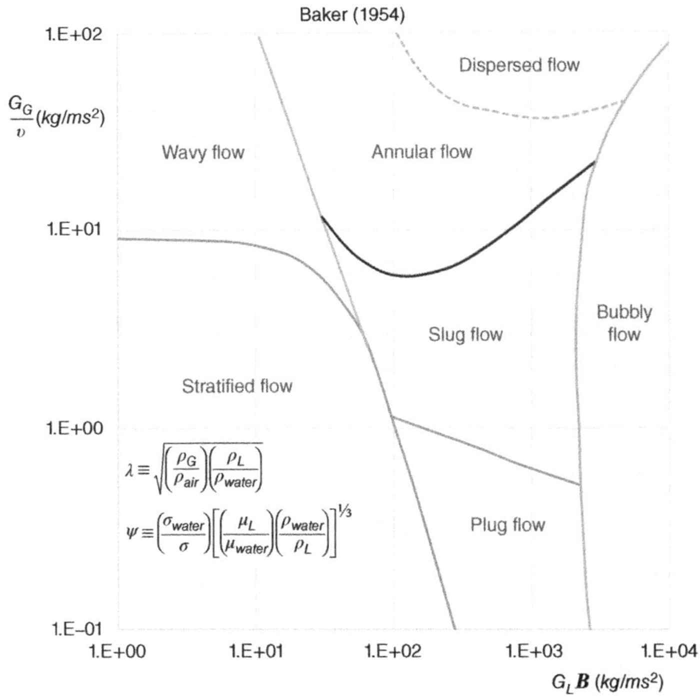

Figure 4.6 Flow pattern map for horizontal pipeline. Source: Adapted from Baker (1954).

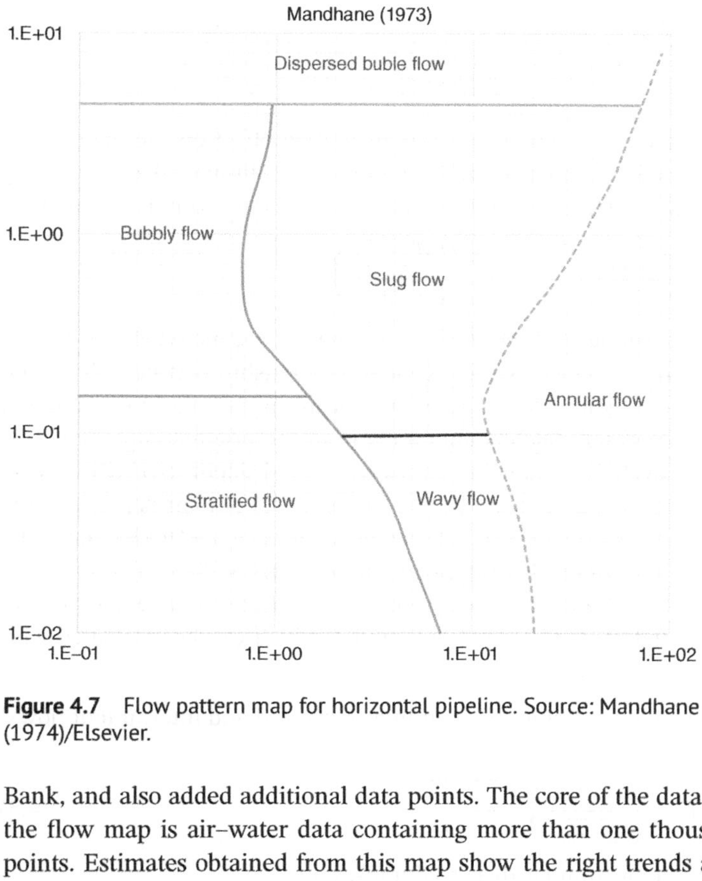

Figure 4.7 Flow pattern map for horizontal pipeline. Source: Mandhane et al. (1974)/Elsevier.

Bank, and also added additional data points. The core of the data used for the flow map is air–water data containing more than one thousand test points. Estimates obtained from this map show the right trends although they are biased toward fluids with properties similar to air and water.

The first mechanistic pattern map for semi-horizontal flow was proposed by Taitel and Dukler (1976), better known as a *T–D* map. The regime transitions of this map, based on physical phenomena principles, provide a solid foundation not found in previous pattern maps empirically derived. Obviously, the map transitions could not be fully developed on principles, but additional experimental data permitted to close the models.

The unifying criterion in this map is that all regimes are converted into an equivalent separated flow model, that is, stratified two-phase flow. The one-dimensional equations of mass and momentum conservation yield a set of algebraic equations in terms of the nondimensional parameters X, Y, and the liquid non-dimensional depth h_L/d. The parameter X is the square root of the Lockhart–Martinelli parameter (Shoham, 2006). The parameter X is calculated from equation (4.3):

$$X \equiv \left(\frac{1-x}{x}\right)^{0.9} \left(\frac{\mu_L}{\mu_G}\right)^{0.1} \left(\frac{\rho_G}{\rho_L}\right)^{0.5} \tag{4.3}$$

The parameter Y accounts for the gravitation effect on the momentum equation, and it is estimated from equation (4.4). This model applies to any pipe inclination θ (i.e. the angle with respect to the horizontal direction).

$$Y = 10.87(\rho_L - \rho_G) g \sin\theta \left(\frac{d^{0.8}\mu_G^{0.2}}{j_G^{1.8}\rho_G^{1.2}}\right) \tag{4.4}$$

Taitel and Dukler (1976), in their original paper, included the nonlinear equation (7), which is the algebraic relationship between X, Y, and the equivalent liquid depth h_L/d. All terms in the algebraic equation are nondimensional. Such equation (7) from the original paper could be solved iteratively to obtain the equivalent liquid depth h_L/d. Otherwise, the authors offer a graphical solution in Figure (2) of their paper with the same objective. Figure (2) was subsequently corrected by Dukler and Taitel (1982) with a factor of 840 multiplying the Y values of Figure (2).

The regimes' transitions, according to the Taitel and Dukler (1976) model, depend on the parameters F and T which can be calculated with equations (4.5) and (4.6).

- *Stratified to non-stratified* (for annular-dispersed and intermittent flows) transition,

$$F = \frac{j_G}{\sqrt{gd\cos\theta}} \sqrt{\frac{\rho_G}{\rho_L - \rho_G}} \tag{4.5}$$

- Intermittent to annular-dispersed transition,

$$T = \sqrt{\frac{2f_{L,S}\rho_L j_L^2}{(\rho_L - \rho_G) gd\cos\theta}} \tag{4.6}$$

Using the parameters X, T, and F, it is possible to identify the flow pattern for gas–liquid flow using the map shown in Figure 4.8.

The void fraction α (or holdup H_L) can be determined by solving equations (4.7) and (4.8). The subtended angle φ corresponds to the equivalent stratified flow regime and defines the interface gas–liquid.

$$\varphi = 2\cos^{-1}\left(\frac{2h_L}{d} - 1\right) \tag{4.7}$$

$$\alpha = \frac{\varphi - \sin\varphi}{2\pi} \tag{4.8}$$

The transition between annular dispersed and dispersed bubble in the map was assumed to occur at $h_L/d \approx 0.35$ ($X \approx 1.61$).

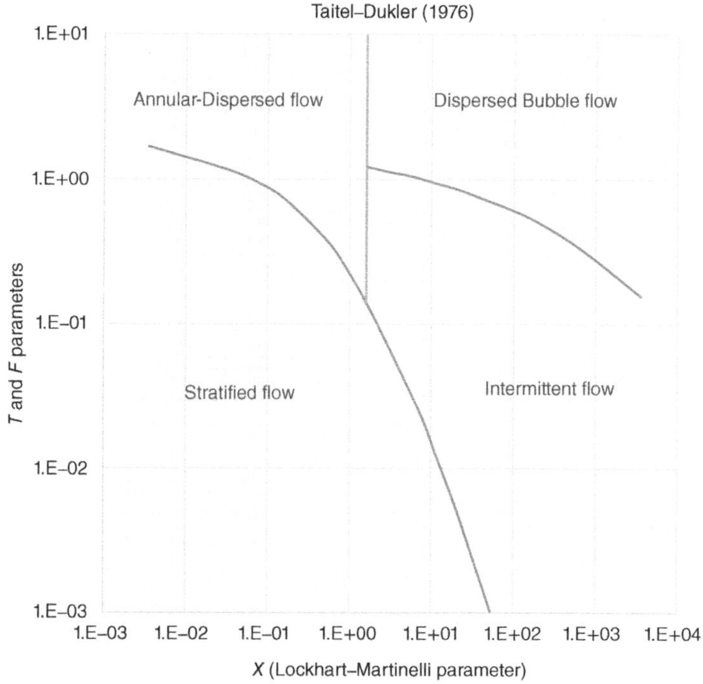

Figure 4.8 Flow pattern map for horizontal pipeline. Source: Adapted from Taitel and Dukler (1976).

4.5 Liquid–Liquid Flow Map

Two-phase liquid–liquid flow has a propensity to form droplets at the interface of stratified flow patterns (Sunder Raj et al., 2005) of immiscible fluids. In two-phase gas–liquid flows, the interface evolves into a larger bubble or plug, or in a developed slug. This performance difference between liquid–liquid and gas–liquid flows prevents the extension of the latter flow pattern maps.

The literature available concerning flow regimes in liquid–liquid flow is limited. A classic and pioneering reference in this area is Govier and Aziz (1972). Presently, such maps have a visualization value rather than being a regular tool supporting mechanistic models for momentum and energy balances. Ibarra et al. (2014) performed experiments and identified nine possible flow patterns in liquid–liquid flow, namely stratified, stratified wavy, dispersions of oil-in-water and water-in-oil with/without continuous layer, dual continuous, intermittent, and annular flow. Sunder Raj et al. (2005) observed regime transitions by increasing the superficial velocity of

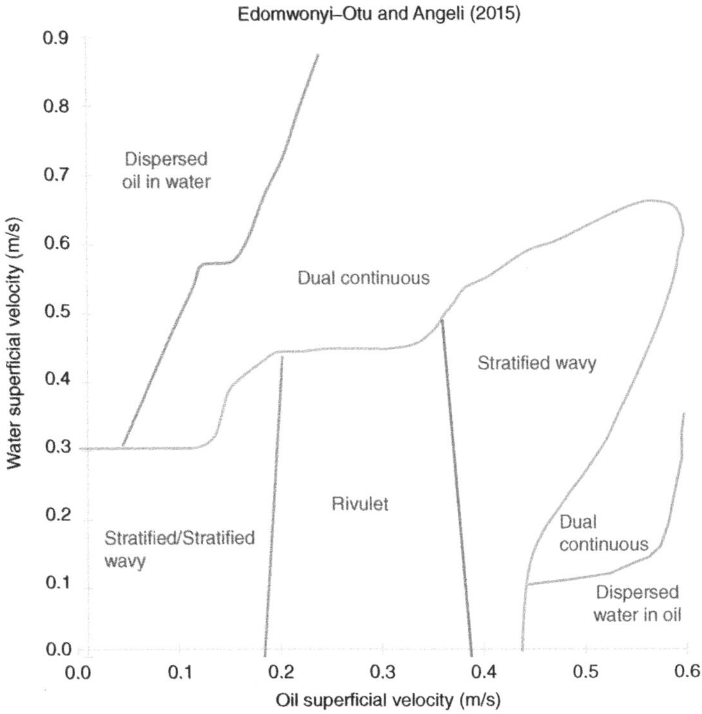

Figure 4.9 Liquid–liquid flow map. Source: Edomwonyi-Otu and Angeli (2015)/Elsevier.

the lighter kerosene liquid. Progressively, the flow pattern changed from smooth-stratified flow, water at the bottom, to stratified wavy. Further increment of the flow rate of kerosene led to form fine drops of water dispersed in the top layer.

Three-layer pattern would form by further increments in the superficial velocities, with a combination of droplets of both phases at the interface. More recently, Edomwonyi-Out and Angeli (2015) published a flow pattern map for the horizontal flow of oil and water based on the superficial velocities of the phases. This plot is shown in Figure 4.9.

4.6 Three-Phase Flow

Three-phase flow in a horizontal pipe is sensitive to the magnitude of the superficial velocities of the phases. Smooth-stratified flow pattern is observed for low superficial liquid velocities and low gas velocities. For increased liquid velocity, there is a probability that the gas becomes

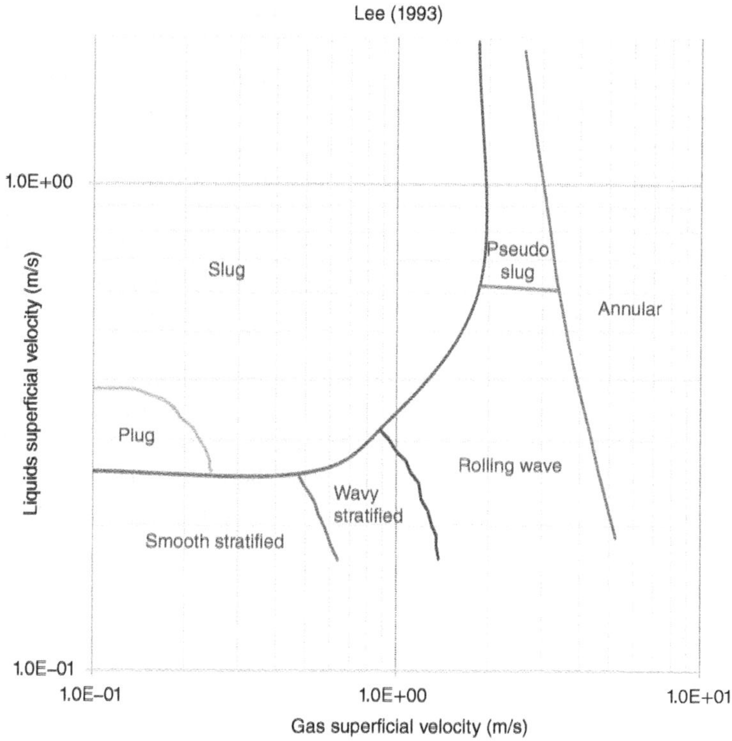

Figure 4.10 Flow pattern map for three-phase flow. Source: Adapted from Lee and Lee (1993).

entrained within the liquid phases, as well as bubbles may flow along the pipe's top. Interface instabilities develop with further increments in liquid flow and slug flow may set in. In general, slug flow becomes the dominating pattern at high liquids' superficial velocities.

There are few three-phase flow patterns available due to the richness of flow regimes (and the means to identify and define the regime transitions). One of these flow maps is by Lee and Lee (1993) herein presented in Figure 4.10.

4.7 Liquid–Solid Flow

There is an extensive list of liquid–solid literature driven by the interest in slurry flow through pipelines. Comprehensive monographs include Nayyar (2000), Crowe (2006), Cheremisinoff (1986), and Govier and Aziz (1972). There are several applications of slurry transport such as minerals in water flows, industrial paints, coal–oil mixtures, and even some rocket fuel applications.

The distinction of liquid–solid slurries is the particle size, which in the case of fine elements enables the particles' suspension in the continuous phase and lessens the settling tendency. In this regard, turbulence level contributes to enhance particle drag while promoting a trend of axisymmetric particle concentration profile. However, liquid flow rate reduction leads to particle settling toward the bottom of horizontal pipelines, named asymmetric suspension flow pattern. The asymmetric flow pattern conduces to particle bouncing off from formed "dunes", agglomerating settled particles.

The complexity of liquid–solid flows cannot be captured in a single-flow pattern map, as it is approximated in the case of gas–liquid flow. Doron and Barnea (1996) demonstrated that flow pattern maps at least depend on the particles' relative density to the carrier liquid, on the particles' average size, on the pipe diameter, and on the concentration of particles.

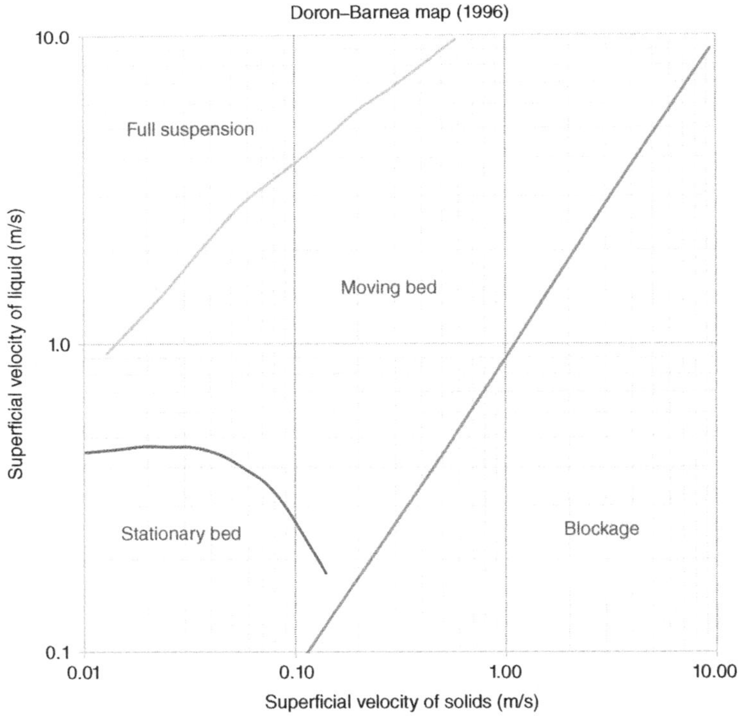

Figure 4.11 Solid–liquid flow map. Source: Adapted from Doron and Barnea (1996).

Concisely, several maps may be required to identify a dominating flow pattern. An example of a flow pattern map describing typical flow regimes is shown in Figure 4.11 taken from Doron and Barnea (1996). The solid particles were 3 mm in diameter and had a density of 1240 kg/m^3, flowing in a water stream.

4.8 Gas–Solid Flow

The flow of solids in gas pipeline streams shares several characteristics with the liquid–solid flows as well. The main difference is the higher gas velocities, and turbulence, required to keep particles suspended in the carrier stream. This holds true because in pneumatic solid transport the relative density between solids and gas is larger than in solid–liquid flows, causing a diminished buoyancy effect. Additionally, drag force of gas carrier upon particles is proportional to its density, which involves larger gas velocity to entrain the solid particles.

As for liquid–solid flows, it is difficult to cast the potential flow patterns of gas–solid flows into a single map. Similar comments apply to the present case of gas–solid flows as in the previous section. A sketch illustrating typical flow patterns encountered in pneumatic solid transport is shown in Figure 4.12.

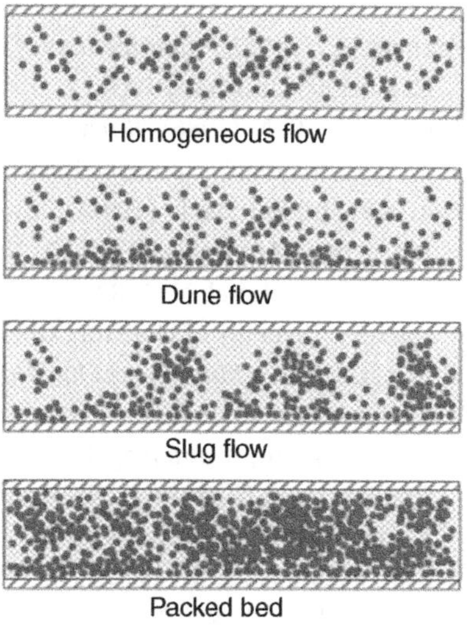

Figure 4.12 Typical gas–solid flow regimes in horizontal flow.

4.9 Conclusions

This chapter summarizes some visualization tools to identify typical multiphase flow structures, that is, flow patterns or regimes, based on simple coordinates related to flow rate or superficial velocity. It is not always possible to develop a single map, as for solid–liquid and solid–gas flows, due to the number of governing parameters that hinder a simple approach. The case of gas–liquid flows has been extensively studied, and there is more clarity as to the possibility of casting several flow regimes onto a 2-D map. However, the factor underpinning the transition among flow regimes is the stability criteria to balance dynamical forces. One of these criteria, if not the most important of all, is the Kelvin–Helmholtz instability conducive to the formation/amplification of interfacial waves. Another important criterion is the balance between surface tension forces and turbulent stresses that determine the size and the number of droplets and bubbles formed at shear layers of the continuous phase.

The value of a representative flow pattern maps cannot be underrated because of the associated issues of idealistic sharp transitions and uncertainty in defining a specific regime; flow pattern maps do help develop valuable mechanistic models, applicable to specific regimes and with better chance of simulating the physics of the flow.

References

Baker, O. (1954). Simultaneous Flow of Oil and Gas. *Oil Gas J.* 53: 185–195.

Brennen, C. (2005). *Fundamentals of Multiphase Flow*. Cambridge, U.K.: Cambridge University Press.

Cheremisinoff, N. (1986). *"Encyclopedia of Fluid Mechanics," Vol. 5, "Slurry Flow Technology"*. Houston, Texas: Gulf Publishing Company.

Doron, P. and Barnea, D. (1996). Flow Pattern Map for Solid-Liquid Flow in Pipes. *Int. J. Multiphase Flow* 22 (2): 273–283.

Dukler, A. and Taitel, Y. (1982). *Lecture Notes of the Two-Phase Gas-Liquid Flow course*. University of Houston, Chemical Engineering Department.

Crowe, C. (2006). *Multiphase Flow Handbook*. Boca Raton, FL, USA: CRC Taylor & Francis.

Edomwonyi-Otu, L. and Angeli, P. (2015). Pressure Drop and Holdup Predictions in Horizontal Oil-Water Flows for Curved and Wavy Interfaces. *Chem. Eng. Res. Des.* 93: 55–65.

Govier, G. and Aziz, K. (1972). *The Flow of Complex Mixtures in Pipes*. New York, USA: Van Nostrand Reinhold Company.

Hewitt, G. and Roberts, D. (1969). "Studies of Two-Phase Flow Patterns," AERE-M2159.

Ibarra, R., Markides, C., and Matar, O. (2014). A Review of Liquid-Liquid Flow Patterns in Horizontal and Slightly Inclined Pipes. *Multiph. Sci. Technol.* 26 (3): 171–198.

Lee, A. and Lee, A.H. (1993). *Study of Flow Regime transitions for Oil-Water-Gas Mixture in Large Diameter Horizontal Pipelines*, MSc thesis. College of Engineering and Technology, Ohio University.

Mandhane, J. et al. (1974). A Flow Pattern Map for Gas-Liquid Flow in Horizontal Pipes. *Int. J. Multiphase Flow* 1: 537–553.

Nayyar, M. (2000). *Piping Handbook*, 7e. New Yor, USA: McGraw-Hill.

Shoham, O. (2006). *Mechanistic Modeling of Gas-Liquid Two-Phase Flow in Pipes*. Richardson, Texas, USA: Society of Petroleum Engineers.

Sunder Raj, T., Chakrabarti, D., and Das, G. (2005). Liquid-Liquid Stratified Flow Through Horizontal Conduits. *Chem. Eng. Technol.* 28 (8): 899–907.

Taitel, Y., Barnea, D., and Dukler, A.E. (1980). Modeling Flow Pattern Transition for Steady Upward Gas-Liquid Flow in Vertical Tubes. *AICHE J.* 26 (3): 345–354.

Taylor, G.I. (1950). The Mechanics of Large Bubbles Rising Through Extended Liquids and Through Liquids in Tubes. *Proc. R. Soc. A* CC: 375–390.

Taitel, Y. and Dukler, A. (1976). A Model of Predicting Flow Regime Transitions in Horizontal and Near Horizontal Gas-Liquid Flow. *AICHE J.* 22: 47–55.

5

Two-Phase Flow in Pipelines

5.1 Introduction

The conservation equations for two-phase flow are covered in this chapter, followed by the boundary and initial conditions necessary to close the system of differential equations. Additional constitutive equations and empirical (field practitioners) or semi-empirical (lab researchers) models can make the system of equations tractable. Purely empirical correlations derive from field measurements and some nondimensional post-processing analysis. Therefore, such empiricism may only succeed when the application operating conditions match the conditions under which the original data were gathered.

The energy conservation equation plays a vital role, jointly with the mass and momentum equations, to link fluid properties, temperature changes, and heat transfer to the surrounding environment. The coverage of this key principle is included in one section of this chapter.

The overarching goal for this chapter is to thoroughly cover the fundamentals of the governing thermal-hydraulic models expressing the changes experienced by a multiphase flow mixture through pipelines.

5.2 Conservation Principles

The foundations of the physical principles applied to multiphase flow were cast in the publications of Delhaye (1968, 1981). Delhaye's work sets the framework for the *local instantaneous equations* addressed in Section 5.2.1. However, the intricacy of the resulting set of equations demands approximations inspired by the experimental work and the empirical data gathered

Multiphase Transport of Hydrocarbons in Pipes, First Edition.
Juan J. Manzano-Ruiz and Jose G. Carballo.

Companion Website: www.wiley.com/go/Manzano-Ruiz/Multiphasetransport

during the past 80 years. The approximations include *space-averaged* and/or *time-averaged*, leading to simplified *composite-averaged* equations. The popular *one-* and *two-dimensional models* implemented in commercial transient codes nowadays become the simulation standard to design and monitor the thermal-hydraulics of hydrocarbons through production gathering systems.

Aside from the space- and time-averaged models another useful approach is to consider two limiting topologies, the disperse and the separated topologies (Brennen, 2005), embracing the variety of spatial distribution flow patterns. A dispersed topology may be either bubbles entrained within a continuum liquid medium or droplets distributed within a gas phase. A separated topology comprises two or more continuum media with distinctive interface(s).

5.2.1 Local Instantaneous Equations

The conservation principles in this section are presented in their *integral form*, applied to a fixed control volume encompassing both the liquid and the gas phases. The instantaneous equations in integral form rely on the identification of the fraction of control volume occupied by each phase (V_g,V_l) and the fraction of the control surface (A_g,A_l) per phase. Figure 5.1 illustrates the mass conservation in integral form, equation (5.1), for a control volume.

$$Mas: \quad \frac{d}{dt}\left[\int_{V_g} \rho_g dV + \int_{V_l} \rho_l dV\right] + \int_{A_g} \rho_g(\vec{u}_g \circ \vec{n}_g)dA + \int_{A_l} \rho_l(\vec{u}_l \circ \vec{n}_l)dA = 0 \tag{5.1}$$

In the momentum conservation principle, equation (5.2), the acting surface forces are represented by the stress tensor (surface pressure and shear stresses), and the volume force is represented by the gravity force.

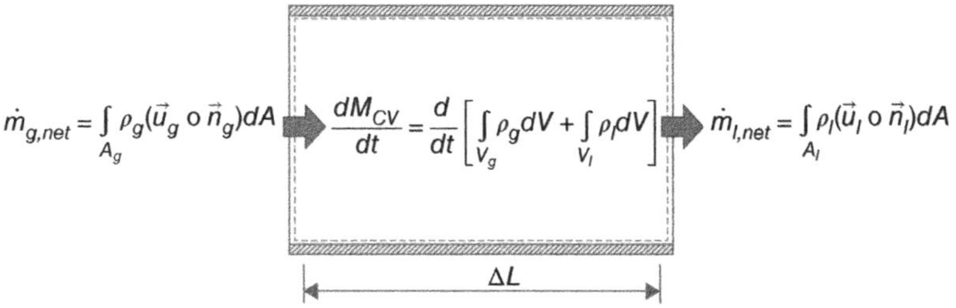

Figure 5.1 Mass conservation equation for a control volume.

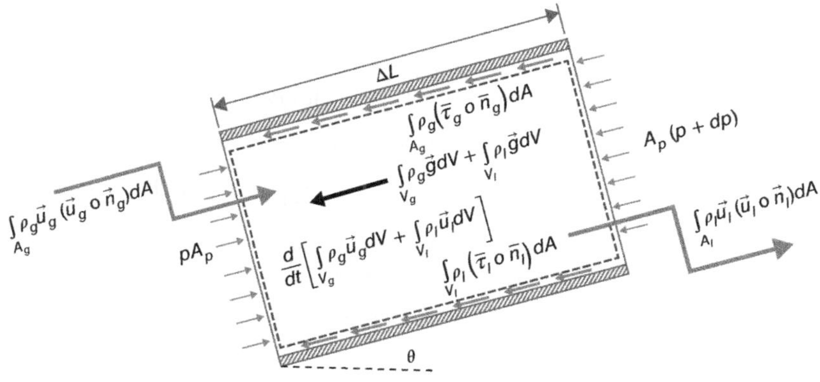

Figure 5.2 Momentum balance on a control volume element.

The summation of the acting forces and the net flux of momentum through the control surface shall be equal to the net change in fluid momentum of both phases within the control volume. The overall balance is shown in Figure 5.2.

$$MM:\ \frac{d}{dt}\left[\int_{V_g}\rho_g\vec{u}_g dV + \int_{V_l}\rho_l\vec{u}_l dV\right] + \int_{A_g}\rho_g\vec{u}_g(\vec{u}_g \circ \vec{n}_g)dA + \int_{A_l}\rho_l\vec{u}_l(\vec{u}_l \circ \vec{n}_l)dA$$
$$= \int_{A_g}\rho_g(\vec{\tau}_g \circ \vec{n}_g)dA + \int_{A_l}\rho_l(\vec{\tau}_l \circ \vec{n}_l)dA + \int_{V_g}\rho_g\vec{g}dV + \int_{V_l}\rho_l\vec{g}dV \qquad (5.2)$$

where the stress tensor has nine components as shown in equation (5.3) for cylindrical coordinates:

$$\vec{\tau} \equiv \begin{pmatrix} -p + \tau_{rr} & \tau_{rz} & \tau_{r\theta} \\ \tau_{zr} & -p + \tau_{zz} & \tau_{z\theta} \\ \tau_{\theta r} & \tau_{\theta z} & -p + \tau_{\theta\theta} \end{pmatrix} \qquad (5.3)$$

The stress tensor for the gas phase and the liquid phase should be determined and introduced into equation (5.2). The shear stress components of the stress tensor relate to the viscous effects through Stokes equations (Brodkey, 1967).

The energy conservation principle in its integral form, equation (5.4), involves work terms associated with gravitational force and shear stress on

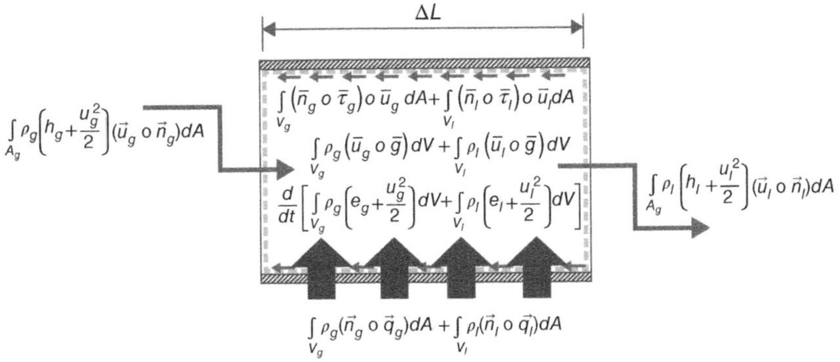

Figure 5.3 Conservation of total energy for a control volume.

the control surface. There are also heat flux terms across the control surface incrementing the total energy inside the control volume, and there is a flow of enthalpy per phase through the control surface. The summation of the previous term's balances with the change of internal and kinetic energy inside the control volume. Such balance is sketched in Figure 5.3.

$$
\begin{aligned}
: \frac{d}{dt}&\left[\int_{V_g} \rho_g\left(e_g + \frac{u_g^2}{2}\right) dV + \int_{V_l} \rho_l\left(e_l + \frac{u_l^2}{2}\right) dV\right] \\
&= \int_{A_g} \rho_g\left(H_g + \frac{u_g^2}{2}\right)(\vec{u}_g \circ \vec{n}_g) dA + \int_{A_l} \rho_l\left(H_l + \frac{u_l^2}{2}\right)(\vec{u}_l \circ \vec{n}_l) dA \\
&+ \int_{V_g} \rho_g(\vec{u}_g \circ \vec{g}) dV + \int_{V_l} \rho_l(\vec{u}_l \circ \vec{g}) dV \\
&+ \int_{V_g} (\vec{n}_g \circ \vec{\tau}_g) \circ \vec{u}_g dA + \int_{V_l} (\vec{n}_l \circ \vec{\tau}_l) \circ \vec{u}_l dA \\
&+ \int_{V_g} \rho_g(\vec{n}_g \circ \vec{q}_g) dA + \int_{V_l} \rho_l(\vec{n}_l \circ \vec{q}_l) dA
\end{aligned} \tag{5.4}
$$

The difficulty of the integral-form approach is accounting for the time and space variations of the *surfaces of discontinuity*, that is, phase interfaces. Such surfaces of discontinuity require additional *jump conditions* to connect the properties of the two phases. Extensive coverage of appropriate jump conditions can be found in Delhaye's (1981) book.

The integral balance principles can be transformed into partial differential equations utilizing the Leibniz rule and the Gauss theorem as follows for each phase:

- Mass conservation (continuity equation)

$$\frac{\partial \rho_g}{\partial t} + \nabla \circ (\rho_g \vec{u}_g) = 0 \tag{5.5}$$

$$\frac{\partial \rho_l}{\partial t} + \nabla \circ (\rho_l \vec{u}_l) = 0 \tag{5.6}$$

- Momentum conservation

$$\frac{\partial \rho_g \vec{u}_g}{\partial t} + \vec{u}_g[\nabla \circ (\rho_g \vec{u}_g)] = \rho_g \vec{g} + \nabla \circ \vec{\tau}_g \tag{5.7}$$

$$\frac{\partial \rho_l \vec{u}_l}{\partial t} + \vec{u}_l[\nabla \circ (\rho_l \vec{u}_l)] = \rho_l \vec{g} + \nabla \circ \vec{\tau}_l \tag{5.8}$$

- Energy conservation

$$\frac{\partial}{\partial t}\left[\rho_g\left(e_g + \frac{u_g^2}{2}\right)\right] + \nabla \circ \left[\rho_g \vec{u}_g\left(h_g + \frac{u_g^2}{2}\right)\right]$$
$$= \rho_g \vec{g} \circ \vec{u}_g + \nabla \circ (\vec{\tau}_g \circ \vec{u}_g) + \nabla \circ \vec{q}_g \tag{5.9}$$

Equations (5.5)–(5.9) are the rigorous backbone of the multiphase flow modeling through pipelines.

5.3 Space-Averaged Equations

The application of equations from the previous section is most challenging due to the space distribution of the phases. The space-averaging process on pipeline geometries rests on the integration of the governing PDEs at cross-sections. Average gas and liquid mass flow, velocities, momentum, and energy represent a compromise between accuracy and simplicity.

A rigorous treatment of the space-averaging processes can be found in Delhaye (1982), supported by the Gauss theorems and the Leibniz rule. The space-averaging technique involves splitting the pipeline cross-section into gas area and liquid area, separated by the interface boundary $C(z,t)$. The *phase cross-sectional averaging operator* is denoted with angle brackets, as defined by equation (5.10) for a generic property "f,"

$$\langle f_g \rangle_g \equiv \frac{1}{A_g} \int_{A_g(z,t)} f_g(r,z,\theta,t)dA$$

$$\langle f_l \rangle_l \equiv \frac{1}{A_l} \int_{A_l(z,t)} f_l(r,z,\theta,t)dA \tag{5.10}$$

The space-averaging technique yields the following set of equations that include integrals that have to be evaluated along the interface gas–liquid, the interface gas–pipe, and the interface liquid–pipe:

- **Mass conservation of gas**

$$\frac{\partial}{\partial t}[A_g\langle\rho_g\rangle] + \frac{\partial}{\partial z}[A_g\langle\rho_g u_g\rangle] = \Gamma_g \tag{5.11}$$

$$\Gamma_g = \text{volumetric mass transfer rate into the gas phase}$$
$$= -\int_{C(z,t)} \frac{\rho_g(\vec{u}_g - \vec{u}_i)\circ\vec{n}_g dC}{\vec{n}_g\circ\vec{n}_{g,C}} \tag{5.12}$$

- **Mass conservation of liquid**

$$\frac{\partial}{\partial t}[A_l\langle\rho_l\rangle] + \frac{\partial}{\partial z}[A_l\langle\rho_l u_l\rangle] = \Gamma_l \tag{5.13}$$

$$\Gamma_l = \text{volumetric mass transfer rate into the liquid phase}$$
$$= -\int_{C(z,t)} \frac{\rho_l(\vec{u}_l - \vec{u}_i)\circ\vec{n}_l dC}{\vec{n}_l\circ\vec{n}_{l,C}} \tag{5.14}$$

and $\Gamma_g + \Gamma_l = 0$ is a boundary condition at the gas–liquid interface, also referred to as a "jump condition."

- Gas-phase momentum

$$\begin{aligned}
&\frac{\partial}{\partial t}[A_g\langle\rho_g\vec{u}_g\rangle] + \frac{\partial}{\partial z}[A_g\langle\rho_g u_g\vec{u}_g\rangle] \\
&= A_g\langle\rho_g\vec{g}\rangle - \frac{\partial}{\partial z}[A_g\langle p_g\vec{n}_z\rangle] + \frac{\partial}{\partial z}[A_g\langle\vec{n}_z\circ\vec{\tau}_g\rangle] \\
&\quad - \int_{C(z,t)} \frac{[\rho_g\vec{u}_g(\vec{u}_g - \vec{u}_i)\circ\vec{n}_g - \vec{n}_g\circ\vec{\tau}_g]dC}{\vec{n}_g\circ\vec{n}_{g,C}} + \int_{C(z,t)} \frac{\vec{n}_g\circ\vec{\tau}_g dC}{\vec{n}_g\circ\vec{n}_{g,C}}
\end{aligned} \tag{5.15}$$

- Liquid-phase momentum

$$\begin{aligned}
&\frac{\partial}{\partial t}[A_l\langle\rho_l\vec{u}_l\rangle] + \frac{\partial}{\partial z}[A_l\langle\rho_l u_l\vec{u}_l\rangle] \\
&= A_l\langle\rho_l\vec{g}\rangle - \frac{\partial}{\partial z}[A_l\langle p_l\vec{n}_z\rangle] + \frac{\partial}{\partial z}[A_l\langle\vec{n}_z\circ\vec{\tau}_l\rangle] \\
&\quad - \int_{C(z,t)} \frac{[\rho_l\vec{u}_l(\vec{u}_l - \vec{u}_i)\circ\vec{n}_l - \vec{n}_l\circ\vec{\tau}_l]dC}{\vec{n}_l\circ\vec{n}_{l,C}} + \int_{C(z,t)} \frac{\vec{n}_l\circ\vec{\tau}_l dC}{\vec{n}_l\circ\vec{n}_{l,C}}
\end{aligned} \tag{5.16}$$

- Gas-phase energy conservation

$$\frac{\partial}{\partial t}\left[A_g\left\langle\rho_g\left(e_g+\frac{u_g^2}{2}\right)\right\rangle\right]+\frac{\partial}{\partial z}\left[A_g\left\langle\rho_g u_g\left(h_g+\frac{u_g^2}{2}\right)\right\rangle\right]$$
$$=A_g\langle\rho_g\vec{g}\circ\vec{u}_g\rangle+\frac{\partial}{\partial z}[A_g\langle(\vec{\tau}_g\circ\vec{u}_g)\circ\vec{n}_z\rangle]-\frac{\partial}{\partial z}[A_g\langle\vec{q}_g\circ\vec{n}_z\rangle]$$
$$-\int_{C(z,t)}\frac{\left[\rho_g(\vec{u}_g-\vec{u}_i)\circ\vec{n}_g\left(e_g+\frac{u_g^2}{2}\right)-(\vec{\tau}_g\circ\vec{u}_g)\circ\vec{n}_g+\vec{q}_g\circ\vec{n}_g\right]dC}{\vec{n}_g\circ\vec{n}_{g,C}}$$
$$+\int_{C_g(z,t)}\frac{\vec{q}_g\circ\vec{n}_g dC}{\vec{n}_g\circ\vec{n}_{g,C}} \tag{5.17}$$

- Liquid-phase energy conservation

$$\frac{\partial}{\partial t}\left[A_l\left\langle\rho_l\left(e_l+\frac{u_l^2}{2}\right)\right\rangle\right]+\frac{\partial}{\partial z}\left[A_l\left\langle\rho_l u_l\left(h_l+\frac{u_l^2}{2}\right)\right\rangle\right]$$
$$=A_l\langle\rho_l\vec{g}\circ\vec{u}_l\rangle+\frac{\partial}{\partial z}[A_l\langle(\vec{\tau}_l\circ\vec{u}_l)\circ\vec{n}_z\rangle]-\frac{\partial}{\partial z}[A_l\langle\vec{q}_l\circ\vec{n}_z\rangle]$$
$$-\int_{C(z,t)}\frac{\left[\rho_l(\vec{u}_l-\vec{u}_i)\circ\vec{n}_l\left(e_l+\frac{u_l^2}{2}\right)-(\vec{\tau}_l\circ\vec{u}_l)\circ\vec{n}_l+\vec{q}_l\circ\vec{n}_l\right]dC}{\vec{n}_l\circ\vec{n}_{l,C}}$$
$$+\int_{C_g(z,t)}\frac{\vec{q}_l\circ\vec{n}_l dC}{\vec{n}_l\circ\vec{n}_{l,C}} \tag{5.18}$$

Evidently, the set of space-averaged equations (5.11)–(5.18) still poses intractable solution challenges related to evaluating the interface integrals, that is, unknown location of the boundaries.

5.4 Time-Averaged Equations

Performing the time averaging of the conservation principles yields additional simplification benefits based on statistics. The *phase time averaging operator* is denoted with an overbar, as defined by equation (5.19) for a generic property "*f*" (Yadigaroglu and Hewitt, 2017),

$$\overline{f_g}\equiv\frac{1}{T_g}\int_{[T_g]}f_g(r,z,\theta,t)dt$$
$$\overline{f_l}\equiv\frac{1}{T_l}\int_{[T_l]}f_l(r,z,\theta,t)dt \tag{5.19}$$

The time integration intervals $[T]$ should be selected in a way that unsteady trends are preserved while eliminating the turbulence fluctuations by the integration process (Yadigaroglu and Hewitt, 2017).

Of particular interest is the time averaging of the *local instantaneous void fraction* $\varepsilon_g(r,z,\theta,t)$, representing the presence of the gas phase at the location at time "*t*." This is characterized by the *gas-phase density* or *gas-presence* function $P_g(r,z,\theta,t)$ defined by

$$P_g(\vec{r},t) \equiv \begin{Bmatrix} 1 \text{ if } \vec{r} \text{ is in the gas phase} \\ 0 \text{ if } \vec{r} \text{ is in the liquid phase} \end{Bmatrix} \tag{5.20}$$

Utilizing definition (5.19) and the Ergodic hypothesis from Statistics allows us to express the local time-averaged void fraction as follows, where the last term represents the fraction of experimental observations of the gas phase at such location:

$$\overline{\varepsilon_g(\vec{r})} = \frac{1}{T}\int_{[T_g]} \varepsilon_g(\vec{r},t)dt = \overline{P_g(\vec{r})} = \frac{T_g}{T} \tag{5.21}$$

Similar time-averaged properties can be derived, analogous to the derivation of the local time-averaged void fraction, to obtain time-smoothed properties for the system of equations (5.5)–(5.9). It is important to note the assumption of considering the phases' interface as discontinuous changes in density (i.e. negligible interphase thickness).

The local time-averaged void fraction can be space-averaged over the cross-sections of the pipelines to estimate the more familiar 1-D void fraction (α) parameter. This illustrates the core of the next section to reduce even further the conservation principles for two-phase flow. The cross-sectional void fraction expression is

$$\alpha = \langle\overline{\varepsilon_g}\rangle = \frac{1}{A}\int_A \overline{\varepsilon_g(\vec{r})}dA = \frac{1}{A}\int_A \frac{1}{T}\int_A P_g(\vec{r},t)dtdA \tag{5.22}$$

5.5 Composite-Averaged Equations

From the previous two sections, it is obvious the advantage of performing time and space averages. In this section, we explore the simultaneous application of both averaging techniques that are conducive to the standard 1-D models in two-phase flow. It is immaterial the order of the two averaging processes (Delhaye, 1982) because both lead to the same set of equations.

One crucial element in performing the averaging processes is the following fundamental relationship reproduced in the Yadigaroglu and Hewitt (2017) monograph, *only* applicable to the product of a property times the void fraction:

$$\langle \overline{\varepsilon_g f_g} \rangle \cong \langle \overline{\varepsilon_g} \rangle \langle \overline{f_g} \rangle \tag{5.23}$$

The resulting set of composite-averaged equations is summarized in Section 1.2.4.3 of Delhaye's (1982) monograph.

5.6 Two-Fluid Model (TFM)

The two-fluid model is a one-dimensional methodology where the properties vary only along the pipeline axis. This model is a simplified application of the composite-averaged equations for a pipeline geometry. As such, the model considers the separate flow of the gas and the liquid phases with interacting boundaries. This type of model blends well with separate types of flow patterns such as annular flow or stratified horizontal flow.

The TFM comprises six conservation equations requiring closure models to match the number of unknowns. The set of conservation principles is as follows:

$$\textit{Continuity (gas phase)}: \quad \frac{\partial}{\partial t}[\rho_g \alpha] + \frac{\partial}{\partial L}[Gx] = \Gamma_g \tag{5.24}$$

$$\textit{Continuity (liquid phase)}: \quad \frac{\partial}{\partial t}[\rho_l(1-\alpha)] + \frac{\partial}{\partial L}[G(1-x)] = \Gamma_l \tag{5.25}$$

$$\textit{Jump condition}: \quad \Gamma_l + \Gamma_g = 0 \tag{5.26}$$

The variables Γ_g and Γ_l are the volumetric mass transfer rates defined by equations (5.24) and (5.25). A positive value of Γ_l means gas condensing into liquid phase, and conversely a positive value of Γ_g means gas liberation from the liquid phase.

Gas phase momentum:

$$\frac{\partial}{\partial t}(Gx) + \frac{\partial}{\partial L}\left(\frac{G^2x^2}{\rho_g \alpha}\right) = \hat{v}\Gamma_g - \alpha\frac{dp_g}{dL} - \frac{\tau_{W,g}P_{W,g}}{A_p} - \frac{\tau_i P_i}{A_p} - \rho_g g \alpha \sin\theta$$

$$\hat{v} = u_g \quad \text{if } \Gamma_g \le 0 \tag{5.27}$$

Liquid phase momentum:

$$\frac{\partial}{\partial t}[G(1-x)] + \frac{\partial}{\partial L}\left[\frac{G^2(1-x)^2}{\rho_l(1-\alpha)}\right] = \hat{v}\Gamma_l - (1-\alpha)\frac{dp_l}{dL}$$

$$
-\frac{\tau_{W,l}P_{W,l}}{A_p}+\frac{\tau_i P_i}{A_p}-\rho_l g(1-\alpha)\sin\theta
$$

$$
\hat{v}=u_l \quad \text{if } \Gamma_l \leq 0 \tag{5.28}
$$

Gas phase energy:

$$
\frac{\partial}{\partial t}\left[\rho_g \alpha E_g^o\right]+\frac{\partial}{\partial L}\left[GxH_g^o\right]
= -h_l^o\Gamma_l+\frac{P_{W,g}q_g^\varepsilon - P_i q_i^\varepsilon}{A_p}-Gxg\sin\theta+p_g\frac{\partial\alpha}{\partial t}-\frac{\tau_i P_i Gx}{A_p\rho_g} \tag{5.29}
$$

Liquid phase energy:

$$
\frac{\partial}{\partial t}\left[\rho_l(1-\alpha)E_l^o\right]+\frac{\partial}{\partial L}\left[G(1-x)H_l^o\right]
= h_l^o\Gamma_l+\frac{P_{W,l}q_l^\varepsilon + P_i q_i^\varepsilon}{A_p}-G(1-x)g\sin\theta+p_l\frac{\partial}{\partial t}(1-\alpha)
+\frac{\tau_i P_i G(1-x)}{A_p\rho_l} \tag{5.30}
$$

It is interesting to point out that equations (5.24)–(5.30) were the original set of equations to develop the well-known OLGA® transient code (Bendiksen et al., 1991) discussed below.

5.6.1 OLGA® Software (SLB)

This section highlights some of the basics of the benchmark code for oil and gas transient and steady-state simulation work in pipelines. OLGA is a software based on the two-fluid model that has undergone numerous improvements in terms of closure equations and numerical schemes. To validate the OLGA code, the experimental data from reputed laboratories have been key for the success of this tool. The laboratories providing experimental data include Norway's SINTEF (Stiftelsen for industriell og teknisk forskning), IFP (Institute of French Petroleum), University of Tulsa, Imperial College, Colorado School of Mines, and the Norwegian University of Science and Technology (NTNU).

The two-fluid model presented in the previous section was extended by Bendiksen et al. (1991) to improve the transient code OLGA. The first version of OLGA was available in 1983 through a joint research program between IFE and SINTEF. At that time only liquid and gas phases were considered, and the pressure drop calculations sometimes reached 50% offset with respect to experimental data. The major contribution of the paper by Bendiksen et al. (1991) was to include a liquid phase entrained within the gas phase in the form of entrained droplets. The result of this addition was successful in diminishing the prediction's uncertainty.

The set of mass conservation balances comprising the core of the OLGA code, including the inclusion of the entrained droplets, is

$$\frac{\partial}{\partial t}[\rho_G V_G] + \frac{\partial}{\partial L}[\rho_G V_G u_G] = \Gamma_G \tag{5.31}$$

$$\frac{\partial}{\partial t}[\rho_L V_L] + \frac{\partial}{\partial L}[\rho_L V_L u_L] = -\Gamma_G\left(\frac{V_L}{V_L + V_D}\right) - \Gamma_E + \Gamma_D \tag{5.32}$$

$$\frac{\partial}{\partial t}[\rho_L V_D] + \frac{\partial}{\partial L}[\rho_L V_D u_L] = -\Gamma_G\left(\frac{V_D}{V_L + V_D}\right) + \Gamma_E - \Gamma_D \tag{5.33}$$

Similarly, momentum conservation balances including the entrained droplets terms are

$$\begin{aligned}\frac{\partial}{\partial t}[\rho_G V_G u_G] &+ \frac{\partial}{\partial L}\left[\rho_G V_G u_G^2\right] + V_G\frac{\partial p}{\partial L} + \frac{f_G\rho_G u_G|u_G|S_G}{2(\pi d^2)}\\ &+ \frac{f_i\rho_G u_r|u_r|S_i}{2(\pi d^2)} - \rho_G g V_G \sin\theta - \Gamma_G(u_L + u_D) + F_D = 0\end{aligned} \tag{5.34}$$

$$\begin{aligned}\frac{\partial}{\partial t}[\rho_L V_D u_D] &+ \frac{\partial}{\partial L}\left[\rho_L V_D u_D^2\right] + V_D\frac{\partial p}{\partial L} - \rho_L g V_D \sin\theta\\ &+ \Gamma_G u_D\left(\frac{V_D}{V_L + V_D}\right) - \Gamma_E u_i + \Gamma_D u_D - F_D = 0\end{aligned} \tag{5.35}$$

$$\begin{aligned}\frac{\partial}{\partial t}[\rho_L V_L u_L] &+ \frac{\partial}{\partial L}\left[\rho_L V_L u_L^2\right] + V_L\frac{\partial p}{\partial L} + \frac{f_L\rho_L u_L|u_L|S_L}{2(\pi d^2)} - \frac{f_i\rho_G u_r|u_r|S_L}{2(\pi d^2)}\\ &- \rho_L g V_L \sin\theta + \Gamma_G\left(\frac{V_L}{V_L + V_D}\right)(u_G + u_D) + \Gamma_E u_i - \Gamma_D u_D = 0\end{aligned} \tag{5.36}$$

The set of equations (5.31)–(5.36) represents a formidable challenge to solve the velocity and density field. The solution option is the application of the finite difference methodology to a discretized domain. A highly effective technique is the control volume (Patankar, 1980) applied to each cell element of the grid.

The momentum equations (5.34)–(5.36) pose a numerical challenge due to the presence of the pressure gradient term. In fact, the velocity field can only be determined when the pressure field is known in advance via an iterative process. An initial guess of the pressure field is the starting point to obtain the velocity field, and velocity corrections are obtained by the residue of the continuity equations (5.31)–(5.33). To obtain the pressure field, the OLGA code uses the following pressure-correction equation derived by

Bendiksen et al. (1991) from the mass conservation equations and assuming that the gas density is a function of temperature, pressure, and gas fraction:

$$\left[\frac{V_G}{\rho_G}\frac{\partial \rho_G}{\partial p} + \frac{(1-V_G)}{\rho_L}\frac{\partial \rho_L}{\partial p}\right]\frac{\partial p}{\partial t} = -\left(\frac{1}{A\rho_G}\right)\frac{\partial}{\partial L}[\rho_G A V_G u_G]$$
$$-\left(\frac{1}{A\rho_L}\right)\frac{\partial}{\partial L}[\rho_L A V_L u_L] - \left(\frac{1}{A\rho_L}\right)\frac{\partial}{\partial L}[\rho_L A V_D u_D] + \Gamma_G\left(\frac{1}{\rho_G} - \frac{1}{\rho_L}\right) = 0 \tag{5.37}$$

The proprietary code OLGA® currently is owned by SLB (formerly known as Schlumberger) and constantly keeps evolving to suit novel applications. Newer versions of OLGA such as the HD (high-definition version) and OLGA 2021.1 have incorporated many improvements that include

- Mass transfer rates equations between fluid phases,
- Up to nine commingled phases:
  - Core gas phase
  - Oil core phase
  - Water core layer
  - Bubbles of gas in the oil core and the water core phases
  - Droplets of oil in the gas core and the water core phases
  - Droplets of water in the oil core and the gas core phases
- A two-dimensional velocity profile for stratified flow yielding a semi-three-dimensional flow approximation,
- Pressure-enthalpy tables for single-component systems,
- Separate specialized modules to complement the basic code, such as
  - Comptracking – to perform node-by-node phase equilibrium and property estimation
  - Corrosion – for internal corrosion estimation
  - Slugtracking – to model hydrodynamic slugging phenomena more accurately
  - FEMTherm – for improved estimation of convection and conduction heat transfer coefficients
  - Complex Fluids – to handle fluids with non-newtonian behavior
  - Multiphase Pumps – to simulate the performance of commingled gas and liquid through a pump
- FieldTwin™ integration to generate a digital copy of the physical system,
- Improved handling of CO_2 property estimation.

5.7 Constitutive Equations

The conservation principles are not sufficient to fully describe the thermal-hydraulic performance of the fluid flow. Constitutive equations are indispensable to "*close*" the set of governing principles and match number of unknowns to number of equations. In the most general sense, two-phase flow problems have 12 unknowns (Hsu and Graham, 1976):

- Void fraction (1)
- Liquid-phase velocity vector (3 components)
- Gas-phase velocity vector (3 components)
- Gas-phase density (1)
- Liquid-phase density (1)
- Pressure (1) – assuming equality of gas- and liquid-phase pressure
- Temperature (2) – non-thermal equilibrium

The set of equations that would match up the set of unknowns is

- Liquid-phase continuity equation (1)
- Gas-phase continuity equation (1)
- Momentum conservation in liquid phase (3 components)
- Momentum conservation in gas phase (3 components)
- Energy conservation in liquid phase (1)
- Energy conservation in gas phase (1)
- EOS for the liquid phase (1)
- EOS for the gas phase (1)

Thus, the equations of state are constitutive equations closing the set of conservation principles. Nevertheless, additional empirical data and models find their way into the problem formulation. Examples of support information required are as follows:

- void fraction correlation as a function of pressure, temperature, and flow pattern
- friction factors to characterize the shear stress of each phase at the wall, and the interfacial shear stress
- heat transfer coefficients needed to estimate heat losses.

5.8 One-Dimensional Modeling

One-dimensional models focusing on property changes along the axial direction of the pipeline are the standard for two-phase flow simulations. Most major commercial and proprietary codes available to date are based on the simplification of the ***1-D model***. Nevertheless, to improve accuracy, some codes superimpose two-dimensional (radial) transport phenomena upon the 1D axial model, with encouraging results. An example of this effective methodology is the combination of the two-fluid model (TFM) with the kinematic drift–flux model (DFM) to simulate steady-state and transient pipeline operations.

5.9 Homogeneous Equilibrium Model (HEM)

The simplest model that can be postulated is the one-dimensional HEM, based on the transformation of the multiphase mixture into an equivalent single-phase fluid with space- or volumetric-averaged properties. One great advantage is that the set of differential equations is amenable to straightforward numerical integration.

The HEM is a good option to obtain a preliminary solution to accelerate the convergence of more complex and accurate models. Also, the HEM can provide fast answers to sensitivity scenarios at an extremely low computing cost when the objective is to explore a trend rather than pursue an accurate answer.

The implicit assumptions of the HEM are

- No relative velocity between the phases (no slippage, $S = 1$),
- Thermodynamic equilibrium between the phases, and
- Pseudo-fluid mixture having mass-weighted density and viscosity.

In general, this model underpredicts pressure drop in pipeline flow, but there are cases where the HEM yields reasonable estimates. The exceptional cases where HEM is an option are

- For dispersed types of two-phase flow such as bubble and mist flow patterns,
- The core of annular flow, and
- When the mixture mass velocity G exceeds 2,000 kg/m^2 s.

The conservation principles for the HEM follow easily from the TFM equations by assuming a single fluid. The HEM equations benefit from using the mass velocity G because of constant-area pipelines and slow changes in mass flow rate. The unsteady-state equations are

$$\text{\textit{Triangular relation:}} \quad \frac{\dot{m}_g}{\dot{m}_l} = \frac{x}{1-x} = \frac{\rho_g}{\rho_l}\left(\frac{\alpha}{1-\alpha}\right) \tag{5.38}$$

$$\text{\textit{Void fraction:}} \quad \alpha = \frac{\rho_l x}{\rho_l x + \rho_g(1-x)} \; ; \; 1-\alpha = H_L = \frac{\rho_g(1-x)}{\rho_l x + \rho_g(1-x)} \tag{5.39}$$

$$\text{\textit{Fluid density:}} \quad \rho_H \equiv (1-\alpha)\rho_l + \alpha\rho_g = \frac{1}{\frac{x}{\rho_g} + \frac{(1-x)}{\rho_l}} \tag{5.40}$$

$$\textit{Mass velocity:}\quad G = \rho_H j = \textit{function}(t) \tag{5.41}$$

$$\textit{Continuity:}\quad \frac{\partial \rho_H}{\partial t} + \frac{\partial}{\partial L} G = 0 \tag{5.42}$$

$$\textit{Momentum:}\quad \frac{\partial G}{\partial t} + \frac{\partial}{\partial L}\left(\frac{G^2}{\rho_H}\right) = -\left[\frac{dp}{dL} + \frac{\tau_W P_W}{A_p} + \rho_H g \sin\theta\right]$$

$$= -\left[\frac{dp}{dL} + \frac{f_{Darcy} G^2}{2D\rho_H} + \rho_H g \sin\theta\right] \tag{5.43}$$

where $\frac{\tau_W P_W}{A_p}$ is the friction pressure gradient, $\rho_H g \sin\theta$ is the gravitation pressure gradient, and $\frac{\partial}{\partial L}\left(\frac{G^2}{\rho_H}\right)$ is the acceleration pressure gradient (usually negligible except at choke points).

$$\textit{Energy:}\quad \left[\frac{\partial}{\partial t}(\rho_H E^o) + \frac{\partial}{\partial L}(\rho_H j H^o)\right] = -\frac{P_W q''}{A_p} \tag{5.44}$$

The wall shear stress in equation (5.43) relates to the friction factor definition through

$$\tau_W = f_F \frac{G^2}{2\rho_H} \tag{5.45}$$

and the Darcy–Weisbach can be approximated for smooth wall piping by

$$f_{Darcy} \approx \frac{0.316}{\mathrm{Re}^{0.2}} \qquad \mathrm{Re} \equiv \frac{\rho_H j d}{\mu_H} \tag{5.46}$$

The viscosity of the homogeneous fluid has been subject to diverse definitions; two of the popular choices to estimate fluid viscosity for the Reynolds number calculation are

$$\mu_H = \mu_g \frac{j_g}{j} + \mu_l \frac{j_l}{j} \tag{5.47}$$

$$\frac{1}{\mu_H} = \frac{x}{\mu_g} + \frac{(1-x)}{\mu_l} \tag{5.48}$$

5.10 Separated Flow Model

The two limiting flow patterns framing two-phase flow in pipelines are *disperse phase* and *separated flows* (Brennen, 2005). A disperse phase flow includes a finite number of either bubbles in a continuous liquid medium or droplets in a continuous gas medium. The identity of each discrete and unconnected element is preserved at any instant of time but may evolve in spatial location, shape, or thermo-mechanical state. The continuous media do transfer mass, momentum, and energy with discrete particles in an extremely complex manner but following the laws of physics.

Currently, there is no accurate description of the full field flow with itemized dispersed particles, even utilizing code based on the *Lagrange* approach. Dispersed flows are dealt with, with an approximated continuous field and the HEM.

Separate flows involve phase identities that enable the *Eulerian* approach and the use of models such as the TFM. However, separate flow is yet challenging and does require simplifications, constitutive equations, and empirical correlations.

The simplification of the TFM can be obtained by lumping the phase equations (Collier and Thome, 1994), leading to the simpler *mixture model* with three conservation equations. Thus, the mixture model does not include the interface shear stress, the volumetric mass rate terms, or the heat transfer flux between the phases. The relevant stresses of the mixture model are only the phases' shear stress with the pipe wall and a single value pressure term common for both phases. The expressions for shear stress are given by

$$\tau_{W,l} = f_{D-W} \frac{\rho_l u_l^2}{8} \tag{5.49}$$

$$\tau_{W,g} = f_{D-W} \frac{\rho_g u_g^2}{8} \tag{5.50}$$

The friction factor depends on the Reynolds number and the nondimensional wall roughness ε/d. The friction factor can be estimated either graphically from the Moody diagram (Figure 5.4) or from empirical correlations. For laminar flow, the Poiseuille equation provides an exact expression of friction factor. For turbulent flow, the Colebrook–White implicit correlation and the Swamee and Jain (Streeter and Wylie, 1979) explicit formula yield excellent friction factor estimates:

$$f_{D-W} = \frac{64}{\mathrm{Re}} \quad \text{(Poiseuielle)} \tag{5.51}$$

$$\frac{1}{\sqrt{f_{D-W}}} = -0.8686 Ln\left(\frac{\varepsilon/d}{3.7} + \frac{2.51}{\mathrm{Re}\sqrt{f_{D-W}}}\right) \quad \text{(Colebrook–White)} \tag{5.52}$$

$$f_{D-W} \cong \frac{1.325}{\left[Ln\left(\frac{\varepsilon/d}{3.7} + \frac{5.74}{\mathrm{Re}^{0.9}}\right)\right]^2} \quad \text{(Swamee–Jain)} \tag{5.53}$$

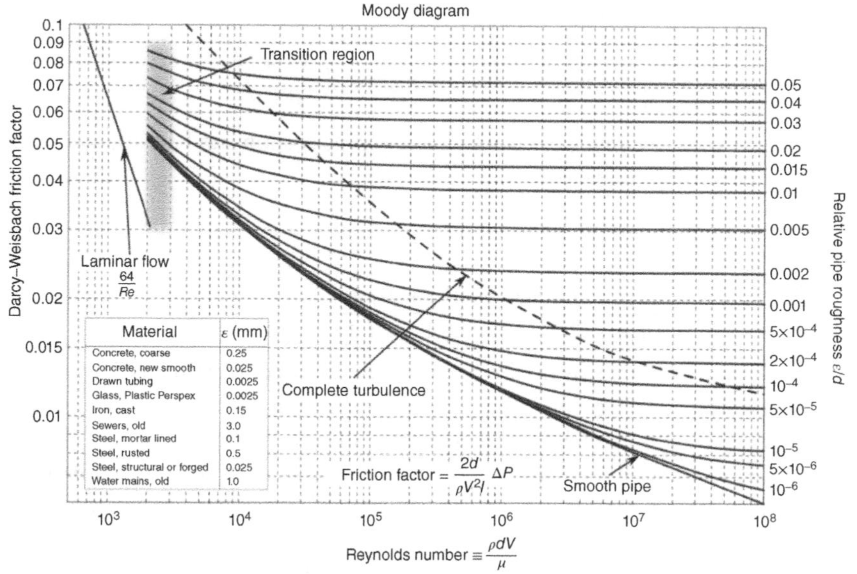

Figure 5.4 Darcy–Weisbach friction factor.

5.11 Kinematic Models

Kinematic is defined as the branch of mechanics concerned with the motion of objects without reference to the forces which cause the motion. In regard to two-phase flow, kinematic models have been developed based solely on the conservation of mass principle. These models were pioneered by the work of Bankoff (1960) and Zuber and Findlay (1965), establishing the foundation for the drift–flux model later consolidated by Wallis (1969).

The kinematic models support the one-dimensional models with cross-sectional corrections for void fraction and velocity profile. Such corrections enhance the accuracy of the 1-D estimates by attempting to incorporate effects of a 3-D flow field.

In this section, we review the Bankoff (1960), the Zuber and Findlay (1965), and the Wallis (1969) kinematic models.

5.11.1 Bankoff Model (1960)

An early attempt to correct 1-D models for two-dimensional (i.e. radial) effects was published by Bankoff (1960) for two-phase steam–water systems. The model postulates that the radial velocity and void fraction profiles can be expressed by the following power functions:

$$\frac{u(r)}{u_{\max,r=0}} = \left(\frac{R-r}{r}\right)^{\frac{1}{m}} \qquad \frac{\alpha(r)}{\alpha_{\max,r=0}} = \left(\frac{R-r}{r}\right)^{\frac{1}{n}} \tag{5.54}$$

The space-average void fraction at the pipe section can be estimated with the power functions as

$$\langle \alpha \rangle = C_a \beta = \frac{C_a j_g}{j} \tag{5.55}$$

$$C_a = \frac{2(m+n+mn)(m+n+2mn)}{(n+1)(2n+1)(2m+1)(m+1)} \tag{5.56}$$

Although there was no formal method proposed to estimate the value of the parameters "m" and "n," there is indication that for one-component two-phase flow, such parameters fall in the ranges $0.5 < n < 1.0$, and $2 < m < 7$. Therefore, the value of the constant C_a is in the range 0.70–0.90. Reported empirical values for the constant C_a by other experimenters are in the range 0.83–0.89.

Bankoff also obtained a relationship between the slip factor S and the cross-sectional void fraction involving a parameter that he called "K," which is a function of pressure, mass flow rate, and mass quality, and which usually falls in the interval 0.5–1.0. The relationship is

$$S = \frac{\langle u_g \rangle}{\langle u_l \rangle} = \frac{1 - \langle \alpha \rangle}{K - \langle \alpha \rangle} \tag{5.57}$$

The Bankoff parameter K can be shown (Zuber, 1960) to relate to the average void fraction and the volumetric quality β by the following equation:

$$K = \frac{\langle \alpha \rangle}{\langle \beta \rangle} \tag{5.58}$$

which combined with the slip factor S produces

$$S = \left[\frac{\langle \beta \rangle}{1 - \langle \beta \rangle}\right]\left[\frac{1 - \langle \alpha \rangle}{\langle \alpha \rangle}\right] \tag{5.59}$$

5.11.2 Zuber–Findlay Model (1965)

Zuber and Findlay (1965) proposed a model applicable to any two-phase flow regime to estimate the cross-sectional void fraction. In this regard, the objective for this model was similar to Bankoff's (1960) objective, to improve the accuracy of one-dimensional models with a correction.

The model incorporates the effect of radial properties variation, that is, non-uniform flow and concentration profiles. It is noteworthy that this paper established the basis for developing the *drift–flux model* (covered in subsection below) and the effect of the slip velocity between the phases.

Zuber and Findlay developed their model utilizing the weighted mean-value definition for the kinematic parameters. This definition applied to the gas velocity yields

$$\overline{u}_g = \frac{\langle \alpha u_g \rangle}{\langle \alpha \rangle} = \frac{\int_A \alpha u_g dA}{\int_A \alpha dA} \tag{5.60}$$

which can be related to the superficial velocity and led the authors to define the parameter C_0:

$$C_o \equiv \frac{\langle \alpha j \rangle}{\langle \alpha \rangle \langle j \rangle} \tag{5.61}$$

$$\overline{u}_g = \frac{\langle j_g \rangle}{\langle \alpha \rangle} = \frac{\langle \alpha j \rangle}{\langle \alpha \rangle} + \frac{\langle \alpha u_{gj} \rangle}{\langle \alpha \rangle} = C_o \langle j \rangle + \frac{\langle \alpha u_{gj} \rangle}{\langle \alpha \rangle} \tag{5.62}$$

enabling to express the space-averaged void fraction as

$$\langle \alpha \rangle = \frac{\langle \beta \rangle}{C_o \langle j \rangle + \frac{\langle \alpha u_{gj} \rangle}{\langle \alpha \rangle \langle j \rangle}} \tag{5.63}$$

Zuber and Findlay (1965) explored a range of potential non-uniform cross-sectional profiles, finding that for those that resemble most empirical data, the parameter C_o lies in the range 1.0–1.2.

5.11.3 Drift–Flux Model (DFM)

The drift–flux model is a kinematic methodology to account for cross-sectional non-uniform profiles of velocity and void fraction. The DFM may complement the TFM, the mixture model, or the mechanistic model to improve prediction accuracy.

The original work based on local relative velocity between the phases was by Behringer (1936), who developed an expression for the gas velocity as the sum of the total volumetric flux and the terminal rise velocity of a bubble. The DFM was shaped by the work of Bankoff (1960), Griffith and Wallis (1961), and Zuber and Findlay (1965).

The book by Wallis (1969) condensed the theory behind the DFM and demonstrated the model's capacity to simulate dispersed bubble and slug flow. Thus, the DFM can support the limiting cases of two-phase flow, disperse and separate flow patterns, adding a relationship between space-averaged void fraction and volumetric mass quality, as if it were an equation of state for two-phase one-dimensional flow.

Some authors call the DFM the void-quality analysis due to the way Zuber and Findlay (1965) developed it. The suitability of the DFM has been distinguished in important nuclear industry simulation codes such as RELAP5/MOD3, and also in the IFP-SimSci code TACITE, see Pauchon and Dhulesia (1994), for oil and gas transport simulations.

Equations (5.60) through (5.63) collect the essence of the DFM. The focus of development for this model is the concentration parameter C_o and the drift velocity term $\langle u_{gj}\rangle$. Two additional useful expressions sometimes used in the DFM (Lahey and Moody, 1993) are

$$\langle \alpha \rangle = \frac{\langle j_g \rangle}{C_o \langle j \rangle + \langle u_{gj} \rangle} = \frac{\langle x \rangle}{C_o\left[\langle x \rangle + \frac{\rho_g}{\rho_l}(1 - \langle x \rangle)\right] + \frac{\rho_g \langle u_{gj} \rangle}{G}} \tag{5.64}$$

$$S = C_o + \left(\frac{\rho_l}{\rho_g}\right)\frac{\langle x \rangle}{(1 - \langle x \rangle)}(C_o - 1) + \frac{\rho_l \langle u_g \rangle}{G(1 - \langle x \rangle)} \tag{5.65}$$

Rassame and Hibiki (2018) presented a summary of correlations for the concentration parameter and the drift velocity that enable to predict the space-averaged void fraction. Hibiki (2019) later issued a corrigendum for the case of horizontal flow. Hibiki's correlations for the concentration parameter and the drift velocity are as follows:

Vertical Upward Flow

$$C_o = \left(1.2 - 0.2\sqrt{\frac{\rho_g}{\rho_l}}\right)\left[1 - \exp\left(-22\frac{d_b}{D}\right)\right] \tag{5.66}$$

- Bubbly flow

$$\langle u_{gj} \rangle = \sqrt{2}(1 - \langle \alpha \rangle)^{1.75}\left[\frac{(\rho_l - \rho_g)g\sigma}{\rho_l^2}\right]^{0.25} \tag{5.67}$$

- Slug flow

$$\langle u_{gj} \rangle = 0.35\left[\frac{(\rho_l - \rho_g)gD}{\rho_l}\right]^{0.5} \tag{5.68}$$

- Churn flow

$$\langle u_{gj} \rangle = \sqrt{2}\left[\frac{(\rho_l - \rho_g)g\sigma}{\rho_l^2}\right]^{0.25} \tag{5.69}$$

- Annular flow

$$\langle u_{gj}\rangle = \frac{1-\langle\alpha\rangle}{\langle\alpha\rangle + \sqrt{\frac{1+75\rho_g(1-\langle\alpha\rangle)}{\rho_l\sqrt{\langle\alpha\rangle}}}}\left[\langle j\rangle + \sqrt{\frac{(\rho_l-\rho_g)gD(1-\langle\alpha\rangle)}{0.015\rho_l}}\right] \tag{5.70}$$

Horizontal Flow

$\langle u_{gj}\rangle$ is negligible because buoyancy forces are perpendicular to the horizontal direction.

$$C_o = 0.8\left(1-\sqrt{\frac{\rho_g}{\rho_l}}\right)\exp\left[0.9545\left(\frac{\langle j_g^+\rangle}{\langle j^+\rangle}\right)^{1.5}\right] + \sqrt{\frac{\rho_g}{\rho_l}} \quad \text{for } 0 \le \frac{j_g^+}{j^+} < 0.9 \tag{5.71}$$

$$C_o = 9.08 - 8.08\left[\frac{\langle j_g^+\rangle}{\langle j^+\rangle} - \left(\frac{\langle j_g^+\rangle}{\langle j^+\rangle} - 1\right)\sqrt{\frac{\rho_g}{\rho_l}}\right] \quad \text{for } 0.9 \le \frac{j_g^+}{j^+} \le 1 \tag{5.72}$$

where the volumetric fluxes are defined by

$$\langle j^+\rangle \equiv \frac{\langle j\rangle}{\left(\frac{g\sigma\Delta\rho}{\rho_l^2}\right)^{0.25}} \tag{5.73}$$

$$\langle j_g^+\rangle \equiv \frac{\langle j_g\rangle}{\left(\frac{g\sigma\Delta\rho}{\rho_l^2}\right)^{0.25}} \tag{5.74}$$

$$\frac{\langle j_g^+\rangle}{\langle j^+\rangle} \simeq \frac{1}{1+\left\langle\frac{1-x}{x}\right\rangle\left\langle\frac{\rho_g}{\rho_l}\right\rangle} \tag{5.75}$$

Research work focused on the drift–flux model started in the 1960s. Bhagwat and Ghajar (2014) published a comprehensible summary of 13 drift–flux correlations (concentration parameter and the drift velocity) spanning four decades of research work.

5.12 Mechanistic/Phenomenological Models

The mechanistic model is a simplification technique to the rigorous conservation principles equations. The assumption is that it is possible to identify in advance the flow pattern before applying a simple model exploiting the characteristics of such flow pattern. Thus, the success (or failure) of the mechanistic models relies on the capacity to pinpoint the flow regime, with a sound flow pattern map, regardless of the previous comment concerning sharp transitions among flow regimes. Sharp boundary transitions are an idealization process; regime transitions are gradual and show characteristics of both regimes involved. Therefore, there is some risk that at certain location of the pipeline flows the wrong regime is identified and an incorrect flow model is applied. One strategy followed in one code to circumvent the uncertainty in the prevailing flow pattern is to perform void fraction and pressure drop calculations using both neighboring regime's models. These two calculations are then weight-averaged to estimate a "transition" regime performance.

Mechanistic models are one-dimensional tools, with emphasis on the patterns depending on the pipe orientation. For horizontal flow, there are stratified, slug-plug, disperse bubbly, and annular flow models. For inclined upward and vertical flow, there are bubbly, churn-turbulent, slug, and annular flow models.

There are two trends of mechanistic models. Older models use the Taitel and Dukler (1976) map or Barnea (1987) map before selecting the corresponding set of equations to calculate the local holdup and the pressure gradient. Representative examples of this trend are Ansari et al. (1994), Shoham (2006) for vertical upward two-phase flows, Xiao et al. (1990) for horizontal two-phase flow, and more recently Petalas and Aziz (2000). Obviously, this generation of older models carries the risk of wrong identification of flow patterns for full-scale field applications, because such flow pattern maps were developed at laboratory scale and conditions.

More recent efforts such as Zhang (2003a, 2003b) and Gomez et al. (2000) have generalized the mechanistic approach evolving toward a "***unified***" methodology, whereby dispersed flow and separate flow regimes are both applied to estimate the two-phase flow parameters, followed by a validation process to verify the flow pattern. The advantage of the unified approach is that right from the onset dynamic analyses determine the flow pattern rather than utilizing a standard map which was created with different operating conditions. Furthermore, the dynamical analyses for transient phenomena (Shirdel and Sepehrnoori, 2012, 2017) may lead to flow pattern identification with a higher probability of success as opposed to traditional flow maps obtained at steady-state laboratory conditions.

Mechanistic models for specific flow patterns are becoming trendy in the literature; for example, models for slug flow have been published by Al-Safran et al. (2005) and Wang et al. (2014). A mechanistic model for stratified flow has been presented by Moradi et al. (2015). The extensive details of the specific mechanistic models are left to the references.

5.13 Empirical Models

At first sight, there seems to be no room for empirical models due to the availability of commercial software developed with the latest technical information. However, sometimes during preliminary design of gathering system or during troubleshooting facilities infrastructure, the objective is not to have an accurate answer but rather a quick and approximate estimate or performance trend. Moreover, empirical methods serve well to double-check results obtained with the commercial tools or to get initial estimates for iterative solutions of advanced codes.

Not surprisingly the two-phase flow literature for the oil and gas industry has numerous legacy empirical models, given the practical need to design and simulate flowlines and the extraordinary complexity of the conservation principles. The earliest models were published in the period between 1940s and 1970s. Examples of the initial models are Lockhart and Martinelli (1949), Baxendell (1955), Flanigan (1958), Eaton (1967), and the well-known Brill and Beggs (1973). Consequently, the empirical approach started to fade away when the mechanistic models started to be published in the mid-1970s and continue to improve to these days.

In this section, we focus on the Flanigan and the Beggs and Brill methods that still find keen users, with the caveat that should be integrated into a discretized pipeline domain jointly with

- An equation of state to predict phase split ratio and phase densities
- Friction factor and viscosities correlations
- The energy conservation balance to estimate temperature changes

5.13.1 Flanigan Model (1958)

This is an empirical and proven method that gives good estimates for gas-dominated flows such as gas-condensate flow with limited loading. The model is inspired by the Panhandle A method (GPSA, 2012) for single-phase gas flow but calibrated with field measurement data. The model can be resolved by hand calculations with average fluid properties determined in advance.

The wet gas flow rate in million standard cubic meters per day for this model is

$$Q_{sc} = 42.12 E_{tp} \left[\frac{d^{4.854} \left(P_1^2 - P_{2,corrected}^2 \right)}{T_m L z_m s_g^{0.8539}} \right]^{0.5394} \tag{5.76}$$

The units in equations (5.76): d [m]; P [kPa]; T_m [K]; L [m]. E_{tp} is a two-phase efficiency parameter given by

$$E_{tp} \cong 4.719 - 0.364\left(\frac{j_g}{R^{0.32}}\right)^{-0.5} \tag{5.77}$$

R represents the local liquid–gas ratio, expressed in bbls/mmscf, and $P_{2,\,corrected}$ is the exit pressure of the segment corrected for gravitational effects,

$$P_{2,corrected} = P_2 + 0.00981\left[\rho_l E_h \sum (head)_{up} - \rho_g \sum (head)_{down}\right] \tag{5.78}$$

$$E_h \cong \frac{1}{1 + 0.33(3.94j_g)^{0.92}} \tag{5.79}$$

This empirical model gives superior results for gas-condensate pipelines with low liquid loading.

5.13.2 Beggs and Brill Model (1973)

This empirical and simple 1-D model has been a popular tool to design oil and gas pipelines for 50 years. Even in our time Beggs and Brill is a popular option to quickly estimate gathering system performance without the hassle of solving intricate systems of PDEs, whenever an approximated rather accurate estimate is sought. The rapid convergence of this method, its applicability to any flow orientation, and the quality of the answers make it an ideal first approximation before attempting more advanced models. Some practitioners apply the Beggs and Brill method throughout the full length of the pipeline for computational economy; this practice leads to substantial errors in the prediction. It is much better to discretize the full pipeline domain into a number of sections and apply the Beggs and Brill model in a piecewise manner.

The model depends on the type of flow regime, namely segregated, intermittent, transition, and distributed. To identify the flow pattern is necessary to estimate the "L" parameters, the volumetric liquid quality, and the mixture Froude number. Knowing the flow regime allows us to predict the holdup for any fluid flow orientation.

The total pressure gradient is the sum of the friction, gravitation, and acceleration (usually negligible compared to the other terms) gradients. The two-phase flow friction factor is calculated from a single-phase Darcy friction factor correlation (e.g. Colebrook, with Reynolds number and nondimensional roughness), and corrected with a normalized factor which is a function of the volumetric liquid quality and the holdup. The method, applied to a subsection of the pipeline, comprises the following steps:

(a) Calculate the nondimensional *Froude* number

$$N_{Fr} = \frac{j^2}{gd} \tag{5.80}$$

(b) Calculate the *Kutateladze* number

$$N_{Ku} = j_l \left(\frac{\rho_l}{g\sigma_l} \right)^{0.25} \tag{5.81}$$

(c) Calculate the volumetric liquid quality

$$\lambda_L = \frac{j_l}{j} = 1 - \beta \tag{5.82}$$

(d) Calculate the "L" parameters:

$$L_1 = 316\lambda_L^{0.302} \tag{5.83}$$

$$L_2 = \frac{0.0009252}{\lambda_L^{2.4684}} \tag{5.84}$$

$$L_3 = \frac{0.1}{\lambda_L^{1.4516}} \tag{5.85}$$

$$L_4 = \frac{0.5}{\lambda_L^{6.738}} \tag{5.86}$$

(e) Estimate the holdup for the pipe section inclined at an angle ϕ with respect to the horizontal using the parameters listed in Table 5.1. The segregated uphill pattern is identified when λ_L is less than 0.01 and N_{Fr} is less L_1, or λ_L is larger than 0.01 and N_{Fr} is less than L_2. The intermittent uphill pattern occurs when λ_L is between 0.01 and 0.4 and when N_{Fr} is between L_3 and L_1, or when λ_L is larger than 0.4 and N_{Fr} is between L_3 and L_4. The distributed uphill pattern is identified when λ_L is less than 0.4 and N_{Fr} is larger than L_1, or when λ_L is larger than 0.4 and N_{Fr} is larger than L_4. The transition regime between the segregated and the intermittent regimes is handled with equation (5.89).

$$h_L = \begin{cases} H_L = \left(\frac{a\lambda_L^b}{N_{Fr}^c} \right) \psi \ \text{if} \ \frac{a\lambda_L^b}{N_{Fr}^c} > \lambda_L \\ H_L = \lambda_L \psi \ \text{if} \ \frac{a\lambda_L^b}{N_{Fr}^c} \leq \lambda_L \end{cases} \tag{5.87}$$

Table 5.1 Parameters for holdup correlation. Source: Adapted from Brill and Beggs (1973).

Regime	*a*	*b*	*c*	*d*	*e*	*f*	*g*
Segregated uphill	0.98	0.4846	0.0868	0.011	−3.77	3.54	−1.61
Intermittent uphill	0.845	0.5351	0.0173	2.96	0.305	−0.447	0.098
Distributed uphill	1.065	0.5824	0.0609			0.124	−0.506
Downhill				4.7	−0.37	0.124	−0.506

$$\psi = 1 + (1 - \lambda_L)Ln\left(d\lambda_L^e N_{Ku}^f N_{Fr}^g\right)\left\{\sin(1.8\phi) - \frac{\sin^3(1.8\phi)}{3}\right\} \tag{5.88}$$

$$H_L(transition) = \frac{(L_3 - N_{Fr})H_L(\text{segregated}) + (N_{Fr} - L_2)H_L(\text{intermittent})}{L_3 - L_2} \tag{5.89}$$

(f) Estimate two-phase and homogeneous densities:

$$\rho = \rho_l H_L + \rho_g(1 - H_L) \tag{5.90}$$

$$\rho_H = \rho_l \lambda_L + \rho_g(1 - \lambda_L) \tag{5.91}$$

(g) Estimate homogeneous and two-phase friction factors:

$$N_{Re} = \frac{\rho_H j d}{\mu_l \lambda_L + \mu_g(1 - \lambda_L)} \tag{5.92}$$

$$f_H = f_{Darcy}\left(N_{Re}, \frac{\varepsilon}{d}\right) \tag{5.93}$$

$$\varpi = Ln\left(\frac{\lambda_L}{H_L^2}\right) \tag{5.94}$$

$$f_{TP} = f_H \exp\left[\frac{\varpi}{0.0185\varpi^4 - 0.873\varpi^2 + 3.18\varpi - 0.0523}\right] \tag{5.95}$$

(h) Evaluate the gravitation and friction pressure gradients. Neglect (for piping) the acceleration pressure gradient.

$$\left(\frac{dp}{dx}\right)_{grav} = \rho g \sin\phi \tag{5.96}$$

$$\left(\frac{dp}{dx}\right)_{fric} = \frac{f_{TP}\rho_H j^2}{2d} \tag{5.97}$$

$$\left(\frac{dp}{dx}\right)_{acc} << \min\left[\left(\frac{dp}{dx}\right)_{fric}, \left(\frac{dp}{dx}\right)_{grav}\right] \tag{5.98}$$

(i) Calculate the change in total pressure, inlet minus outlet pressure, for the piping section and advance the calculations for the next one (p_{out}):

$$p_{out} \cong p_{in} - \Delta x \left[\left(\frac{dp}{dx} \right)_{grav} + \left(\frac{dp}{dx} \right)_{fric} \right] \tag{5.99}$$

It is important to emphasize that gas and liquid properties such as density and viscosity are a function of pressure and temperature. Thus, the analysis of each subsection requires compositional information (thermodynamic equilibrium) and the solution of the energy conservation equation coupled to the mass and momentum conservation equations.

5.13.3 Void Fraction Empirical Models

There are some good references summarizing the state of the art in empirical void fraction correlations, such as Garcia et al. (2005), Woldesemayat and Ghajar (2007), and Bhagwat and Ghajar (2014). The latter reference is centered on the DFM distribution parameter and the drift velocity.

In the nuclear industry, extensive experimental and theoretical studies have been carried out by the Electric Power Research Institute (Chexal et al., 1997) or EPRI for short. The Chexal–Lellouche void fraction model described in this reference is independent of flow patterns, and it is included in the RELAP5 and RETRAN-3D thermal-hydraulic models.

5.14 Computational Fluid Dynamics (CFD)

CFD is simulating physical fluid flow and heat transfer phenomena with mathematical models solved with numerical methods using computer resources.

With the developments in computer power and numerical methods, the field CFD found several applications in two-phase flow. The power of CFD lies in the discretization of the domain with small-size and adaptive mesh to follow specific surface contours. Full three-dimensional domain modeling became possible for pipeline accessory geometries opening the door to solutions not viable with the 1-D models. However, CFD comes at the price of computing time and memory size to describe the physical phenomena with better resolution.

Due to the high computing time involved, CFD is left to solve small segments or accessories in the gathering system of the hydrocarbon production.

Full pipeline modeling by CFD is untenable at the current time, and practical use of this tool is limited to coupled solutions with other transient tools such as Schlumberger's OLGA or Kongsberg's LEDAFLOW applied to the vast majority of the pipeline domain.

The advanced CFD code available nowadays can be set up to adopt either a Lagrangian or an Eulerian (control volume) approach. The Lagrangian approach following the particles is particularly useful for dispersed types of flows. Furthermore, the majority of two-phase flows are turbulent, and the CFD codes offer a selection of turbulence models such as *k-ε*, *DNS, LES, RANS*, and Langevin, among others.

The system of equations in CFD codes is aimed at continuum media and comprises continuity, Navier–Stokes (momentum), and energy conservation equations. Realistic and practical boundary conditions, interface jump conditions, and initial conditions for the fluid flow are the challenges faced by the CFD user. Local fluid properties require either tabular pre-estimated data, as a function of pressure and temperature, or compositional subroutines supported by equations of state.

Some examples of the use of CFD tools in two-phase flow are

- Thermal analysis of materials (e.g. steel) subject to extremely low temperatures due to Joule–Thomson rapid gas expansion. The analysis allows us to identify cold spots, thermal gradients, and verify design strength limits.
- Modeling of two-phase flow through production wells choke valve.
- Choked two-phase flow through orifice plates (blowdown safety operation).
- Thermal modeling of partly buried, backfilled, rock dump, and mattressed pipelines.
- Simulation of entrained solids causing pipeline erosion and identification of hot spots. Typical application being simulation of sand fines entrainment and accumulation at points of low velocity.
- Study the phenomenon of two-phase maldistribution at tees and manifolds.
- Simulation of slug catchers (vessels, finger-type).
- Transient water hammer local effects during quick closure of shutoff valves in two-phase flow.

References

Al-Safran, E., Sarica, C., Zhang, H., and Brill, J. (2005). *Probabilistic/Mechanistic Modeling of Slug-Length Distribution in a Horizontal Pipeline*, 160–172. SPE Production and Facilities.

Ansari, A., Sylvester, N., Sarica, C. et al. (1994). *A Comprehensive Mechanistic Model for Upward Two-phase Flow in Wellbores*, 143–152. SPE Production and Facilities.

Bankoff, S. (1960). A Variable Density Single-Fluid Model for Two-Phase Flow with Particular Reference to Steam-Water Flow. *J. Heat Trans.* 82: 265.

Barnea, D. (1987). A Unified Model for Predicting Flow Pattern Transitions for the Whole Range of Pipe Inclinations. *Int. J. Multiphase Flow* 13: 1.

Baxendell, P. (1955) Fourth World Petroleum Congress, Section 2/E, Paper No. 4, Rome, Italy.

Behringer, H. (1936). The Flow of Liquid-Gas Mixtures in Vertical Tubes. *Zeit. Ges. Kalte.- Ind.* 43: 55–58.

Bendiksen, K., Malnes, D., Moe, R., and Nuland, S., (1991) "The Dynamic Two-Fluid Model OLGA: Theory and Application," SPE Paper 19451.

Bhagwat, A. and Ghajar, A. (2014). A Flow Pattern Independent Drift Flux Model Based Void Fraction Correlation for a Wide Range of Gas-Liquid Two-Phase Flow. *Int. J. Multiphase Flow* 59: 186–205.

Brill, J. and Beggs, H. (1973). A Study of Two-Phase Flow in Inclined Pipes. *J. Pet. Technol.* 25: 607.

Brennen, C. (2005). *Fundamentals of Multiphase Flow*. Cambridge, U.K.: Cambridge University Press.

Brodkey, R. (1967). *The Phenomena of Fluid Motions*. Reading, MA, USA: Addison-Wesley.

Chexal, B., Merilo, M., Maulbetsch, J. et al. (1997) "Void fraction technology for design and analysis," EPRI TR-106326, California, USA

Collier, J. and Thome, J. (1994). *Convective Boiling and Condensation*, 3e. Oxford, U.K.: Oxford Science Publications.

Delhaye, J. (1968). "Equations fondamentales des écoulements diphasiques," Commissariat a L'energie Atomique, Rapport CEA-R-3429.

Delhaye, J. (1981). *Two-Phase Flow and Heat Transfer in the Power and Process Industries*," Chapter 2 (ed. A. Bergles). New York, NY, USA: Hemisphere Publishing Corporation.

Delhaye, J. (1982). *Handbook of Multiphase Systems* (ed. G. Hestroni). New York, NY, USA: Hemisphere Publishing Corporation Chapter 1, Section 1.2.

Eaton, B. (1967). The Prediction of Flow Patterns, Liquid Holdup, and Pressure Losses Occurring During Continuous Two-Phase Flow in Horizontal. *J. Pet. Technol.* 815–828.

Flanigan, O. (1958). Effect of Uphill Flow on Pressure Drop in Design of Two-Phase Gathering Systems. *Oil Gas J.* 132.

Garcia, F., Garcia, R., and Joseph, D. (2005). Composite Power Law Holdup Correlations in Horizontal Pipes. *Int. J. Multiphase Flow* 31: 1276–1303.

Gomez, L., Shoham, O., Schmidt, Z. et al. (2000). Unified Mechanistic Model for Steady-State Two-Phase Flow: Horizontal to Vertical Upward Flow. *SPE J.* 5 (3): 339–350.

(2012). *GPSA Engineering Data Book*, 13e. Tulsa, OK, USA: Gas Processors Suppliers Association (GPSA).

Griffith, P. and Wallis, G. (1961). Two-phase Slug Flow. *J. Heat Transfer, Trans. ASME, Ser. C* 83: 307.

Hibiki, T. (2019). Personal Communication (2021), Corrigendum to "One-dimensional drift-flux correlations for two-phase flow in medium-size channels". *Exp. Comput. Multiphase Flow* 1 (2): 85–100.

Hsu, Y. and Graham, R. (1976). *Transport Processes in Boiling and Two-Phase Systems.* Washington DC, USA: Hemisphere Publishing Company.

Lahey, R. and Moody, F. (1993). *The Thermal Hydraulics of a Boiling Water Nuclear Reactor*, 2e. The American Nuclear Society.

Lockhart, R. and Martinelli, R. (1949). Proposed Correlation of Data for Isothermal Two-Phase Two-Component Flow in Pipes. *Chem. Eng. Prog.* 45: 39–48.

Moradi, B., Hossain, M., and Oluyemi, G. (2015). Mechanistic Model for Four-Phase Sand/Water/Oil/Gas Stratified Flow in Horizontal Pipes. In: *Computational Methods in Multiphase Flow VIII*, vol. 89, 335–346. WIT Press.

Patankar, S. (1980). *Numerical Heat Transfer and Fluid Flow.* New York, NY: Hemisphere Publishing Corporation.

Pauchon, C. and Dhulesia, H. (1994). "TACITE: A Transient Tool for Multiphase Pipeline and Well Simulation," SPE Paper No. 28545, Dallas, TX, USA.

Petalas, N. and Aziz, K. (2000). A Mechanistic Model for Multiphase Flow in Pipes. *J. Can. Pet. Technol.* 39: 43–55.

Rassame, S. and Hibiki, T. (2018). Drift-flux Correlation for Gas-Liquid Two-Phase Flow in a Horizontal Pipe. *Int. J. Heat Fluid Flow* 69: 33–42.

Shirdel, M. and Sepehrnoori, K. (2012). Development of a Transient Mechanistic Two-Phase Flow Model for Wellbores. *SPE J.* 942–955.

Shirdel, M. and Sepehrnoori, K. (2017). Development of a Transient Mechanistic Three-phase Flow Model for Wellbores. *SPE J.* 374–388.

Shoham, O., (2006). Mechanistic Modeling of Gas-Liquid Two-Phase Flow in Pipes," SPE, Dallas, TX, USA.

Streeter, V. and Wylie, E. (1979). *Fluid Mechanics*, 7e. New York, NY, USA: McGraw Hill Book Company.

Taitel, Y. and Dukler, A. (1976). A Model of Predicting Flow Regime Transitions in Horizontal and Near Horizontal Gas-Liquid Flow. *AICHE J.* 22: 47–55.

Wallis, G. (1969). *One-Dimensional Two-Phase Flow.* New York, NY, USA: McGraw Hill Book Company.

Wang, S., Zhang, H., Sarica, C., and Pereyra, E. (2014). *A Mechanistic Slug-Liquid-Holdup Model for Different Oil Viscosities and Pipe-Inclination Angles*, 329–336. SPE Production and Operations.

Woldesemayat, M. and Ghajar, A. (2007). Comparison of Void Fraction Correlations for Different Flow Patterns in Horizontal and Upward Inclined Pipes. *Int. J. Multiphase Flow* 33: 347–370.

Xiao, J., Shoham, O., Brill, J., (1990). "A Comprehensive Mechanistic Model for Two-phase Flow in Pipelines," SPE paper 20631-MS, Dallas, TX, USA.

Yadigaroglu, G. and Hewitt, G. (2017). *Introduction to Multiphase Flow*. Switzerland: Springer.

Zhang, H., Wang, Q., Sarica, C. et al. (2003a). A Unified Mechanistic Model for Slug Liquid Holdup and Transition Between Slug and Dispersed Bubble Flows. *Int. J. Multiphase Flow* 29: 97–107.

Zhang, H., Wang, Q., Sarica, C. et al. (2003b). Unified Model for Gas-Liquid Pipe Flow via Slug Dynamics. *ASME J. Energy Res. Technol.* 125: 266–283.

Zuber, N. (1960). On the Variable-Density Single-Fluid Model for Two-Phase Flow. *J. Heat Transfer, Trans. ASME, Ser. C* 82: 255–258.

Zuber, N. and Findlay, J. (1965). Average Volumetric Concentration in Two-Phase Flow Systems," Trans. ASME, Ser. C. 87: 453.

6

Two-Phase Flow in Wells

6.1 Introduction

This chapter covers the dominating factor of gravity on two-phase flow in quasi-vertical piping. Three-phase flow (O/W/G) is customarily modeled as a two-phase liquid–gas flow. Some reference is also made to horizontal well flow despite not being a quasi-vertical flow.

A review of legacy models developed for well flow is included in Section 6.6 because some of these models are still in significant use in commercial software, spreadsheets, or even hand calculations. Newer mechanistic models, supported by flow pattern predictions (see Chapter 4), provide a firmer ground to predict two-phase flow performance in wells.

The well flow model serves to determine the production operating point by superimposing the inflow production relationship (IPR) jointly in a graph with a well flow model. This useful methodology is introduced in Section 6.3, better known as *nodal analysis*.

During the life cycle, the production stream changes composition. One detrimental phenomenon affecting the wells' production process is water coning, introduced in Section 6.5. Another interesting effect is the GOR for an oil well which also covered in Section 6.5.

Intermittent well flow, that is, slugging, may cause perturbation on surface facilities. The phenomenon of slugging is discussed in Section 6.9, with an aim at production rate.

For well production affected by increment in fluid density, that is, produced water, it is necessary to assist the well with artificial lift. Section 6.10 touches on gas lift, one of the common and economical ways to assist production by means of downhole gas injection.

A brilliant monograph covering all the topics in this chapter is the book by Brill and Mukherjee (1999), where the reader can find further details on topics skipped herein in pursuit of a streamlined chapter.

Multiphase Transport of Hydrocarbons in Pipes, First Edition.
Juan J. Manzano-Ruiz and Jose G. Carballo.

Companion Website: www.wiley.com/go/Manzano-Ruiz/Multiphasetransport

6.2 Reservoir Boundary Conditions

The reservoir conditions of pressure, temperature, and fluid composition represent the upstream boundary conditions for the system of dynamic equations describing the production processes. During the production life of the project, the hydrocarbon reserves are depleted (Craft and Hawkins, 2015) due to the hydrocarbon withdrawal. This depletion process translates into lower pressure at the upstream end of the production model as time goes by. In some reservoirs, there may be a large underlying aquifer supporting pressure for a long time and enabling a prolonged life-cycle operation. Without an artificial lift, the only source of energy to deliver the hydrocarbon stream comes from the reservoir, and its decline has a profound effect on production.

Although reservoir temperature remains constant during the life cycle, the produced fluid composition will change, for example, increasing amounts of water produced, and increased amounts of gas are released. All these changes will determine the reservoir's true recovery factor.

The logical approach to simulate hydrocarbon production is to consider an integrated production model, from the sand face, through the well string, followed by the gathering system toward the processing facilities. For offshore fields, the gathering system is known as the SURF (subsea umbilical, risers, and flowlines) infrastructure, delivering the production at the topside facility for preliminary processing.

6.2.1 Well Production String: Vertical, Inclined, and Horizontal Wells

The production well joins the gap between the reservoir and the gathering system infrastructure. The well itself could also be viewed as a restriction for the stream flow when the gravitational and frictional effects are accounted for. Wells are designed, drilled, and completed to meet the expected production rates at high pressure and temperature. The production string is the tubing enclosed by layers of casing piping that protects the integrity of the hole. Aside from the different casing sizes and depths, the tubing may be of a single size or telescopic.

The effects of friction and gravity relate to two key factors in every well, namely the *measured depth* (MD) and the *true vertical depth* (TVD). The effect of gravity manifests through the TVD. The effect of friction relates to MD. The latter should be better understood as *measured length* to also apply to horizontal wells. Both terms are normally referenced with respect to the rotating table of the drilling rig.

Very seldom a well is truly vertical but has some degree of inclination in pursuit of the reservoir's objective. An inclined well deviates from the vertical direction below the so-called *kick-off point* (KOP). In some instances, the deviation angle is quite pronounced, and the well is labeled *highly deviated* or *extended reach wells* due to the step-out. The purpose of drilling extended wells sideways is to reach far objectives from a specific pad or cluster enabling farther objectives from a given location. These locations are known as *drill centers* where the rig is stationed, eliminating the need to mobilize the rig to several other drill locations. The savings in rig mobilization are, however, offset by drilling time associated with longer drilling paths; thus, there is a Capex tradeoff between drilling time and rig mobilization costs. Additionally, drilling risks must be evaluated too for the drill center option. A sketch of a subsea well completion including casings piping and production tubing is shown in Figure 6.1.

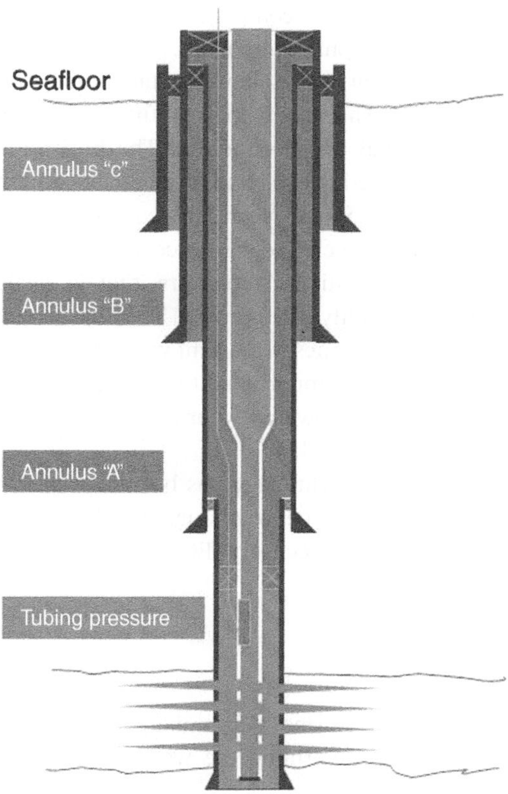

Figure 6.1 Well completion sketch. Source: This photo by unknown author is licensed under CC BY-ND.

The simulation of the production system requires starting from the upstream reservoir boundary condition. Therefore, the wells must be integrated into the flow model, with their completion details (e.g. tubing string, perforation intervals, kick-off point, SSCSV, layers of casings, and packers' fluid). This information is crucial to simulate the well performance (thermal and hydraulic).

The well hydraulic performance is driven by gravitational effects. The gravitational component of pressure drop is proportional to the two-phase flow density. Thus, it is vital to have a good estimate of the liquid holdup distribution along the well to obtain the best estimate of the local two-phase flow density. It is not unusual to realize, for large-size tubing strings, that the frictional component is about two orders of magnitude smaller than the gravitational component of pressure drop. Conversely, there are high-performing wells (e.g. gas wells with extraordinary potential) that must be provided with extra-large production string (i.e. completed with casing pipe) to handle the large frictional component of pressure drop which may be comparable to the gravitational component.

The technology of drilling horizontal wells allows to tap production zones with larger production zones. This technology matured with the advent of geo-steering multidirectional drill technology. The horizontal well has greater contact with the production sand and therefore boosts the productivity index of the well. For the past 15 years, the combination of horizontal well and reservoir-fracturing technologies successfully enabled the great productivity of tight unconventional reservoirs, that is, shale reservoirs, with extremely low permeability and porosity. The objective of the horizontal well is to tap multiple pay zones, with semi-vertical bedding. The productivity of the horizontal well is proportional to its length, but at additional drilling time, risks, and costs. Nowadays, horizontal wells have been drilled with a record lateral of about 5.5 km.

Aside from horizontal producers, the technology has been exported to injector wells for improving well injectivity due to the augmented reservoir contact, making them best for enhanced-oil-recovery (EOR) applications.

6.3 Nodal™ Analysis

Nodal analysis is the trademark of SBL (former Schlumberger company), and it is a methodology employed to determine the well performance.

Herein, well performance means the achievable standard flow rate of the production fluid and the pressure at the nodal point.

The nodal analysis is the balance between the reservoir energy, represented by the reservoir inflow relationship, and the lifting requirements of the production stream through the well completion. Implicit in this lift requirement is the wellhead pressure (P_{wh}) imposed by the gathering system at the surface. In other words, the well performance is subject to the performance of the overall production system. The IPR is the function (Economides et al., 2013) connecting the production flow rate with the available bottom-hole flowing pressure (P_{wf}). The oil stream production rate is stated in terms of the equivalent stock-tank flow rate to avoid the uncertainty of the volumetric flow rate at other conditions. Similarly, for gas wells, the flow rate is expressed at standard conditions (i.e. 1.01 bara and 15.6 °C). This practice of expressing flow at standard conditions is equivalent to specifying the mass flow rate which is conserved.

The lift performance of the well is referred to as the vertical lift performance (*VLP*) relationship or sometimes labeled tubing performance relationship (*TPR*). The nodal analysis does the matching of IPR against VLP to determine the resulting flow rate and the bottom-hole pressure P_{wf}. The nodal graph offers the possibility for sensitivity analyses to explore different well completions and also the effect of well choking at the tree level.

Figure 6.2 illustrates a nodal analysis for an oil well during various stages of the life cycle. The IPR curves decline from early life to late life, while the VLP remains constant (e.g. constant fluid composition, constant wellhead pressure, same well completion). It is obvious from the IPR curves that the reservoir is depleted, and the reservoir pressure declines appreciably. The expected well performance drops from an initial 456 sm^3/d (2,868 stb/d) to 255 sm^3/d (1,604 stb/d).

6.4 Wells Cluster and Manifolds

When multiple wells are drilled to deplete a hydrocarbon reservoir, it makes sense to minimize the cost and time of rig mob/demob (i.e. mobilization). Thus, several wells may be drilled from a small footprint or pad, forming a cluster of wells feeding into a production manifold. The manifold is a structure assembled with shutoff valves to isolate each individually well. For subsea manifolds (Figure 6.3), the valves are remotely operated either

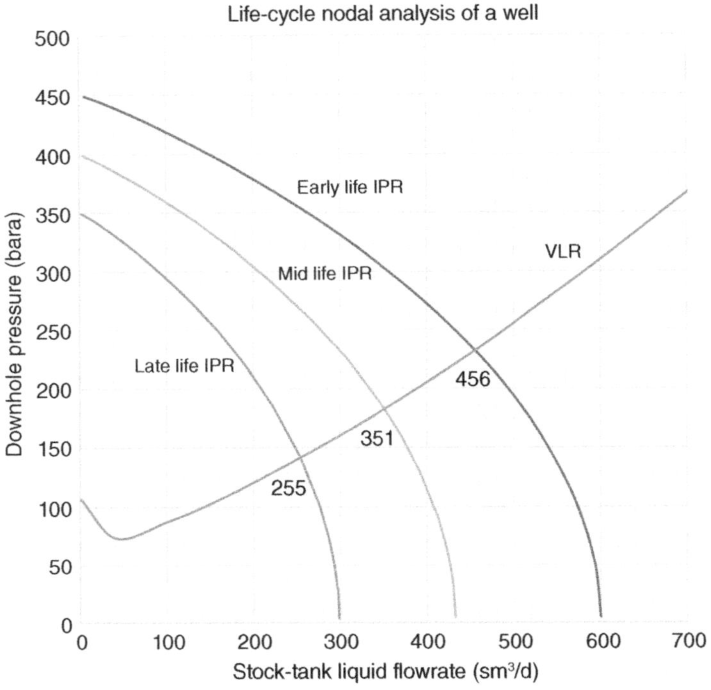

Figure 6.2 Example of a Nodal Analysis™.

Figure 6.3 Subsea manifold. Source: Jason Woodhead/Flickr/CC BY 2.0.

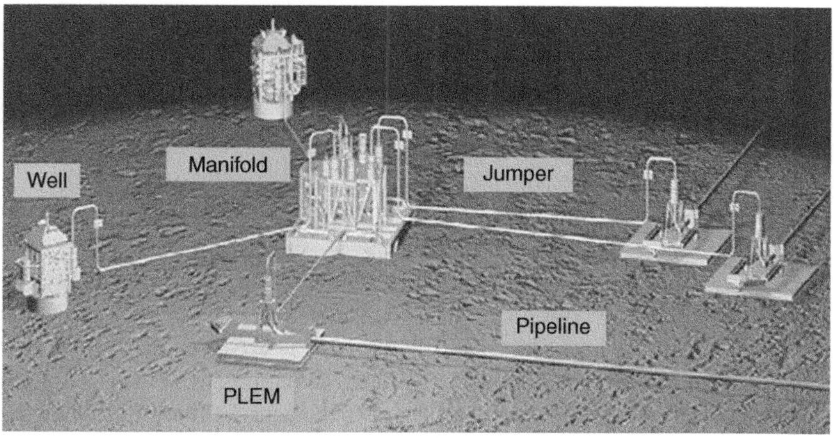

Figure 6.4 Subsea wells tie into the manifold. Source: Rui et al. (2023)/John Wiley & Sons/CC BY 4.0.

hydraulically or electrically. Usually, the production manifold has spare slots to connect at a future time when additional local drilling may be performed.

Typically, subsea trees are connected via a jumper structure (see Figure 6.4) directly to a manifold or with a jumper and a short flowline tie-in to a production manifold. A multiphase flowline conveys the production from the manifold to the processing facilities or taps another multiphase flow trunk line. The production manifold normally connects to an independent test flowline that could be selectively aligned with a single well to be tested. This test line enables the operator to execute the regular testing of each well to determine allocation factors and perform reservoir management in general.

The production manifold structure facilitates workover operations of several cluster wells if needed, without rig mobilization. Often, for onshore applications, a gas–liquid production separator and/or a test separator may be present at the cluster pad. For subsea applications, there are a handful of fields that have subsea processing and boosting. In both cases, the objective is the same – to separate gas from liquids as early as possible to eliminate the higher hydraulic losses caused by multiphase flow in pipes.

The comparison between onshore fields and subsea field developments leads to a striking difference in the number of wells and drilling and completion costs. Onshore wells overwhelm the number of subsea wells. Subsea wells carry an exceedingly high price tag as compared to onshore drilling; for this reason, subsea developments are justified when the materiality (i.e. reserves) of the reservoir, the potential productivity index of the wells, and the smaller number of wells can deliver a profitable production profile for a long life cycle. In this case, a small footprint of subsea manifold(s) minimizes investment and future maintenance costs.

6.5 Water and Gas Coning

Water and/or gas coning refers to the fluid migration overriding the oil production in the reservoir formation. The water coning problem arises when free water in the lower layers of the reservoir is driven to the wellbore by the drawdown pressure gradient. Similarly, gas coning from the gas cap gas encroaches through the perforation zone in the near-wellbore area when the remaining oil in place is saturated.

Produced water will break through during the life cycle because most hydrocarbon traps are supported by aquifers. In general, the production of water will increase requiring a handling and disposal management program (Arnold and Stewart, 2008).

When water encroaches, the near-wellbore area creates several problems diminishing fluid production. The operational issues arising from water coning are as follows:

- *Decreased lift capacity*. The reservoir will face a larger back pressure because the average density of the production fluid within the tubing string will be higher.
- *Risk of hydrate formation*. The incremental presence of water, the presence of gas in the solution released, and the higher pressure within the tubing might create the conditions to form hydrate and potentially plug the well.
- *Emulsions*. The presence of water and the shear stresses (turbulence effects) on the oil phase might enhance the chance of creating some tight emulsions if there are natural surfactants at play.
- *Scale*. Water breakthrough is the signal of scaling issues setting in, due to the dissolved salts that may come out of the solution depending on the local temperature and *pH*.

Gas coning in oil wells may or may not enhance oil production. The additional amount of gas pulled from the reservoir gas cap may serve as a gas lift mechanism favoring liquid lifting. This beneficial effect applies to low rates of liquid flow produced, but for high rates, the contrary takes place because of the higher friction losses caused by the faster gas phase velocity.

6.6 Legacy Models

Older models published in the period 1940–1970 relied on the basic momentum equation to estimate the pressure drop components (always neglecting the acceleration component) and were well supported with empirical field data correlations (e.g. holdup, friction coefficient). An outstanding summary of methodologies before the 1980s was compiled in the course by Dukler and Taitel (1982).

In general, all models share the same pressure gradient calculation procedure, characterized by equations (6.1)–(6.6). Even the mechanistic models also split the total pressure gradient into three terms. The total pressure gradient along the tubing axis ℓ is

$$\left(\frac{dp}{d\ell}\right)_{total} \cong \left(\frac{dp}{d\ell}\right)_{fric} + \left(\frac{dp}{d\ell}\right)_{grav} \tag{6.1}$$

after neglecting the acceleration term with respect to the other terms:

$$\left(\frac{dp}{d\ell}\right)_{acc} \ll \left(\frac{dp}{d\ell}\right)_{grav} \text{or/and} \left(\frac{dp}{d\ell}\right)_{fric} \tag{6.2}$$

The friction term is expressed in terms of the tubing's internal diameter d, a characteristic two-phase flow density ρ_m and velocity u_m depending on the model considered

$$\left(\frac{dp}{d\ell}\right)_{fric} = \frac{f\rho_m u_m^2}{2d} \tag{6.3}$$

The Darcy–Weisbach friction factor is obtained from either Figure 5.4 or from a correlation using the Reynolds number (equation (6.5)) and the nondimensional roughness parameter:

$$f = function\left(\text{Re}, \frac{\varepsilon}{d}\right) \tag{6.4}$$

$$Re = \frac{\rho_m u_m d}{\mu} \tag{6.5}$$

The gravitation pressure gradient term is also calculated with a characteristic density ρ_m depending on the model considered:

$$\left(\frac{dp}{d\ell}\right)_{grav} = \rho_m g \tag{6.6}$$

To date, some of the models covered in this chapter are still popular, successfully supporting the nodal analysis for well producers. Given the strong influence of the field database used to develop these correlations, it is important to select the models that historically have given satisfactory results depending on the type of fluid or the type of well completion.

6.6.1 Poettmann–Carpenter Model (1952)

This pioneering model is the longest-standing legacy one, with some use by field practitioners due to its simplicity lending to hand calculations (Poettmann and Carpenter 1952).

The model is expressed in *FPS units* as originally published, with the caveat that our correlation of the friction factor has been included below. Densities are expressed in *lbm/ft*^{3}, mixture velocity j is expressed in *ft/s*, tubing internal diameter is expressed in *ft*, and the total pressure gradient (neglecting acceleration effects) is expressed in *psi/ft*.

The characteristic density for this model is evaluated with the volumetric quality (equation 2.28):

$$\rho_m = \rho_n = \rho_G \beta + \rho_L (1 - \beta) \tag{6.7}$$

And the liquid density is evaluated with the water cut (WC),

$$\rho_L = \rho_W WC + \rho_O (1 - WC) \tag{6.8}$$

The characteristic friction factor was presented by the authors in graphical form as a function of the parameter ϖ (expressed in *lbm/ft-s*):

$$\varpi \equiv \rho_n jd \tag{6.9}$$

$$f_{P-C} \simeq \frac{1}{0.163\varpi^{1.5} - 1.45 + 5.99\frac{Ln(\varpi)}{\varpi^2}} \tag{6.10}$$

The total pressure gradient equation becomes

$$\frac{dp}{d\ell} \cong 0.00694\rho_n + \frac{f_{P-C}\rho_n j^2}{9{,}265d} \tag{6.11}$$

As pointed out before, for vertical or slanted wells, the first term above (gravitation term) dominates over the friction term (for regular-size tubing design). Therefore, the accuracy of equation (6.11) is as good as the mixture density estimate. In this particular model, the void fraction parameter has been approximated by the volumetric gas quality β for simplicity, which is not a good assessment for the ample spectrum of flow patterns in vertical upward flow. Order-of-magnitude estimates for pressure gradients can be expected from the Poettmann and Carpenter model.

6.6.2 Griffith–Wallis Model (1961)

Griffith and Wallis (1961) published a study on slug and bubble flow in vertical upward flow, which became a standard for vertical upward flow modeling and training (Griffith, 1978) in two-phase flow. The original work was amply covered in the monograph by Wallis (1969). This work was imported into later models such as Hagedorn and Brown (1965) and Orkiszewski (1967).

According to this method, the boundary between bubbly and slug flow is (in *SI units*)

$$j_G = \left(1.071 - 0.06762\frac{j^2}{d}\right)j, \quad \text{with } j < \sqrt{13.9d} \tag{6.12}$$

In other words, when the actual j_G is less than the value obtained from equation (6.12), the flow pattern would be bubbly. The liquid holdup can then be estimated using

$$H_L = \frac{(u_S - j) + \sqrt{(u_S - j)^2 + 4u_S j_L}}{2u_S} \tag{6.13}$$

In this model, the slip velocity u_S is approximated by the constant value 0.2439 m/s. The Reynolds number is calculated with the liquid density ρ_L and the liquid velocity j_L/H_L as characteristic parameters. The equations for estimating the friction pressure gradients are

$$\text{Re} = \frac{\rho_L j_L d}{H_L \mu_L} \tag{6.14}$$

$$\left(\frac{dp}{d\ell}\right)_f = \frac{f\rho_L}{2d}\left(\frac{j_L}{H_L}\right)^2 \tag{6.15}$$

The gravitation pressure gradient is evaluated with the characteristic density ρ_s in the case of bubbly flow:

$$\rho_S = [\rho_L H_L + \rho_G(1 - H_L)] \tag{6.16}$$

When the gas superficial velocity exceeds the value obtained from equation (6.12), the flow pattern would be slug flow. This regime is characterized by the bubble rise velocity u_b, and the relevant Reynolds numbers for slug flow can be computed from

$$u_b = C_1 C_2 \sqrt{gd} \tag{6.17}$$

$$N_{\text{Re},b} = \frac{\rho_L u_b d}{\mu_L} \tag{6.18}$$

$$N_{\text{Re},L} = \frac{\rho_L j d}{\mu_L} \tag{6.19}$$

and the parameters C_1 and C_2 are read from Figures 6.5 and 6.6.

These parameters allow us to calculate the friction component of the pressure gradient, in combination with equations (6.4), (6.6), and (6.13) through (6.19), after an iterative solution for equations is attained jointly with Figures 6.5 and 6.6.

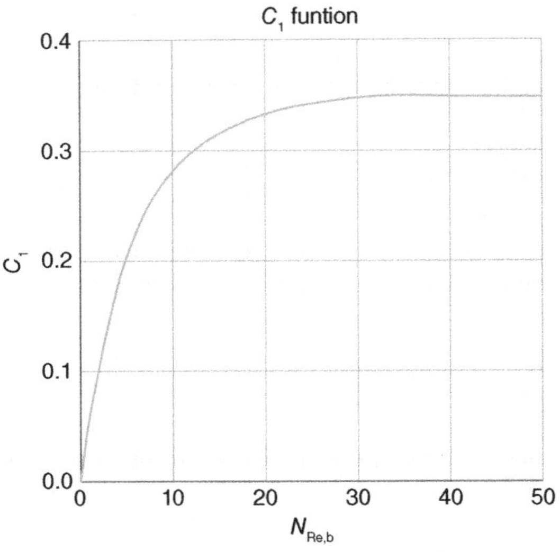

Figure 6.5 C_1 function for Griffith–Wallis model.

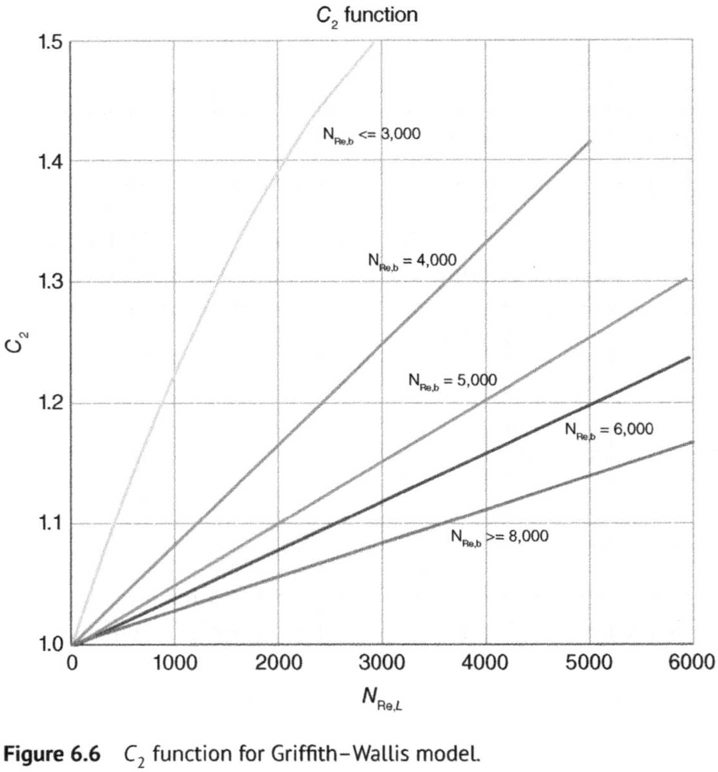

Figure 6.6 C_2 function for Griffith–Wallis model.

The two-phase flow density in the gravitation term shall be evaluated with the following density expression to be used in equation (6.6) in the case of slug flow:

$$\rho_{slug} = \frac{\rho_L(j_L + u_b) + \rho_G j_G}{j + u_b} \tag{6.20}$$

6.6.3 Duns–Ros Model (1963)

This is another model based on empirical field measurements which has been extensively used and still is in vogue in some commercial codes applied for oil wells.

The basis of the model is the flow pattern identification followed by choosing the corresponding correlations. The proposed flow pattern map is illustrated in Figure 6.7 and comprises bubbly/plug, churn/slug, annular-mist, and transition between slug and mist flow.

The methodology starts by calculating three nondimensional parameters – the liquid velocity number N_{LV}, the gas velocity number N_{GV}, and the pipe diameter number N_d,

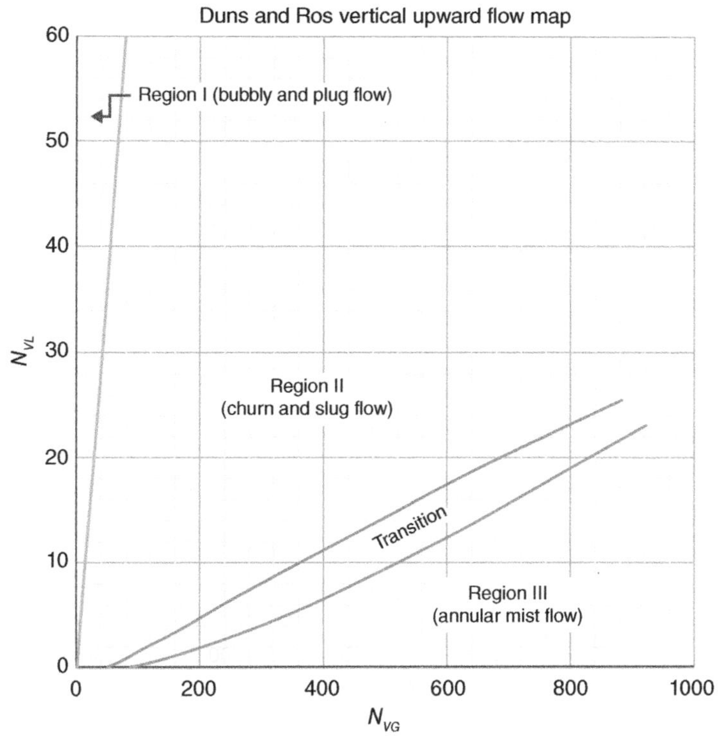

Figure 6.7 Duns and Ros flow pattern map.

- Liquid velocity number:

$$N_{LV} = j_L \left(\frac{\rho_L}{g\sigma_L} \right)^{1/4} \tag{6.21}$$

- Gas velocity number:

$$N_{GV} = j_G \left(\frac{\rho_L}{g\sigma_L} \right)^{1/4} \tag{6.22}$$

- Pipe diameter number:

$$N_d = d \left(\frac{g\rho_L}{\sigma_L} \right)^{1/2} \tag{6.23}$$

The flow pattern is identified based on the parameters L_1 and L_2, from Figure 6.8, and the following criteria:

- *Bubbly and plug flow:*

$$N_{GV} < L_1 + L_2 N_{LV} \tag{6.24}$$

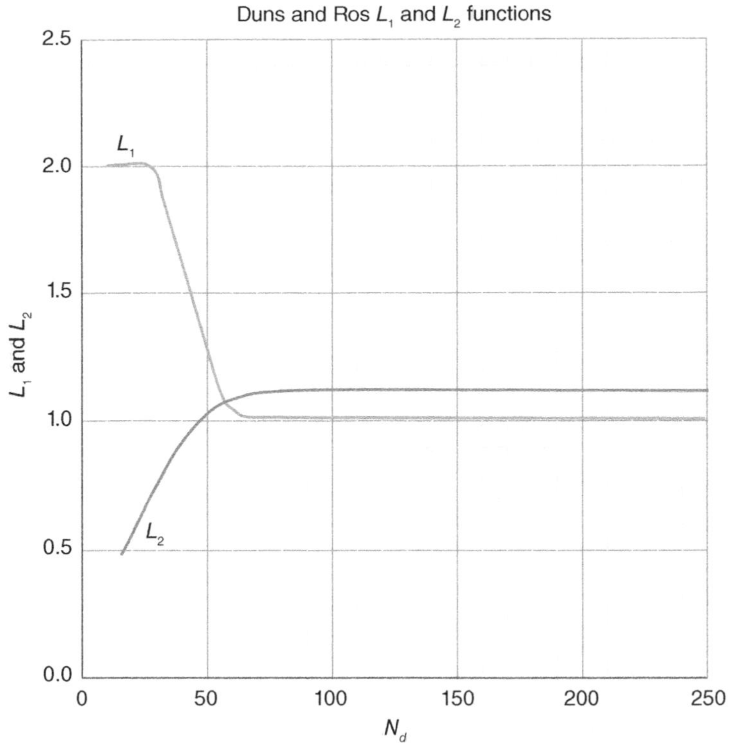

Figure 6.8 Functions L_1 and L_2 to identify flow pattern.

- *Churn and slug flow:*

$$L_1 + L_2 N_{LV} < N_{GV} < 50 + 36N_{LV} \tag{6.25}$$

- *Annular-mist flow:*

$$N_{GV} > 75 + 84N_{LV}^{0.75} \tag{6.26}$$

- *Transition:*

$$50 + 36N_{LV} < N_{GV} < 75 + 84N_{LV}^{0.75} \tag{6.27}$$

The next step in the method is to estimate the slip factor S corresponding to the flow pattern as summarized by the following equations:

- *Bubbly and plug flow:*

$$S = F_1 + F_2 N_{LV} + \left(F_3 - \frac{F_4}{N_d}\right)\left(\frac{N_{GV}}{1 + N_{LV}}\right)^2 \tag{6.28}$$

- *Churn and slug flow:*

$$S = \frac{(1 + F_5)\left(N_{GV}^{0.982} + F_6 + 0.029N_d\right)}{(1 + F_7 N_{LV})^2} \tag{6.29}$$

- *Annular-mist flow:*

$$S = 0 \tag{6.30}$$

The parameters F_1 through F_7 were empirically determined and are presented in Figures 6.9 and 6.10, using the liquid viscosity number N_L from equation (6.38). The liquid holdup can be estimated with equation (6.13), where the slip velocity u_S is calculated by

$$u_S = S\left(\frac{g\sigma_L}{\rho_L}\right)^{1/4} \tag{6.31}$$

For annular-mist flow, the slip velocity is assumed to be zero and the holdup is equal to $1 - \beta$.

Once the liquid holdup is estimated, the characteristic density for this model is ρ_S (equation (6.16)). The two-phase flow friction factor "f" depends on the three parameters f_1, f_2, and f_3, obtained by

- Estimate Reynolds number

$$\mathrm{Re}_L = \frac{\rho_L j_L d}{\mu_L} \tag{6.32}$$

- Read friction factor f_1 Moody diagram (Figure 5.4) using Reynolds number

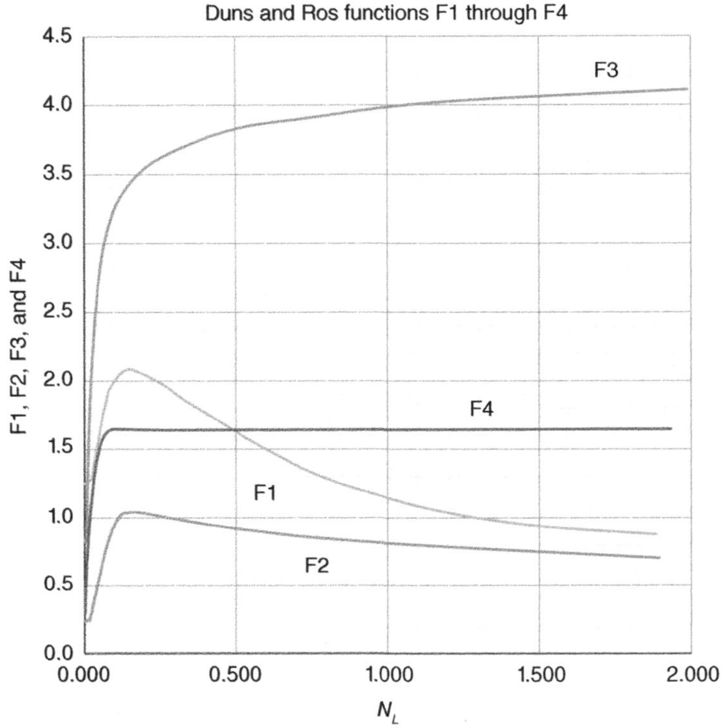

Figure 6.9 Functions F1, F2, F3, and F4.

- Read f_2 from Figure 6.11 with the parameter

$$f_1 \left(\frac{j_G}{j_L} \right) N_d^{0.66} \tag{6.33}$$

- Calculate f_3 from

$$f_3 = 1 + \frac{f_1}{4} \sqrt{\frac{j_G}{50 j_L}} \tag{6.34}$$

- Compute friction factor

$$f = \frac{f_1 f_2}{f_3} \tag{6.35}$$

Finally, the pressure gradients are evaluated with the following equations:

- Friction:

$$\left(\frac{dp}{d\ell} \right)_f = \frac{f \rho_L j_L j}{2d} \tag{6.36}$$

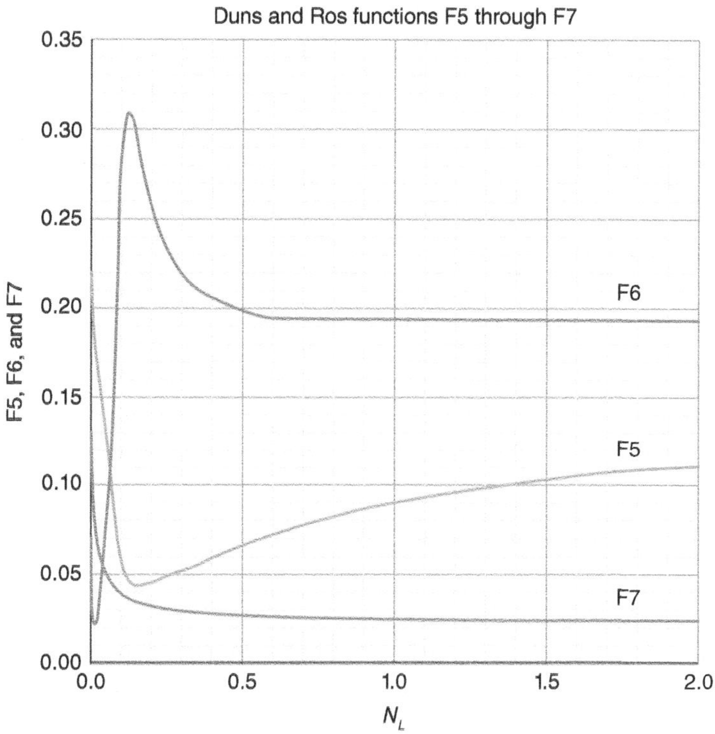

Figure 6.10 Functions F5, F6, and F7.

- Gravitation:

$$\left(\frac{dp}{d\ell}\right)_g = \rho_S g \tag{6.37}$$

6.6.4 Hagedorn–Brown Model (1965)

According to a study done by Ansari et al. (1994), the empirical model of Hagedorn and Brown (*H–B*) produces as good estimates of pressure drop in two-phase flow through oil wells as the mechanistic models available at that date. Currently, this model stands as a regular choice in commercial codes to simulate well performance via TPR.

Aside from the three nondimensional parameters defined by equations (6.21)–(6.23), the *H–G* model needs the liquid viscosity number N_L,

$$N_L = \mu_L \left(\frac{g}{\rho_L \sigma_L^3}\right)^{1/4} \tag{6.38}$$

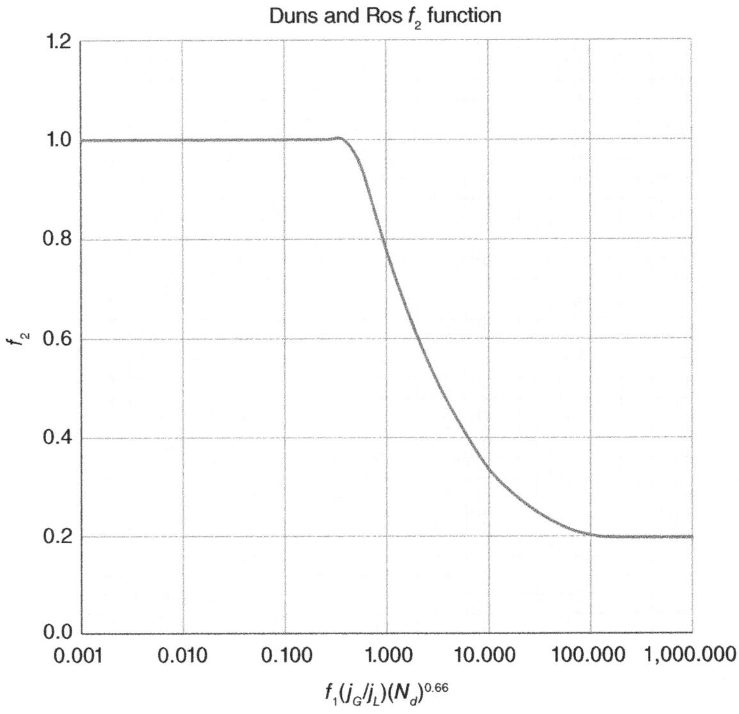

Figure 6.11 Function f_2.

These parameters serve to read the variables N_{LC}, ψ, and H_L/ψ, from Figures 6.12, 6.13, and 6.14.

Once the parameters N_{LC}, ψ, and H_L/ψ are read, the liquid holdup H_L will follow, and the two-phase flow density can be obtained from equation (6.16).

The two-phase flow viscosity can be estimated with the liquid holdup H_L and the following equation:

$$\mu_s = \mu_L^{H_L} \mu_G^{(1-H_L)} \tag{6.39}$$

and the Reynolds number is obtained from equation (6.40) using equation (6.7):

$$\text{Re} = \frac{\rho_n jd}{\mu_s} \tag{6.40}$$

to read the friction factor f from Figure 5.4. Then, the frictional pressure gradient can be obtained from equation (6.41):

$$\left(\frac{dp}{d\ell}\right)_f = \frac{f \rho_n^2 j^2}{2d\rho_S} \tag{6.41}$$

and the gravitation pressure gradient can be calculated using equation (6.10).

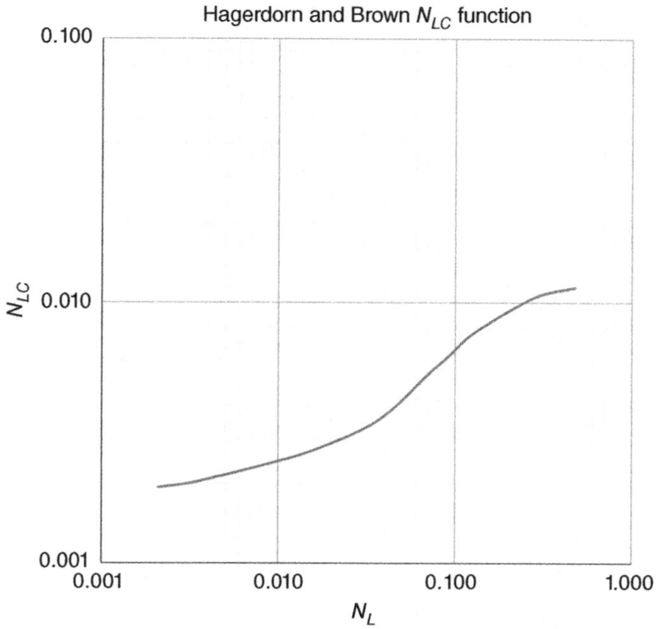

Figure 6.12 Hagedorn–Brown model function N_{LC}.

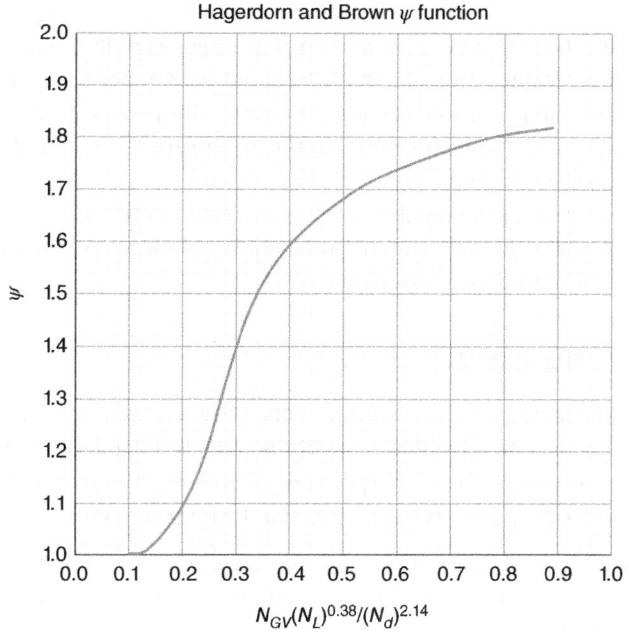

Figure 6.13 Hagedorn–Brown model function ψ.

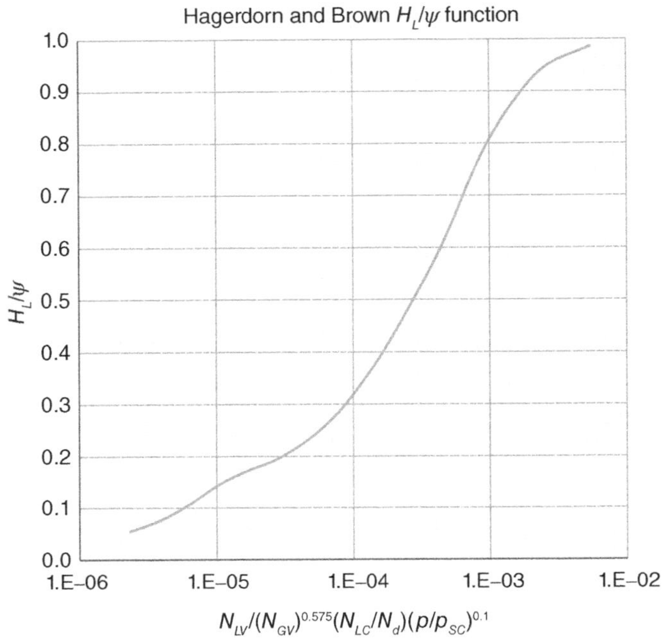

Figure 6.14 Hagedorn–Brown model function H_L/ψ.

The Hagedorn–Brown model can sometimes predict incorrect liquid holdup for the slug flow regime. It is known that under certain circumstances the holdup obtained from Figure 6.14 could be smaller than the value predicted assuming a no-slip condition among the phases, that is, $H_L = 1 - \beta$. In such case, the recommendation is to substitute the holdup factor with the no-slip estimate.

Another improvement in this methodology is to apply Griffith and Wallis's (1961) model when this other model predicts bubbly flow. A similar approach has been followed by Orkiszewski (1967).

6.6.5 Orkiszeski Model (1967)

Orkiszewski (1967) compared the accuracy of the previous well flow correlations available and concluded that none of the models did a decent job predicting the performance of the full spectrum of flow regimes in a well. From the previous models, Orkiszewski borrowed the following models:

- Griffith and Wallis (1961) for bubbly/plug upward flow. The same flow pattern map from this model is used to differentiate between bubbly/plug and slug/churn regimes.

- A modified Hagedorn and Brown (1965) model for slug/churn flow and,
- Duns and Ros (1963) for annular/mist flow. The same flow pattern map from this model is used to differentiate between annular/mist and slug/churn regimes.

In this section, we focus on the modified *H–B* model for slug flow because the other flow regimes have been covered in previous sections.

Once the pattern identification indicates slug flow, the characteristic density is a modification of equation (6.16) to include the effects of slipping liquid film surrounding slugging the Taylor's bubbles and the entrained droplets within the bubbles themselves. This led to add the liquid distribution coefficient Γ,

$$\rho_{slug} = \frac{\rho_L(j_L + u_b) + \rho_G j_G}{j + u_b} + \rho_L \Gamma \tag{6.42}$$

Orkiszewski (1967) provided correlations for the liquid distribution coefficient Γ based on the identification of the continuous phase (either oil or water).

The friction pressure gradient is calculated with

$$\left(\frac{dp}{d\ell}\right)_f = \frac{f \rho_L j^2}{2d} \left[\frac{j_L + u_b}{j + u_b} + \Gamma\right] \tag{6.43}$$

6.6.6 Beggs–Brill Model (1973)

The popular Beggs and Brill (1973) model applies to the inclined and vertically upward flow over a discretized domain, with the support of a compositional fluid model (e.g. phase equilibrium). The results obtained with this model are within +30/−20% compared to mechanistic models or two-fluid models. Despite this level of uncertainty, the model is useful for a first approximation or to quickly discern among design options.

Section 5.13.2 covered the Beggs and Brill model for flowlines, but the same methodology can be applied to wells by appropriate adoption of angle of inclination ϕ.

6.6.7 Gray Model (1978)

Gray (1978) method finds its application in vertical gas-condensate wells, including water production. The restriction of applying this method involves gas velocity up to 15 m/s, tubing ID larger than 89 mm, and liquid loadings exceeding 280 m^3 (condensate) per mmsm^3 (millions of standard cubic meters of gas), and 28 m^3 (water) per mmsm^3 (gas).

This model evaluates the liquid holdup as

$$H_L = 1 - \frac{1 - \exp\left\{-2.314\left[N_V\left(1 + \frac{205}{N_D}\right)\right]^B\right\}}{\left(1 + \frac{j_L}{j_G}\right)} \tag{6.44}$$

where the parameters N_V, B, and N_d are evaluated with

$$N_V = \frac{\rho_n^2 j^4}{g\sigma_L(\rho_L - \rho_G)} \tag{6.45}$$

$$N_d = \frac{g(\rho_L - \rho_G)d^2}{\sigma_L} \tag{6.46}$$

$$B = 0.0814\left[1 - 0.0554Ln\left(1 + \frac{730}{1 + \frac{j_G}{j_L}}\right)\right] \tag{6.47}$$

The liquid holdup is used to calculate the two-phase flow density (equation (6.16)), and the latter allows us to estimate the gravitation term (equation (6.46)).

The friction factor can be read from Figure 5.4 with the following Reynolds number and the modified nondimensional roughness parameter ε/d:

$$\text{Re} = \frac{\rho_s jd}{\mu_s} \tag{6.48}$$

$$\text{for } \frac{j_L}{j_G} > 0.007 \quad \frac{\varepsilon}{d} = \frac{28.5\sigma_L}{\rho_n j^2 d} \tag{6.49}$$

$$\text{for } \frac{j_L}{j_G} < 0.007 \quad \frac{\varepsilon}{d} = \frac{\varepsilon_{single-phase\ gas}}{d} \times\left[1 - \left(\frac{j_L}{0.007 j_G}\right)\right] + \left(\frac{28.5\sigma_L}{\rho_n j^2 d}\right)\left(\frac{j_L}{0.007 j_G}\right) \tag{6.50}$$

6.6.8 Mukherjee and Brill Model (1985)

This model is an improvement on the earlier Beggs and Brill (1973) model, by incorporating newer data from an enlarged database. As such this model is another empirical endeavor aimed to apply to any pipeline orientation, that is, vertical up, inclined, downward, and horizontal as well.

The model strategy is similar to the earlier Duns and Ros (1963) approach of first recognizing the flow pattern and following with an empirical model specific to the identified regime. The next step consists of estimating the liquid holdup of such flow pattern to evaluate the two-phase flow density and the Reynolds number. With these results, the friction factor can be read from the Moody graph, and the gravitation and friction pressure drop can be computed.

The two boundaries separating the bubbly from the slug flow, and the slug from the annular-mist flow are expressed in terms of the tubing inclination θ with respect to the horizon. These boundaries are as follows:

- *Bubbly-Slug flow:*

$$j_{L,bubbly-slug} = \left(\frac{g\sigma_L}{\rho_L}\right)^{1/4} 10^{(Log_{10}N_{GV} + 0.94 + 0.074\ \sin\theta - 0.855\sin^2\theta + 3.695N_L)} \tag{6.51}$$

- *Slug-Annular flow:*

$$j_{G,slug-annular} = \left(\frac{g\sigma_L}{\rho_L}\right)^{1/4} 10^{[1.401 - 2.694N_L + 0.521(N_{LV})^{0.329}]} \tag{6.52}$$

The parameters required in equations (6.51) and (6.52) are obtained with equations (6.21), (6.22), and (6.38). The liquid holdup for slanted wells is computed from

$$H_L = \exp\left[(-0.380 + 0.130\sin\theta - 0.120\sin^2\theta + 2.34N_L^2)\left(\frac{N_{GV}^{0.476}}{N_{LV}^{0.289}}\right)\right] \tag{6.53}$$

The friction pressure drop estimate depends on the flow regime identified with expressions (6.51) or (6.52). The corresponding equation

$$\left(\frac{dp}{d\ell}\right)_{fric,bubbly/slug} = \frac{f\left(N_{\mathrm{Re}}, \frac{\varepsilon}{d}\right)[\rho_L H_L + \rho_G(1 - H_L)]\,j^2}{2d(1 - E_k)} \tag{6.54}$$

$$\left(\frac{dp}{d\ell}\right)_{fric,annular} = \frac{f\left(N_{\mathrm{Re}}, \frac{\varepsilon}{d}\right) f_R[\rho_G\beta + \rho_L(1 - \beta)\,]\,j^2}{2d(1 - E_k)} \tag{6.55}$$

where

$$N_{\mathrm{Re}} = jd\left[\frac{\rho_G\beta + \rho_L(1 - \beta)}{\mu_G\beta + \mu_L(1 - \beta)}\right] \tag{6.56}$$

$$E_k = \frac{j_G\, j[\rho_L H_L + \rho_G(1 - H_L)]}{p} \tag{6.57}$$

$$f_R = 1 \quad \text{for } H_R \equiv \frac{1 - \beta}{H_L} < 0.2 \text{ and } H_R > 1 \tag{6.58}$$

$$f_R = -7.832H_R^4 + 20.13H_R^3 - 19.84H_R^2 + 8.646H_R - 1.008 \quad \text{for } 0.2 \le H_R \le 1 \tag{6.59}$$

Lastly, the gravitation pressure drop is found in equations (6.6) and (6.16).

Example 6.1 ***Legacy Models***
Consider an oil reservoir with an original pressure of 450 bara and temperature of 98 °C. The reservoir will be drilled with a vertical well down to 2900 m, completed with a vertical tubing of 114 mm internal diameter, and casing perforated at 2800 m TVD. The reservoir has been simulated with Fetkovich's model (Economides et al., 2013), assuming an open potential of 600 sm^3/d, and an exponent equal to 0.89. The well has a choke at the tree with a bean size of 50 mm. ***Plot*** the well performance in terms of equivalent oil production (sm^3/d) in the abscissa, and downhole flowing pressure (bara) in the ordinate, for the following models: Duns and Ros (Section 6.6.3), Hagedorn and Brown (Section 6.6.4), Orkiszeski (Section 6.6.5), Beggs and Brill (Section 6.6.6), and Mukherjee and Brill (Section 6.6.8).

The solution to this problem is captured in Figure 6.15, superimposing the five TPR models with the IPR curve.

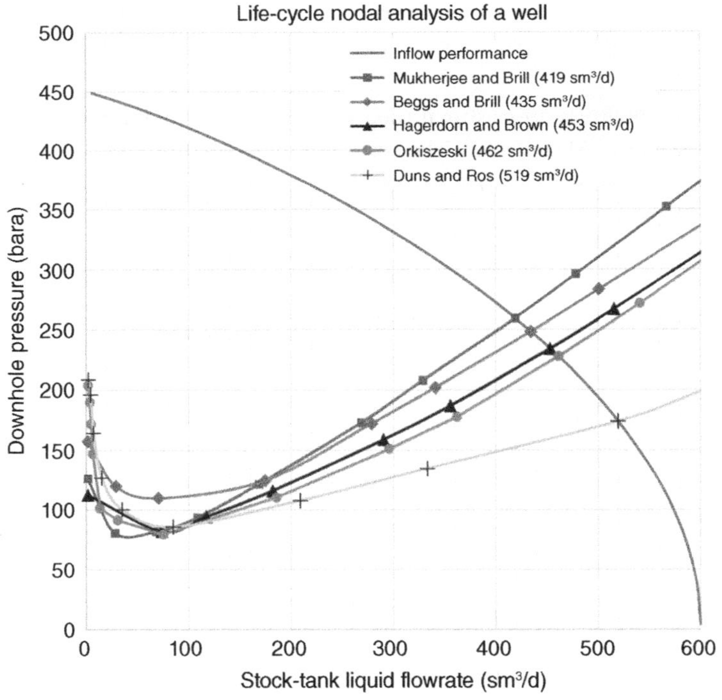

Figure 6.15 Example 6.1 – Comparison of legacy models.

The first conclusion from the previous graph is that the potential oil production may lie within the range of 420–460 sm^3/d, and P_{wf} may be in the range of 230–260 bara. The second conclusion is that Duns and Ros model far overpredicts (+18%) the potential production rate and the well drawdown.

Although there are visible differences among four of the TPR models, the uncertainty in flowrate is ±4.5% which is surprisingly good for modeling two-phase flow for this specific reservoir-well case.

6.7 Mechanistic Models for Well Flow

Mechanistic models refer to involving conservation principles to represent the physical phenomena for each specific flow pattern. Broader and more dependable models (Shoham 2006) have been developed using physical principles with the help of experimental data. Validation of such models has been done at different scales but only to estimate the necessary closure parameters that underpin the core of fundamentals. One of the best keynote lectures on multiphase flow is Brill (2012) who summarized the state of the art in the scope of mechanistic models available to date.

The strategy behind the mechanistic models is the application of fundamentals to particular flow regimes after the proper identification of the pattern has been confirmed. This strategy is aimed at both the hydrodynamic and the heat transfer phenomena involved.

Three mechanistic models are reviewed in this section with a minimum of details due to the added complexity of the models. The avid reader seeking more details can refer to the original publications. Although at first sight the equations presented below might look as simple empirical correlations, as those presented for the legacy models, it is important to keep in mind that mechanistic equations are supported by mass and momentum conservation principles with a minimum of experimental data to close the system of equations.

6.7.1 Ansari et al. Model (1994)

Like all other mechanistic models, the flow pattern map takes a vital role, with distinctive hydrodynamic models for each flow pattern. The model applies to vertical and slanted wells.

To differentiate between bubbly, dispersed, slug/churn, and annular flow, Ansari et al. (1994) selected the flow map of Barnea (1987) which is a unified improvement of the mechanistic model of Taitel et al. (1980).

The flow regime boundaries for the unified model Barnea (1987) are as follows:

- *Bubble-Slug:*

$$j_G = 0.383\left[\frac{g\sigma_L(\rho_L - \rho_G)}{\rho_L^2}\right]^{0.25} + 0.333 j_L \tag{6.60}$$

- *Bubble-Disperse:*

$$j_G = j\left[0.482 j^{1.2}\sqrt{\frac{0.4\sigma_L}{g(\rho_L - \rho_G)}} + \left(\frac{\rho_L}{\sigma_L}\right)^{0.6}\left(\frac{f_{TP}\left(N_{\mathrm{Re}}, \frac{\varepsilon}{d}\right)}{2d}\right)^{0.4} - 0.175\right]^2 \tag{6.61}$$

- *Slug-Annular:*

$$j_G = 3.1\left[\frac{g\sigma_L(\rho_L - \rho_G)}{\rho_G^2}\right]^{0.25} \tag{6.62}$$

- *Dispersed-Slug:*

$$j_G = 3.17 j_L \tag{6.63}$$

This model also uses equations (6.1)–(6.6) to calculate the pressure gradients specific for each flow regime as follows:

- *Bubble and disperse flow*

$$\rho_{TP} = \rho_G\beta + \rho_L(1 - \beta) \tag{6.64}$$

$$\mu_{TP} = \mu_G\beta + \mu_L(1 - \beta) \tag{6.65}$$

$$N_{\mathrm{Re},TP} = \frac{\rho_{TP} j d}{\mu_{TP}} \tag{6.66}$$

$$f_{TP}\left(N_{\mathrm{Re},TP}, \frac{\varepsilon}{d}\right) \tag{6.67}$$

- *Slug and churn flow*

These flow regimes require to estimate the length ratio (ζ) between the Taylor bubble (L_{TB}) and the whole slug unit (L_{SU}) and the liquid holdup in the liquid slug volume (H_{LLS}) – see original paper by Ansari et al. (1994). The slug/churn model estimates the total pressure drop with the following set of equations:

$$\zeta = \frac{L_{TB}}{L_{SU}} \tag{6.68}$$

$$\rho_{LS} = \rho_G(1 - H_{LLS}) + \rho_L H_{LLS} \tag{6.69}$$

$$\mu_{LS} = \mu_G(1 - H_{LLS}) + \mu_L H_{LLS} \tag{6.70}$$

$$N_{\text{Re},TP} = \frac{\rho_{LS}\, jd}{\mu_{LS}} \tag{6.71}$$

$$\left(\frac{dp}{d\ell}\right)_{fric} = \frac{f_{TP}\left(N_{\text{Re},TP}, \frac{\varepsilon}{d}\right)\rho_{LS}\, j^2}{2d}(1-\zeta) \tag{6.72}$$

$$\left(\frac{dp}{d\ell}\right)_{grav} = \rho_{LS} g \sin\theta \tag{6.73}$$

- *Annular-Mist flow*

The model for these flow regimes requires an estimate of the nondimensional liquid film (δ/d) by means of an iterative solution – see original paper by Ansari et al. (1994).

$$u_{crit} = 10{,}000\left(\frac{j_G\mu_G}{\sigma_L}\right)\sqrt{\frac{\rho_G}{\rho_L}} \tag{6.74}$$

$$F_E = 1 - \exp\left[-\frac{u_{crit} - 1.5}{8}\right] \tag{6.75}$$

$$\lambda_{LC} = \frac{1}{1 + \left(\frac{j_G}{F_E\, j_L}\right)} \tag{6.76}$$

$$\rho_C = \rho_G(1 - \lambda_{LC}) + \rho_L\lambda_{LC} \tag{6.77}$$

$$\mu_{SC} = \mu_G(1 - \lambda_{LC}) + \mu_L\lambda_{LC} \tag{6.78}$$

$$u_{SC} = F_E\, j_L + j_G \tag{6.79}$$

$$N_{\text{Re},TP} = \frac{\rho_C u_{SC} d}{\mu_{SC}} \tag{6.80}$$

$$\delta \equiv \frac{\delta}{d} \tag{6.81}$$

$$Z = 1 + 300\delta \quad \text{for} \quad F_E < 0.9, \text{or } Z = 1 + 24\delta\left(\frac{\rho_L}{\rho_G}\right)^{1/3} \text{for } F_E > 0.9 \tag{6.82}$$

$$\left(\frac{dp}{d\ell}\right)_{fric} = \left[\frac{f_{TP}\left(N_{\text{Re},TP}, \frac{\varepsilon}{d}\right)\rho_C v_{SC}^2}{2d}\right]\left[\frac{Z}{(1-2\delta)^5}\right] \tag{6.83}$$

$$\left(\frac{dp}{d\ell}\right)_{grav} = \rho_C g \sin\theta \tag{6.84}$$

6.7.2 Unified Mechanistic Models

Barnea (1986) and Barnea (1987) work initiated the trend to develop unified two-phase flow models for pipe flow of any spatial orientation, including flow pattern transitions. These papers were followed by other unified models by Felizola and Shoham (1995), Gomez et al. (2000a), and Zhang et al. (2003a, 2003b). The idea of a unified model found its cradle at Tulsa University with the *TUFFP* (Tulsa University Fluid Flow Project) since the mid-1980s until now. The objective of the unified model is to capture all potential flow regimes and their hydrodynamic features under a single methodology based on first principles.

The complexity of the unified models goes beyond the intent of this chapter, which is to present streamlined methodologies to simulate vertical and inclined to upflow in pipes, covering the range from fully empirical to phenomenological models. The references listed should be consulted for the minute details of the methodologies described herein.

Gomez et al. (2000b) proposed one valuable contribution to the unified models to improve Ansari et al. (1994) slug/churn model by simplifying the estimation of liquid slug volume (H_{LLS}) with the following expression:

$$H_{LLS} = \exp\left(-0.00785\theta - 0.00000248\frac{\rho_L jd}{\mu_L}\right) \tag{6.85}$$

valid for well inclinations θ (in degrees measured with respect to horizontal) ranging from vertical to horizontal.

6.7.3 Hasan and Kabir Model (2002)

The Hasan and Kabir (2002) model includes its own regime map and holdup correlations specific per flow pattern. The model comprises the following:

Flow pattern map

- *Bubble-Slug flow transition*

$$j_G = \sin\theta\left(\frac{C_0 j_L + 1.53\left[\frac{g\sigma_L(\rho_L-\rho_G)}{\rho_L^2}\right]^{0.25}}{4 - C_0}\right) \tag{6.86}$$

where the parameter C_o is

$$C_0 = 1.2 \text{ if } d < 0.12\,\text{m or if } j_L > 0.02\,\text{m/s}$$
$$C_0 = 2.0 \text{ if } d > 0.12\,\text{m or if } j_L < 0.02\,\text{m/s} \tag{6.87}$$

with the understanding that if the gas superficial velocity is less than the value from equation (6.86), the flow pattern is bubble, and the converse is slug flow.

- *Bubble-Disperse flow transition*

$$j = 3.967 d^{0.4286}\left[\frac{g(\rho_L - \rho_G)}{\sigma_L}\right]^{0.4464}\left(\frac{\sigma_L}{\rho_L}\right)^{0.5357}\left(\frac{\rho_L}{\mu_L}\right)^{0.0714} \tag{6.88}$$

Again, if the mixture velocity is less than the value from equation (6.88), the flow pattern is bubble, and the converse is dispersed flow.

- *Slug-Annular flow transition.* This transition is the same as equation (6.62), derived from the Ansari et al. (1994) model.

Pressure drop

- *Bubble and dispersed flow*

$$H_L = 1 - \frac{j_G}{C_0 j + 1.53\left[\frac{g\sigma_L(\rho_L - \rho_G)}{\rho_L^2}\right]^{0.25}} \tag{6.89}$$

$$N_{\mathrm{Re}_m} = \frac{\rho_L j d}{\mu_L} \tag{6.90}$$

$$\rho_S = \rho_L H_L + \rho_G(1 - H_L) \tag{6.91}$$

$$\left(\frac{dp}{d\ell}\right)_{grav} = \rho_S g \sin\theta \tag{6.92}$$

$$\left(\frac{dp}{d\ell}\right)_{fric} = \frac{f_{TP}\left(N_{\mathrm{Re}_m}, \frac{\varepsilon}{d}\right)\rho_S j^2}{2d} \tag{6.93}$$

- *Slug and churn flow*

$$H_L = 1 - \frac{j_G}{1.2j + 0.35\sqrt{\sin\theta}(1 + \cos\theta)^{1.2}\left[\frac{gd(\rho_L - \rho_G)}{\rho_L}\right]^{0.5}} \tag{6.94}$$

This regime uses equations (6.92)–(6.93) to compute the pressure drop terms.

- *Annular-Mist Flow*

The model starts by calculating the root square of the *Lockhart–Martinelli parameter*:

$$\chi \equiv \left(\frac{1-x}{x}\right)^{0.9}\left(\frac{\rho_G}{\rho_L}\right)^{0.5}\left(\frac{\mu_L}{\mu_G}\right)^{0.1} \tag{6.95}$$

and follows with the next set of equations and equation (6.74):

$$\lambda_{LC} = (1+\chi^{0.8})^{-0.378} \tag{6.96}$$

$$f_C = \frac{0.046}{\left(\frac{\rho_G j_G d}{\mu_G}\right)^{0.2}}(1+75\lambda_{LC}) \tag{6.97}$$

$$F_E = 0.0055 u_{crit}^{2.86} \text{ if } u_{crit} < 4 \tag{6.98}$$

$$F_E = 0.857(\log_{10} u_{crit}) - 0.2 \text{ if } u_{crit} > 4 \tag{6.99}$$

$$\rho_C = \frac{\rho_G j_G + \rho_L j_L F_E}{j_G + j_L F_E} \tag{6.100}$$

$$\left(\frac{dp}{d\ell}\right)_{grav} = \rho_C g \sin\theta \tag{6.101}$$

$$\left(\frac{dp}{d\ell}\right)_{fric} = \frac{f_C \rho_C}{2d}\left(\frac{j_G}{1-\lambda_{LC}}\right)^2 \tag{6.102}$$

Example 6.2 *Mechanistic Models*
Consider again the same oil reservoir and well from Example 6.1, Section 6.8. ***Plot*** the well performance, in terms of equivalent oil production (sm^3/d) in the abscissa, and downhole flowing pressure (bara) in the ordinate, for the Ansari et al. model (Section 6.7.1), and the Unified TUFFP model (Section 6.7.2).

The solution with the mechanistic models is shown in Figure 6.16, superimposing the two TPR models with the same IPR curve from Example 6.1.

The mechanistic models provide remarkably close results with uncertainty of ±1.3%, and P_{wf} in the range of 235–244 bara. This type of result builds confidence in the flowrate achievable and the necessary well drawdown.

Comparing the results of Examples 6.1 and 6.2 leads to conclude that the legacy models provide a similar estimate with a bit more uncertainty. However, the simulation effort to produce results with legacy models is less than the effort of running mechanistic models. Thus, there is a tradeoff of accuracy versus speed of computation.

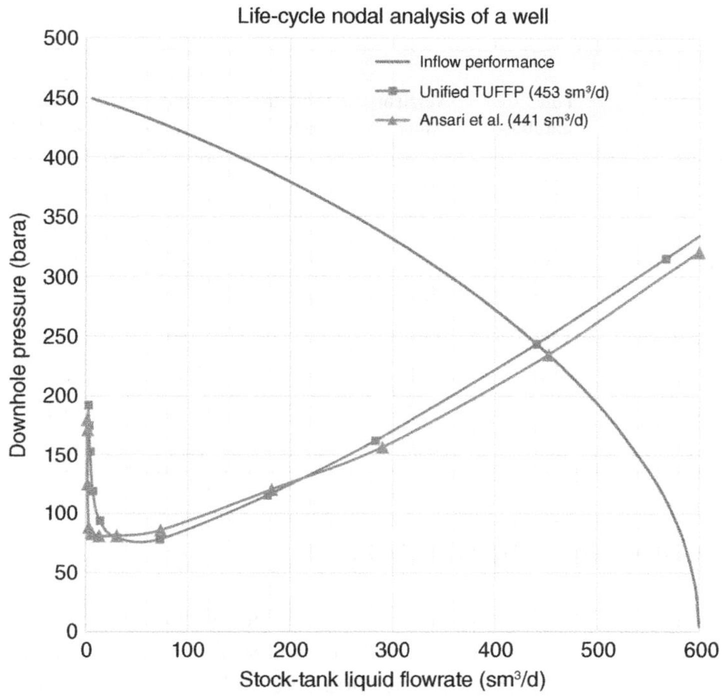

Figure 6.16 Example 6.2 – Comparison of mechanistic models.

6.8 Comparison of Flow Models

Shoham (2006) presents a comprehensive comparison (Table 6.1) among seven well simulation methods. To compare the relative merits of each model, the performance factor FRP was calculated, ranging between "0" (perfect score) and "6" (worst case). The comparison was done with a full database of 1712 experiments, comprising 1086 vertical wells and the rest were deviated wells. The database was evaluated with and without the data from the Hagedorn and Brown (H–B) database.

The results indicate that the two methods that do best are Hagedorn and Brown (legacy) and Ansari et al. (mechanistic). However, 30% of the vertical wells and 19% of the full database were experiments performed by Hagedorn and Brown which tilted the FRP index in favor of their correlation. The last two columns of Table 6.1 show the best performance by Ansari et al. after eliminating the experiments by Hagedorn and Brown from the full database.

It is obvious from the data above that Hagedorn–Brown is a dependable model comparable in performance to a mechanistic model.

Table 6.1 Comparison of seven models with database.

	Full database	Vertical wells	Deviated wells	Vertical wells w/o H-B	Deviated wells w/o H-B
No. experiments	1712	1086	626	755	1381
Ansari et al.	0.7	1.12	1.38	0.08	0
Hagedorn–Brown	0.59	0.6	0.92	0.88	0.77
Duns–Ros	1.72	1.68	1.68	1.71	1.79
Hasan–Kabir	1.94	2.01	2.2	1.84	1.78
Beggs–Brill	2.98	2.91	3.45	3.32	3.41
Orkiszeski	4.26	5.27	2.32	5.84	4.69
Mukherjee–Brill	4.88	4.65	6	3.91	4.6

6.9 Severe Slugging Phenomenon

The mechanistic models presented in the previous section address the phenomenon of slugging in vertical and inclined well geometries, on the basis of a time average. Such averaging should be understood as the statistical performance of multiple slugs with different lengths and frequencies. However, aside from the regular slug flow regime, there is another more harmful condition known as severe slugging.

Severe slugging is an intermittent flow pattern having exceedingly long liquid slugs that form at the last section of a near-horizontal flowline, at the entrance of a riser pipeline. A liquid slug is generated when the incoming flow is larger than the exit liquid flow into the riser. This phenomenon gets enhanced when the flowline has a negative slope toward the base of the riser.

There are four recognized phases in severe slugging that follow a cyclic mechanism. These phases are slug generation, slug production, bubble penetration, and gas blowdown, as described by Schmidt et al. (1980) and Fabre et al. (1990). The pressure build-up in the flowline at some point overcomes the forces acting on the liquid slug and pushes it up through the riser. Once most of the liquid slugs leave at the top, a gas bubble penetrates the riser, and the process ends with gas blowdown at the top. The cycle then reinitiates with more liquid storage at the base of the riser. Figure 6.17 shows a sketch depicting these four phases with an indication of the relative duration of each of these.

The severe slugging hydrodynamic phenomenon is quite complex to model. Numerous laboratory and field tests have been conducted to shed

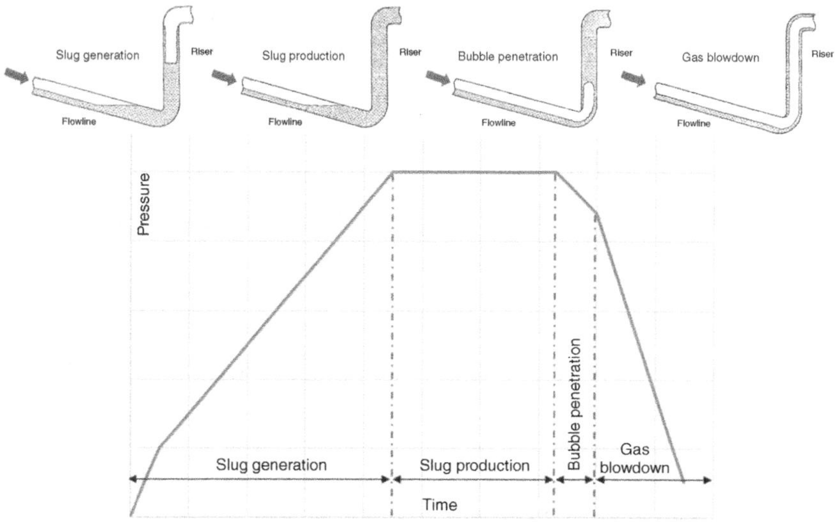

Figure 6.17 Phenomenon of severe slugging.

some light and develop some empirical correlations. The two-phase flow literature is rich in reporting slugging phenomena in vertical pipes and risers. Some other important publications highlighting semi-theoretical and empirical approaches to understand the slugging phenomena include Bøe (1981), Schmidt et al. (1985), Taitel (1986), Pots et al. (1987), Taitel et al. (1990), Sarica and Shoham (1991), Cai et.al. (1999), Shoham (2006), and Al-Safran and Brill (2017).

Pots et al. (1987) offered an important criterion to predict the onset of severe slugging. This criterion is the *severe slugging group* Π_{SS} defined as

$$\Pi_{SS} \equiv \left(\frac{m_G}{m_L}\right)\left[\frac{ZRT}{MW_G L_F g(1-H_L)}\right] \tag{6.103}$$

In this expression, Z is the gas compressibility and T is the absolute temperature, both evaluated at the conditions at the base of the riser. MW_G is the gas molecular weight, H_L is an average value for the liquid holdup along the whole flowline, and L_F is the effective length of the flowline, calculated as the product of the real flowline length times H_L.

According to Pots et al., severe slugging takes place when Π_{SS} is less than one. This criterion leads to the following inequality to identify the severe slugging condition:

$$j_L > j_G \frac{\left(\frac{p}{\rho_L g L_F}\right)}{(1-H_L)} \tag{6.104}$$

Solving for the effective flowline length L_F from the inequality (6.104), we obtain an estimate for the liquid slug length that would trigger severe slugging in the riser:

$$L_F > \left(\frac{j_G}{j_L}\right)\frac{\left(\frac{p}{\rho_L g}\right)}{(1-H_L)} \tag{6.105}$$

In other words, an order of magnitude for the slug length during severe slugging can be estimated as

$$L_{slug} \approx \frac{L_{riser}}{\Pi_{SS}} \tag{6.106}$$

The slug's frequency during severe slugging can be assessed from the following equation:

$$\omega_{slug} \approx \frac{L_{slug}}{j} \tag{6.107}$$

It is important to highlight that the analysis of severe slugging is related to the maximum length of liquid slugs in the flowline upstream of the riser, which has been amply researched for horizontal flow. Slugging phenomenon in flowlines is implicit in the derivation of the *severe slugging group* previously addressed. Brill et al. (1981) conducted a pioneering field-level (12 inch and 16 inch in size) two-phase flow study in 3-mile-long pipelines, finding that the steady-state liquid slug lengths follow a log-normal distribution. These field tests performed at Prudhoe Bay, Alaska, had slug lengths in the range of 80–800 internal diameters, as measured with gamma densitometers. Scott et al. (1987) developed a model that predicted well the Prudhoe Bay slug length data. Scott et al. (1989) presented a correlation for the slug length as a function of flowline's internal diameter (equation (6.108)). Barnea and Taitel (1993) reported laboratory-scale air/water experiments performed by other researchers, determining that L_{slug}/d is in the range 15–40 for horizontal pipes, and L_{slug}/d is in the range 8–25 for vertical pipes.

$$L_{slug} \simeq \exp\{28.5[Ln(d)]^{0.1} - 25.4\} \quad d \text{ in inch and } L_{slug} \text{ in feet} \tag{6.108}$$

Scott et al. (1989) equation (6.108) yields values of L_{slug}/d in the range 280–330 for typical size flowlines in the range 8 inch to 24 inch.

Recently, Al-Safran et al. (2005) published a mechanistic model to estimate the slug length in horizontal pipelines based on probabilistic analyses.

In conclusion, there is significant spread regarding the various ranges of L_{slug}/d observed at laboratory and field scales. Thus, it is important to realize that the *severe slugging group* Π_{SS} was inspired by the works of Schmidt et al. (1980) and Schmidt et al. (1985).

6.10 Gas Lift

A common production technique for ailing oil wells with high water cuts is the injection of compressed gas. Conceptually, the hydrodynamic phenomena taking place within the tubing are no different than the subject addressed in this chapter, and therefore can also be described by the methods listed.

Gas lift refers to a technique, Hernandez (2016) and Guo et al. (2017), aimed to reduce the back pressure on the reservoir by lowering the average density of the produced liquids inside the tubing string. Gas lift (G/L) facilities at the surface treat (dehydrate the gas) and compress the gas to be injected at spaced intervals of the production string. Dehydrated natural gas is necessary to avoid tubing plugging by hydrate formation at the G/L valves.

The gas injected is pumped through the annular space of the well tubing at some pre-determined locations, where pre-installed mandrels will serve to fit gas lift valves. The mandrels are deployed during well completion onto the tubing string.

There are several types of G/L valves for conventional gas lift systems, either continuous or intermittent flow conditions service. The most common valves are injection-pressure-operated (IPO), production-pressure-operated (PPO), pilot-operated gas, and single-point injection orifice valves. Typically, pressure-operated valves open by differential pressure due to the higher gas pressure in the well annulus from the compressed gas at the surface.

Gas lift, aside from its typical application to oil wells, it is frequently applied to deliquefy gas wells when water coning sets in becoming a "liquid-loading" issue.

An additional application for gas lift is to inject gas at the base of a long riser to minimize severe slugging originating from terrain conditions. The objective of this technique is to reduce the liquid accumulations at the base of the riser and the slug sizes traveling up the riser. This practice helps avoid separator flooding at the first stage on the topsides.

References

Al-Safran, E., Sarica, C., Zhang, H., and Brill, J. (2005). *Probabilistic/ Mechanistic Modeling of Slug-length Distribution in a Horizontal Pipeline*, 160–172. SPE Production and Facilities.

Al-Safran, E. and Brill, J. (2017). *Applied Multiphase Flow in Pipes and Flow Assurance*. Dallas, TX, USA: SPE.

Ansari, A., Sylvester, N., and Sarica, C. (1994). A Comprehensive Mechanistic Model for Upward Two-Phase Flow in Wellbores. *SPE Prod. Eng.* 9: 143–152.

Arnold, K. and Stewart, M. (2008). *Surface Production Operations – Design of Oil Handling Systems and Facilities*, 3e. Burlington, MA, USA: Gulf Professional Publishing.

Barnea, D. (1986). Transition from Annular Flow and From Dispersed-Bubble Flow-Unified Models for the Whole Range of Pipe Inclination. *Int. J. Multiphase Flow* 12: 733.

Barnea, D. (1987). A Unified Model for Predicting Flow Pattern Transitions for the Whole Range of Pipe Inclinations. *Int. J. Multiphase Flow* 13: 1.

Barnea, D. and Taitel, Y. (1993). A Model for Slug Length Distribution in Gas-Liquid Slug Flow. *Int. J. Multiphase Flow* 19 (5): 829–838.

Beggs, H. and Brill, J. (1973). A Study of Two-Phase Flow in Inclined Pipes. *J. Pet. Technol.* 255: 607.

Bøe, A., "Severe Slugging Characteristics – Part 1 and 2", Selected Topics in Two-Phase Flow, Trondheim, Norway (1981).

Brill, J., Schmidt, Z., Coberly, W. et al. (1981). Analysis of Two-Phase Tests in Large-Diameter Flow Lines in Prudhoe Bay field. *Soc. Pet. Eng. J.* 21: 363–378.

Brill, J. and Mukherjee, H. (1999). *Multiphase Flow in Wells*, vol. Vol. 17. Dallas, TX, USA: SPE, Monograph.

Brill, J., "Multiphase Flow Technology – Past, Present and Future", PIPESIM User Group Meeting, Houston, Texas, USA (2012).

Dukler, A. and Taitel, Y. (1982). *Lecture Notes of the Two-Phase Gas-Liquid Flow course*. University of Houston, Chemical Engineering Department.

Cai, J., Wang, H., Hong, T., and Jepson, W., "Slug frequency and length in inclined large diameter multiphase pipeline." Multiphase Flow and Heat Transfer Proceedings of the Fourth International Symposium, pp. 195–202, Xi'an, China (August) (1999).

Craft, B.C. and Hawkins, M. (2015). *Applied Petroleum Reservoir Engineering*. New Jersey: Pearson Education Inc.

Duns, H., and Ros, N., "Vertical Flow of Gas and Liquid Mixtures in Wells," Proceedings Sixth World Petroleum Congress, p. 451, Tokio, (1963).

Economides, M.J., Hill, A.D., Ehlig-Economides, C., and Zhu, D. (2013). *Petroleum Production Systems*, 2e. Hoboken, NJ, USA: Pearson Education Inc.

Fabre, J., Peresson, L., Corteville, J. et al. (1990). Severe Slugging in Pipeline/Riser Systems. *SPE Prod. Eng.* 5: 299–305.

Felizola, H. and Shoham, O. (1995). A Unified Model for Slug Flow in Directional Wells. *ASME J. Energy Res. Technol.* 117: 7.

Gomez, L., Shoham, O., Schmidt, Z. et al. (2000a). Unified Mechanistic Model for Steady-State Two-Phase Flow: Horizontal to Vertical Upward Flow. *SPE J.* 5 (3): 339–350.

Gomez, L., Shoham, O., and Taitel, Y. (2000b). Prediction of Slug Liquid Holdup-Horizontal to Upward Vertical Flow. *Int. J. Multiphase Flow* 26 (3): 517.

Gray, H. (1978). *User's Manual for API 14B, SSCSV Sizing Computer Program*, 2e, 38–41. API, Appendix B.

Griffith, P. (1978). *Class Notes Subject 2.57J/22.36J, "Two-phase Flow and Boling and Heat Transfer"*. Cambridge, Massachusetts: MIT.

Griffith, P. and Wallis, G. (1961). Two-phase slug flow. *J. Heat Transfer Trans. ASME, Ser. C* 83: 307–320.

Guo, B., Liu, X., and Tan, X. (2017). *Petroleum Production Engineering*, 2e. Cambridge, Massachusetts, USA: Gulf Professional Publishing.

Hagedorn, A. and Brown, K. (1965). Experimental Study of Pressure Gradients Occurring During Continuous Two-Phase Flow in Small-Diameter Vertical Conduits. *J. Pet. Technol.* 234: 475.

Hasan, A. and Kabir, C. (2002). *Fluid Flow and Heat Transfer in Wellbores*. Dallas, TX, USA: SPE.

Hernandez, A. (2016). *Fundamentals of Gas Lift Engineering*. Cambridge, Massachusetts, USA: Gulf Professional Publishing.

Orkiszewski, J. (1967). Predicting Two-Phase Pressure Drops in Vertical Pipes. *J. Pet. Technol.* 240: 829.

Poettmann, F. and Carpenter, P. (1952). *The Multiphase Flow of Gas, Oil and Water through Vertical Flow Strings with Application to the Design and Gas-Lift Installations*, 257. American Petroleum Institute, Drilling and Production Practice.

Pots, B., Bromlow, I., and Konijn, M. (1987). Severe Slug Flow in Offshore Flowline/Riser System. *SPE Prod. Eng.* 2: 319–324.

Rui, S., Zhang, H., Xu, H. et al. (2023). Seabed Structures and Foundations Related to Deep-Sea Resource Development: A Review Based on Design and Research. *Deep Underground Sci. Eng.* DOI 10.1002/dug2.12042.

Sarica, C. and Shoham, O. (1991). A Simplified Transient Model for Pipeline-Riser Systems. *Chem. Eng. Sci.* 46: 2167–2179.

Schmidt, Z., Brill, J., and Beggs, H. (1980). Experimental Study of Severe Slugging in a Two-Phase-Flow Pipeline-Riser Pipe System. *SPE J.* 20: 407–414.

Schmidt, Z., Doty, D., and Dutta-Roy, K. (1985). Severe Slugging in Offshore Pipeline-Riser Pipe Systems. *SPE J.* 25: 27–38.

Scott, S., Shoham, O., and Brill, J., "Modeling Slug Growth inn Large Diameter Pipes," 3rd International Conference on Multiphase Flow, The Hague, Netherlands pp. 55-63 (1987).

Scott, S., Shoham, O., and Brill, J. (1989). Prediction of Slug Length in Horizontal, Large-Diameter Pipes. *SPE Prod. Eng.* 4: 335–340.

Shoham, O. (2006). *Mechanistic Modeling of Gas-Liquid Two-Phase Flow in Pipes*. Dallas, TX, USA: SPE.

Taitel, Y., Barnea, D., and Dukler, A. (1980). Modelling Flow Pattern Transitions for Steady Upward Gas-Liquid Flow in Vertical Tubes. *AICHE J.* 26: 345.

Taitel, Y. (1986). Stability of Severe Slugging. *Int. J. Multiphase Flow* 12: 203–217.

Taitel, Y., Vierkandt, S., Shoham, O., and Brill, J. (1990). Severe Slugging in a Riser System: Experiments and Modeling. *Int. J. Multiphase Flow* 16: 57–68.

Wallis, G. (1969). *One-Dimensional Two-Phase Flow*. New York, USA: McGraw-Hill Book Company.

Zhang, H., Wang, Q., Sarica, C. et al. (2003a). A Unified Mechanistic Model for Slug Liquid Holdup and Transition Between Slug and Dispersed Bubble Flows. *Int. J. Multiphase Flow* 29: 97–107.

Zhang, H., Wang, Q., Sarica, C. et al. (2003b). Unified Model for Gas-Liquid Pipe Flow via Slug Dynamics. *ASME J. Energy Res. Technol.* 125: 266–283. 350 (2000).

7

Two-Phase Flow Through Restrictions and Piping Components

7.1 Introduction

Section 7.2 covers the main flow restrictions found in the hydrocarbon production systems. Section 7.3 covers the important subject of critical flow, also known as choked flow. The following sections cover specific types of fittings found in production systems.

Most of the information presented in this chapter originated from studies performed for the nuclear industry, in pursuit of mitigating safety concerns before potential catastrophic scenarios such as loss-of-coolant accidents (LOCA). The majority of the models and conclusions drawn for such scenarios do apply to hydrocarbon two-phase flow in pipelines, despite the fact that the technical results were developed for steam–water mixtures.

7.2 Flow Through Restrictions

Generally speaking, there are two groups of fittings regarding the relevance of the associated pressure drop. One group comprises elements that modify the system's deliverability. In this group, we find

- Wellhead choke
- Delivery-point choke
- Flow orifice for system blowdown/depressurization
- Control flow valves

The second group includes flow elements that either ensure the safe operation of the production system or are key components of the gathering system architecture. In general, such elements cause a local pressure drop that is orders of magnitude smaller than the pressure difference between the

Multiphase Transport of Hydrocarbons in Pipes, First Edition.
Juan J. Manzano-Ruiz and Jose G. Carballo.

Companion Website: www.wiley.com/go/Manzano-Ruiz/Multiphasetransport

flowing wellhead pressure and the arrival pressure at the delivery point. This second group includes elements such as,

- Isolation valves operating fully open
- Pressure safety valves (PSVs)
- Check or non-return valves
- Chemical injection metering valves (CIMVs)
- Reducers and expanders
- Elbows
- Splitting-tee junction
- Manifolds and headers with valves fully open
- Combining-tee junction

In this chapter, we review the two-phase flow characteristics across fittings with more emphasis on the first group. The second group of fittings causes negligible effects on the production system deliverability and some impact the two-phase flow at the production processing facilities.

The most comprehensive coverage of the fundamentals of two-phase flow-through restrictions can be found in Delhaye (1981).

7.3 Critical Flow

The *critical flow* phenomenon takes place when the flow rate cannot be increased despite changes in operating conditions downstream from a constriction or throttle section. This is a thermodynamic effect related to the transmission velocity of small-magnitude waves (i.e. fluid compressibility), also referred to as the *speed of sound* in the medium. Perturbations occurring downstream from the throttle cannot travel upstream because the local wave velocity cannot be exceeded. In this situation, the flow is tagged as *critical* or *choked* (Hsu and Graham, 1976).

Critical flow is an important subject for two-phase flow because some fittings such as choke valves and orifices operate under this condition when the downstream-to-upstream pressure ratio adopts some minimum value. In a way, we should look at choked flow as a means to control flow rate, although it can operate either at critical conditions or at subcritical conditions. Moreover, there are other two-phase flow scenarios where choked flow may occur, for instance, during voluntary system blowdown or during a loss-of-containment accident after a piping rupture.

The fundamentals of critical flow for single-phase fluid are very well understood and covered in classical references such as Shapiro (1953) and Liepmann and Roshko (1957). The key parameter for the theory of compressible fluid flow is the *speed of sound*. The estimation of the speed of sound for many single-phase substances can nowadays be easily computed by equations of state with great confidence. Unfortunately, this is not the case for two-phase critical flow where there is ample debate on how to define speed of sound in two compressible media interwoven in complex flow patterns upstream of the constriction.

The intricacy of two-phase critical flow has driven the proposal of several approximate models, namely isenthalpic, isothermal, isentropic, polytropic, and frozen-flow, Levy (1999). Also, models with/without interphase slip or with/without thermal equilibrium have been suggested. The common ground among all these models is the reference state defined as the *stagnation state*, symbolized by the subindex "stag," applied to properties such as pressure, temperature, and density. The term stagnation refers to the fluid(s) brought to rest.

There are two categories of models, either homogeneous or nonhomogeneous flow. The homogeneous model (Griffith, 1978) is the simplest and most straightforward model, supported by the fact that under large acceleration pressure gradients, the flow pattern has a large probability of being dispersed flow. The assumptions for homogeneous flow have been addressed in Section 5.9. For instance, to develop an equation for the maximum (choked) flow with the HEM, the density is expressed by equation (5.40), the dynamic viscosity is given by equation (5.46) and the friction factor is obtained from equation (5.46). Equivalent expressions can be drawn for the homogeneous mixture enthalpy and entropy based on the mass quality x:

$$h_H = h_L(1-x) + h_G x \tag{7.1}$$

$$s_H = s_L(1-x) + s_G x \tag{7.2}$$

Starting with the HEM momentum equation (5.43), and assuming *isentropic* flow, the *speed of sound* (Wallis, 1969) for the homogeneous mixture can be expressed in terms of the single-phase speed of sound for gas and liquid, c_L and c_G,

$$\frac{c_{mix}}{c_L} = \frac{1}{\sqrt{1-\alpha\left(1-\frac{\rho_G}{\rho_L}\right)}\sqrt{\left\{1-\alpha\left[1-\left(\frac{\rho_L}{\rho_G}\right)\left(\frac{c_L}{c_G}\right)^2\right]\right\}}} \tag{7.3}$$

The assumption of constant entropy is an adequate approximation for those fittings which, due to their reduced size, sustain negligible heat losses (adiabatic flow) and friction. The major change in pressure drop in a fitting is usually the acceleration term that dominates over friction and gravity terms. Thus, the isentropic assumption is compatible with the low relevance of friction and heat transfer in fittings.

The speed of sound for the liquid phase c_L can be computed with the following expression (Riazi, 2005):

$$c_L \equiv \sqrt{-\left(\frac{v_L^2}{MW_L}\right)\left(\frac{\partial p}{\partial v_L}\right)_s} = \sqrt{\gamma_L\left(\frac{\partial p}{\partial \rho_L}\right)_T} \tag{7.4}$$

where v_L is the molar volume, and ρ_L is the mass density. The derivative term should be evaluated with an equation of state. As a reference, typical values of sonic speed for live oil vary in the range of 940–990 m/s at the temperature of 60 °C within the pressure range of 1–10 bara. The speed of sound for the gas phase follows from

$$c_G = \sqrt{Z\bar{R}T\gamma_G} \tag{7.5}$$

The homogeneous isentropic model can be further developed into "*frozen*" and "*equilibrium*" models Whalley (1987). The *frozen* model assumes that the mass quality remains constant across the fitting, implying that there is no change of phase during this process. In other words, the very brief residence time of the flow within the fitting, compared to the thermodynamic relaxation time of phase change, justifies the assumption of nonequilibrium and constant mass quality. The *equilibrium* model, on the contrary, assumes thermal equilibrium during the flow process with phase and mass quality changes. Whalley (1987) indicates that the frozen model provides the best estimates for the critical mass flux G_{crit}, whereas the equilibrium model gives the best results for the critical pressure ratio.

The frozen model critical mass flux is given by the following expression:

$$G_{crit} = \sqrt{\frac{\left[\frac{2\xi x_1 p_O}{\rho_{G,O}(\xi+1)}\right]}{\left[\frac{1-x_1}{\rho_{L,O}} + \frac{x_1}{\rho_{G,O}}\left(\frac{2}{\xi+1}\right)^{\frac{1}{1-\xi}}\right]^2}} \tag{7.6}$$

The mass quality x_1 is the value upstream of the fitting where critical flow occurs, and the stagnation properties should be evaluated with an EOS to obtain the parameters through an isentropic path to where the fluid is stagnant. The isentropic expansion index ξ for the mixture should also be evaluated with EOS calculations at the upstream conditions.

The critical pressure ratio for the frozen model is given by the following expression which is equivalent to the single-phase ideal gas case. However, the difference lies in the usage of an isentropic expansion index γ for the ideal-gas mixture case instead of the ξ index.

$$\frac{p_{crit}}{p_O} = \left(\frac{2}{\xi+1}\right)^{\frac{\xi}{\xi-1}} \tag{7.7}$$

The homogeneous equilibrium model can be used to estimate the critical mass flux G_{crit} using the following equation (Levy, 1999):

$$G_{crit} = \frac{\sqrt{2[h_{stag} - (1-x)h_L(p_{crit}) - xh_G(p_{crit})]}}{\left(\frac{1-x}{\rho_L(p_{crit})} + \frac{x}{\rho_G(p_{crit})}\right)} \tag{7.8}$$

In contrast to the frozen model, the equilibrium model requires an iterative solution. The critical pressure at the constriction area must be guessed to evaluate the enthalpies, the fluid densities, and the mass quality from

$$x = \frac{s_{stag} - s_L(p_{crit})}{s_G(p_{crit}) - s_L(p_{crit})} \tag{7.9}$$

The iterative process proceeds with several assumed values of critical pressure until determining the maximum value for the critical mass flux G_{crit} from equation (7.8). This way the critical pressure ratio is obtained too.

7.4 Choke Valves

The choke valve is a key component of every production system. These valves may be placed downstream of the well's tree or at the topsides or process plant downstream of the gathering system.

The choke valves carry several duties, namely

- Gradual start-up of production wells to achieve soft reservoir drawdown while protecting tree valves,
- Gradual shut-in of production wells prior to master and wing valves shutdown,
- Setting the well production rate,
- Pressure maintenance (balancing well clusters) and flow control at manifolds,
- Sustain erosion and corrosion damage from powerful wells,
- Maintain the desired backpressure upon the production system at the arrival point,
- Separator pressure letdown,
- Heater bypass to control gas pressure,
- Backpressure control on well annulus during drilling and cleanup,
- Water injection flow rate control,
- CO_2 well injection flow control for EOR programs,
- Pressure or flow control for gas lift feed to wells,
- Control blowdown rate delivered to flare/vent.

A choke valve may be intended to limit, but not modulate, the flow rate through it. This type of choke is referred to as a valve with *positive or fixed bean*, usually made out of material to withstand high-velocity erosion. A typical body material is AISI 410 stainless steel, lined with tungsten carbide or ceramic to protect it from corrosive and abrasive wear. The fixed choke bean has typical bean size increments of 0.40–1.60 mm (1/64- or 1/16-inch) screwed to the choke box. A choke bean can be as large as 76 mm (192/64 inches).

Flow modulation can be performed with different choke valve types such as *plug and cage*, *multistage trim*, or *external sleeve*.

The flow-regulating choke valves are tested by the manufacturer, and the *choke flow capacity coefficient* C_V (customarily in FPS units, gallons per minute of water at 16°C, for a pressure drop of 1 psi across the valve) is reported to verify the selection adequacy for the expected range of flow rates during production. Figure 7.1 shows a typical shape of the flow coefficient as a function of the opening fraction, usually expressed in 1/64-inch (0.397 mm) intervals.

To determine the flow rate through a choke we have to distinguish between single-phase or two-phase flow. Also, we need to verify whether the choke would be under subcritical or critical conditions.

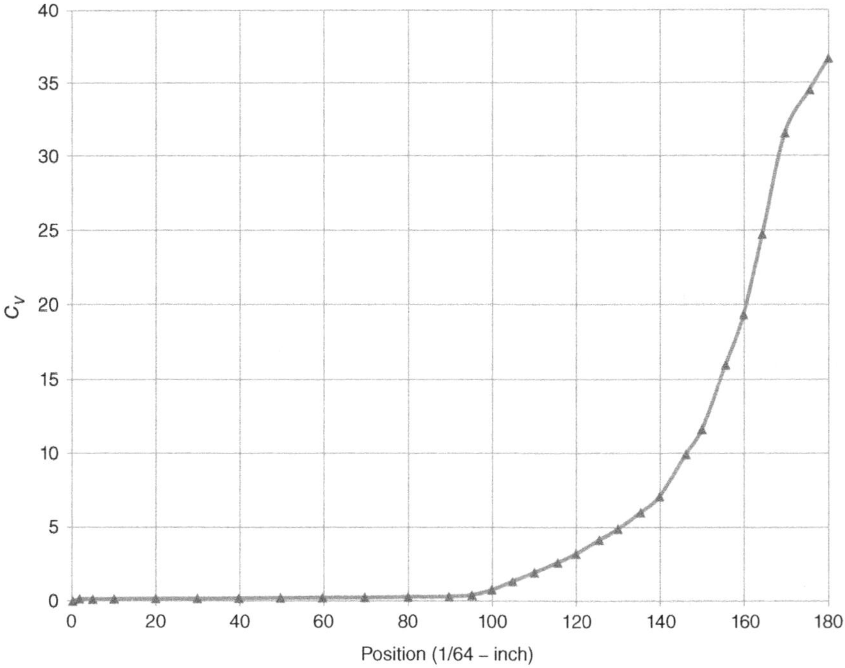

Figure 7.1 C_V coefficient as a function of stem position.

The fundamentals of control valve flow rates apply to choke flow as well with the appropriate empirically determined flow coefficient. The fundamentals for *single-phase flow-through* valves have been condensed in the standard ANSI/ISA-75.01.01 (2012). The concepts addressed by this standard rigorously apply to turbulent, single-phase single-component fluid. For hydrocarbon multicomponent mixtures, the models from ANSI/ISA-75.01.01 (2012) are still a particularly good approximation, if the mixture is homogeneous, in single-phase state, and in thermodynamic equilibrium.

The first step for single-phase choke valve calculation is to ascertain the type of flow, subcritical, or critical (choked flow). ANSI/ISA-75.01.01 (2012) recommends estimating the pressure-drop ratio, the ratio of specific heat factors, and the liquid pressure-recovery factor using the following equations:

$$x = \left(1 - \frac{p_2}{p_1}\right) \tag{7.10}$$

$$F_\gamma = \left(\frac{\xi}{1.4}\right) \tag{7.11}$$

$$F_F = 0.96 - 0.28\sqrt{\frac{p_{vap}}{p_{crit}}} \tag{7.12}$$

F_γ is a factor that compensates for the way valves are tested with air (IEC60534-2-3, 2015), and the results can be extrapolated to other fluids with equation (7.12).

The pressure-drop ratio factor at choked flow condition, x_T, is measured and supplied by the choke manufacturer. An approximate correlation of representative x_T values for different choke sizes (Emerson-Fisher, 2019) is equation (7.13):

$$x_T = 0.657 + 0.0134 d_{ch.,nom.} \tag{7.13}$$

Likewise, the F_L factor is also supplied by the valve manufacturer. A correlation of representative values for different choke sizes (Emerson-Fisher, 2019) is

$$F_L = \frac{0.872}{d_{ch.,nom.}^{0.05}} \tag{7.14}$$

Based on ANSI/ISA-75.01.01 (2012), the criterion to identify subcritical flow becomes

$$\frac{p_2 - p_1}{F_L^2[p_1 - F_F p_{vap}]} < 1 \tag{7.15}$$

For single-phase undersaturated oil flow, the approximate volumetric flow rate for subcritical conditions is ANSI/ISA-75.01.01 (2012),

$$Q_L = 0.865C_V\sqrt{\frac{p_1 - p_2}{S_L}} \tag{7.16}$$

For single-phase undersaturated oil flow, the approximate volumetric flow rate for critical conditions (ANSI/ISA-75.01.01, 2012) is given by the following equation, although it is unlikely that the flow becomes critical due to the exceedingly high speed of sound in the liquid state:

$$Q_L = 0.865C_V\sqrt{\frac{F_L^2[p_1 - F_F p_{vap}]}{S_L}} \tag{7.17}$$

The units for the terms in equations (7.15) and (7.16) are pressure [bar absolute], C_V [US GPM], S_L is the dimensionless specific gravity of the oil, $d_{ch.,nom.}$ is the nominal size of the choke unit [inches], and the oil flow rate Q_L (m^3/s). An EOS is needed to estimate the oil critical pressure P_{crit} and the bubble point P_b on an isentropic path from the upstream condition p_1 and T_1 to the choked pressure p_2.

For gas flows in chokes, there is another criterion quite common to identify critical flow in chokes. This second criterion derives from the isentropic flow assumption for ideal gases:

$$\frac{p_{2,crit}}{p_1} = \left(\frac{2}{\xi + 1}\right)^{\frac{\xi}{\xi-1}} \tag{7.18}$$

where

$$\xi \equiv \frac{c_p}{c_v} \tag{7.19}$$

The heat capacity ratio ξ for gas hydrocarbons normally lies in the range 1.25–1.32, indicating that the critical pressure ratio for gases is of the order of 0.54–0.56. In other words, critical flow in gases takes place when the choke discharge pressure is about half the inlet pressure. For compressible gas single-phase flow, the *subcritical flow* rate equation is

$$Q_G = 2,250C_V p_1\left[1 - \frac{x}{2.143\xi x_T}\right]\sqrt{\frac{x}{T_1 Z_1 MW}} \tag{7.20}$$

and the *critical flow* rate equation is

$$Q_G = 1,268C_V p_1\sqrt{\frac{\xi x_T}{T_1 Z_1 MW}} \tag{7.21}$$

The gas flow rate in the last two equations is expressed in (sm^3/min). S_G is the nondimensional *specific gravity* of the gas, and Y is the *expansion factor* which has a limit value of 2/3 for choked flow (ANSI/ISA-75.01.01, 2012).

Single-phase flow fundamentals are very well supported by accurate experimental work, and the theory behind it has been compiled in standards. Unfortunately, this is not the case for two-phase flow choke flow. Modeling two-phase choke flow is much more challenging and empirical, fully dependent on data supplied by the manufacturers.

There are several models available for two-phase choke flow and in this section, we cover the highlights of the pioneering homogeneous model by Ashford and Pierce (1975) and a more recent comprehensive non-homogeneous model by Al-safran and Kelkar (2007).

Ashford and Pearce (1975) published a criterion to identify critical flow. This criterion requires trial-and-error to solve the implicit equation (7.22) for the critical pressure ratio y_c,

$$\frac{\left(\frac{2R}{\xi}\right)\left[\left(\frac{R\xi}{\xi-1}\right)\left\{1-y_c^{\left(\frac{\xi-1}{\xi}\right)}\right\}+1-y_c\right]}{\left[1+\frac{R}{y_c^{\left(\frac{1}{\xi}\right)}}\right]y_c^{\left(\frac{\xi+1}{\xi}\right)}}=1 \tag{7.22}$$

$$y_c=\frac{p_{2,crit}}{p_1} \tag{7.23}$$

$$R=\frac{j_G}{j_L} \tag{7.24}$$

Al-safran and Kelkar (2007) also proposed another criterion to detect critical flow, requiring the calculation of the *polytropic index* n, the *slip-factor* S, the parameter η, and the iterative solution of the following implicit equation in y_c:

$$n=\frac{x\xi c_{V,G}+(1-x)c_L}{xc_{V,G}+(1-x)c_L} \tag{7.25}$$

$$S=(1+0.6e^{-5x})\sqrt{1+x\left(\frac{\rho_L}{\rho_G}-1\right)} \tag{7.26}$$

$$\eta\equiv\frac{S(1-x)\rho_G}{x\rho_L} \tag{7.27}$$

$$\frac{\eta(1-y_c)+\frac{n}{n-1}}{\left[y_c^{\left(\frac{n-1}{n}\right)}\right]\left[\left(\frac{n}{n-1}\right)+\left(\frac{n}{2}\right)\left(1+\eta y_c^{\left(\frac{1}{n}\right)}\right)^2\right]}=1 \tag{7.28}$$

Generally, the estimation of the critical pressure ratio by equations (7.22) and (7.28) yields similar results within the accuracy of both models.

Additionally, Al-safran and Kelkar (2007) included equations to estimate the mass flow rate of two-phase flow-through subcritical and critical chokes.

For *subcritical chokes*, Al-safran and Kelkar (2007) recommend the following set of equations:

$$S = \left(\frac{\rho_L}{\rho_G}\right)^{0.17} \tag{7.29}$$

$$\eta \equiv \frac{S(1-x)\rho_G}{x\rho_L} \tag{7.30}$$

$$y = \frac{p_2}{p_1} \tag{7.31}$$

$$\dot{m} = \sqrt{\frac{0.75A_{choke}^2 p_1 \left\{\eta(1-y) + \frac{n}{n-1}\left[1 - y^{\left(\frac{n-1}{n}\right)}\right]\right\}}{\frac{x_1}{\rho_{G,1}}\left[\eta + \frac{1}{y^{\left(\frac{1}{n}\right)}}\right]^2 \left[x_1 + \frac{1-x_1}{S}\right]}} \tag{7.32}$$

And for *critical chokes* Al-safran and Kelkar (2007) recommend this other set of equations:

$$S = (1 + 0.6e^{-5x})\sqrt{1 + x\left(\frac{\rho_L}{\rho_G} - 1\right)} \tag{7.33}$$

$$y_c = \frac{p_{2,crit}}{p_1} \tag{7.34}$$

$$\dot{m} = \sqrt{\frac{0.75A_{choke}^2 p_1 \left\{\eta(1-y_c) + \frac{n}{n-1}\left[1 - y_c^{\left(\frac{n-1}{n}\right)}\right]\right\}}{\frac{x_1}{\rho_{G,1}}\left[\eta + \frac{1}{y_c^{\left(\frac{1}{n}\right)}}\right]^2 \left[x_1 + \frac{1-x_1}{S}\right]}} \tag{7.35}$$

From the previous equations it is evident that two-phase critical flow through a production choke is more difficult to estimate than single-phase critical flow.

A comprehensive review of choke flow can be found in Haug (2012) thesis.

7.5 Sudden Pipe Enlargement

A sudden expansion in pipe diameter, although it is not common in hydrocarbon production systems, slows down the fluids' velocity and augments the local pressure. Figure 7.2 shows a sketch of a pipe expansion.

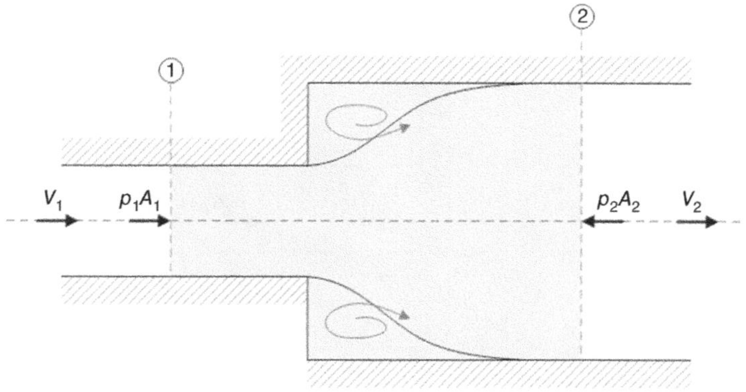

Figure 7.2 Sudden pipe expansion.

The sudden expansion creates a recirculation zone where turbulent eddies dissipate mechanical energy. Thus, the increment in pressure associated with the momentum change is diminished by local losses due to turbulence.

The pressure increment across an abrupt diameter enlargement in the system can be estimated using Chisholm (1986) model:

$$\frac{p_2 - p_1}{\frac{G_1^2}{\rho_L}} \cong \left(\frac{d_1}{d_2}\right)^2 \left[1 - \left(\frac{d_1}{d_2}\right)^2\right] (1-x)^2 \left[1 + \frac{C}{\chi} + \frac{1}{\chi^2}\right] \tag{7.36}$$

where χ is the parameter calculated (Collier and Thome, 1994) with equation (7.37):

$$\chi = \left(\frac{1-x}{x}\right)\left(\frac{\rho_G}{\rho_L}\right)^{0.5} \tag{7.37}$$

and the constant "C" can be estimated (Collier and Thome, 1994) from

$$C = \left\{1 + \left[\left(\frac{\rho_G}{\rho_L}\right)^{0.167} - 1\right]\sqrt{\frac{\rho_G}{\rho_L - \rho_G}}\right\}\left[\sqrt{\frac{\rho_G}{\rho_L}} + \sqrt{\frac{\rho_L}{\rho_G}}\right] \tag{7.38}$$

Additional coverage about flow-through pipe enlargement can be found in Salcudean (1986).

7.6 Sudden Pipe Contraction

A sudden contraction in pipe diameter accelerates the fluids' velocity and reduces the local pressure. Figure 7.3 shows an outline of a pipe contraction.

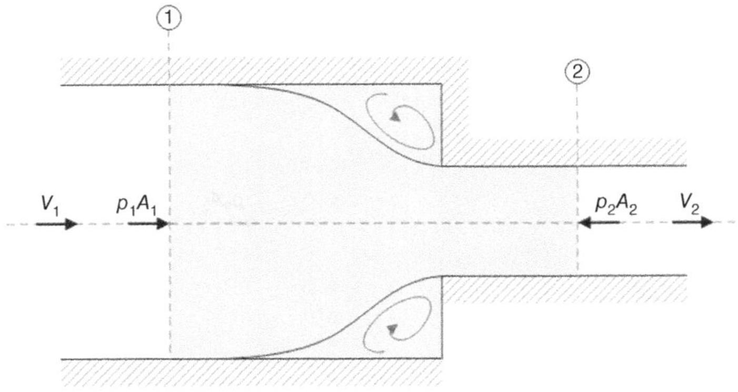

Figure 7.3 Sudden pipe contraction.

The pressure drop across an abrupt diameter reduction in the production system can be computed with equation (7.39), (Collier and Thome, 1994) and (Lahey and Moody, 1993):

$$\frac{p_1 - p_2}{\frac{G_2^2}{\rho_L}} \cong \frac{1}{2}\left(\frac{d_1}{d_2}\right)^2 \left[1 + \left(\frac{1}{C_C} - 1\right)^2 - \left(\frac{d_2}{d_1}\right)^2\right] \left[1 + x\left(\frac{\rho_L}{\rho_L - \rho_G}\right)\right] \tag{7.39}$$

and the constriction coefficient C_C can be approximated (Collier and Thome, 1994) by

$$C_C \cong 0.573\left(\frac{d_2}{d_1}\right)^8 - 0.588\left(\frac{d_2}{d_1}\right)^6 + 0.447\left(\frac{d_2}{d_1}\right)^4 - 0.018\left(\frac{d_2}{d_1}\right)^2 + 0.586 \tag{7.40}$$

Further material on pipe contraction flow can be found in Salcudean (1986).

7.7 Flow Orifice

An orifice plate is a fitting commonly used to fix the flow rate of the production stream. Thus, it is quite normal that the flow will be choked at this fitting. The choked flow estimation is based on the methodology addressed in Section 7.3.

Fluid flows through an orifice of area A_0 undergo a deformation of the streamlines forming a neck (i.e. "*vena contracta*") downstream of the plate, usually about one orifice's radius from the plate. The cross-section of the vena contracta is labeled A_c, and the ratio A_c/A_o is slightly less than 1.

To estimate the pressure drop across an orifice plate in *two-phase flow*, first evaluate the parameters ζ and κ with the equations (7.41), (7.42), and (7.43):

$$\zeta = \left(\frac{1-x}{x}\right)\left(\frac{\rho_G}{\rho_L}\right)^{0.5} \tag{7.41}$$

$$\text{For } \zeta \geq 1 \quad \kappa = \sqrt{1 + x\left(\frac{\rho_L}{\rho_L - \rho_G}\right)} \tag{7.42}$$

$$\text{For } \zeta < 1 \quad \kappa = \left(\frac{\rho_L}{\rho_G}\right)^{0.25} \tag{7.43}$$

Next evaluate the constant C from (7.44) at the vena contracta flow area,

$$C = \kappa\sqrt{\frac{\rho_G}{\rho_L}} + \frac{1}{\kappa}\sqrt{\frac{\rho_L}{\rho_G}} \tag{7.44}$$

$$A_c = C_D A_c \tag{7.45}$$

The *discharge coefficient* C_D can be estimated from Appendix A, page A-20 of Crane Corp (1988), as a function of the Reynolds number and the area ratio (orifice area to approach pipe area). For Reynolds' number above 10,000, the discharge coefficient varies in the range 0.61–0.64 depending on the area ratio. Finally, the pressure drop across the orifice follows from equation (7.46):

$$p_1 - p_2 \cong \left(\frac{1}{2\rho_L}\right)\left[1 + \frac{C}{\zeta} + \frac{1}{\zeta^2}\right]\left[\frac{G_1(1-x_1)A_1}{A_c}\right]^2 \tag{7.46}$$

For further details of two-phase flow-through orifices, please refer to the publication by Lin (1986).

7.8 Gate and Globe Valves

Gate valves are isolating fittings which have two operating positions: fully open or shut off. For an open gate valve, the constant C_c in equation (7.39) is

$$\textit{gate valve} \quad C_c = \left\{1 + 0.5\sqrt{\frac{\rho_G}{\rho_L - \rho_G}}\right\}\left[\sqrt{\frac{\rho_G}{\rho_L}} + \sqrt{\frac{\rho_L}{\rho_G}}\right] \tag{7.47}$$

Globe valves are used to control fluid flow rate and can operate between fully open or partly shut off. For an open globe valve, the constant C in

Chisholm's equation (7.39) is

$$\text{globe valve} \quad C_c = \left\{1 + 1.3\sqrt{\frac{\rho_G}{\rho_L - \rho_G}}\right\}\left[\sqrt{\frac{\rho_G}{\rho_L}} + \sqrt{\frac{\rho_L}{\rho_G}}\right] \tag{7.48}$$

The two-phase flow pressure drop across open gate and globe valves can then be estimated using equation (7.49):

$$p_1 - p_2 \cong \left(\frac{1}{2\rho_L}\right)\left[1 + \frac{C_c}{\zeta} + \frac{1}{\zeta^2}\right]\left[\frac{G_1(1 - x_1)A_1}{A_c}\right]^2 \tag{7.49}$$

For a globe valve partly open, the pressure-drop calculation is similar to the orifice plate procedure with the caveat that the discharge coefficient has to be determined empirically or obtained from the valve manufacturer.

7.9 Elbows and Bends

The two-phase flow-through bends can be estimated with Chisholm's (1986) model and a correlation for the constant C. The constant equation for bends is

$$C = \left\{1 + \left[\frac{2.2}{\left(2 + \frac{R}{d}\right)k_{fo}}\right]\sqrt{\frac{\rho_G}{\rho_L - \rho_G}}\right\}\left[\sqrt{\frac{\rho_G}{\rho_L}} + \sqrt{\frac{\rho_L}{\rho_G}}\right] \tag{7.50}$$

where the loss coefficient for bends is approximated by

$$k_{fo} \cong 0.675 d_{nom.}^{-0.208} \tag{7.51}$$

The two-phase flow pressure drop across elbows can then be estimated relative to the single-phase liquid flow with equation (7.52) and equation (7.37).

$$\frac{(p_1 - p_2)_{Two-phase}}{(p_1 - p_2)_{SP,liquid}} = 1 + \frac{C}{\chi} + \frac{1}{\chi^2} \tag{7.52}$$

7.10 Tee Junctions and Manifolds

Mixing and splitting tees are common elements of gathering systems. These elements serve to commingle production or to divide the flow. Figure 7.4 shows a sketch of a mixing tee.

Applying a simple homogeneous model (Azzopardi, 1986) across the boundary of the control volume of Figure 7.4, and setting up steady-state

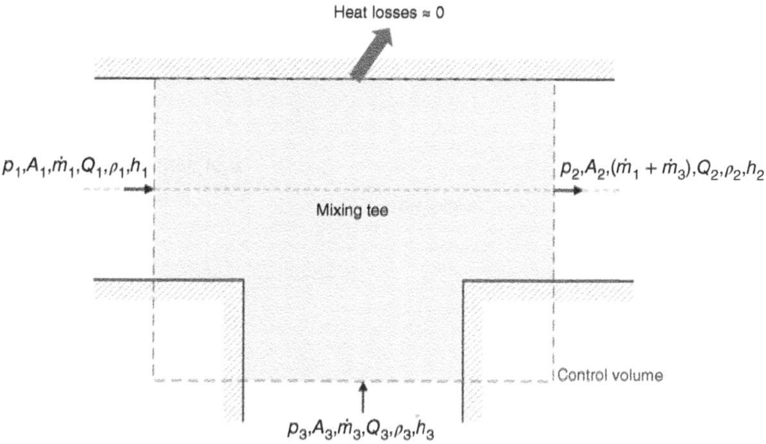

Figure 7.4 Schematics of a mixing/joining adiabatic tee.

momentum balances (neglecting gravitational effects) yields

$$p_1 - p_2 = \frac{1}{2}\left(\rho_2 j_2^2 - \rho_1 j_1^2\right) + \frac{1}{2}\rho_2 j_2^2 \left[0.045 + 1.38\left(\frac{j_3}{j_2}\right)\right] \quad (7.53)$$

$$p_3 - p_2 = \frac{1}{2}\left(\rho_2 j_2^2 - \rho_3 j_3^2\right) + \frac{1}{2}\rho_2 j_2^2 \left[1.09 - 0.53\left(\frac{j_1}{j_2}\right)\right] \quad (7.54)$$

The last terms of equations (7.53) and (7.54) (Blevins, 1992), respectively, represent the mechanical-energy losses of blending stream 1 with stream 3. Therefore, the pressure drops from these equations account for the necessary incremental static pressure upstream of the mixing tee. Also, due to mass conservation, equation (7.55) holds for the two-phase mixing junction tee, as long as the overall flow rate is high enough to be approximated by a homogeneous model:

$$j_2 = \left(\frac{\rho_1}{\rho_2}\right) j_1 + \left(\frac{\rho_3}{\rho_2}\right)\left(\frac{d_3}{d_1}\right)^2 j_3 \quad (7.55)$$

From a practical point of view, the mixture densities at the three ports of the control volume may vary slightly, enabling to approximate equation (7.55) by equation (7.56):

$$j_2 = j_1 + j_3\left(\frac{d_3}{d_1}\right)^2 \quad (7.56)$$

Equations (7.53) to (7.56) can support the simulation of mixing tees on flowlines (e.g. daisy-chain tie-ins) and trunklines, or for the simulation of a manifold collecting the flow from several wells from a tie-back.

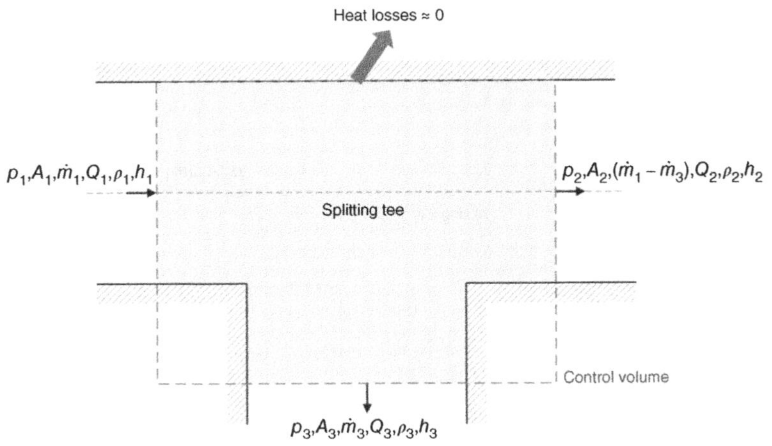

Figure 7.5 Splitting/dividing adiabatic tee.

Mixing tees in general may pose a simulation challenge regarding the estimation of fluid properties downstream of the fitting. When streams 1 and 3 (Figure 7.4) have different compositions, it is necessary to perform compositional analysis invoking phase equilibrium computations with an equation of state. Not only a local thermodynamic analysis is required, but mass balances (per species) and an energy balance across the tee depend upon the flow rates of streams 1 and 3. In conclusion, the presence of mixing tees in the production system becomes an expensive item in terms of simulation computing time, due to the need of including compositional tracking if a rigorous job is intended.

Splitting or dividing tees can be present when "system looping" is utilized to reduce the system backpressure on the production sources. Figure 7.5 illustrates a splitting tee.

One of the research challenges in two-phase flow-through diving tees is the resulting flow patterns at the exiting streams and the length of pipe necessary to develop a stable regime downstream. For further details about flow regimes, consult the reference by Azzopardi (1986).

The main characteristic of the splitting tee is that the fluid properties of the inlet stream 1 are shared by the two exiting streams 2 and 3, eliminating the need to run compositional analysis.

Similar models can be drawn for a splitting tee, invoking the homogeneous flow model to represent high flow rate two-phase flows. These are

$$p_1 - p_2 = \frac{1}{2}\left(\rho_2 j_2^2 - \rho_1 j_1^2\right) + \frac{1}{2}\rho_1 j_1^2 \left[1.55\left(0.22 - \frac{j_3}{j_1}\right)^2 - 0.03\right]$$

$$\text{if } \frac{j_3}{j_1} \leq 0.22 \tag{7.57}$$

$$p_1 - p_2 = \frac{1}{2}\left(\rho_2 j_2^2 - \rho_1 j_1^2\right) + \frac{1}{2}\rho_1 j_1^2 \left[0.65\left(0.22 - \frac{j_3}{j_1}\right)^2 - 0.03\right]$$

$$\text{if } 0.22 < \frac{j_3}{j_1} \leq 1 \tag{7.58}$$

$$p_1 - p_3 = \frac{1}{2}\left(\rho_3 j_3^2 - \rho_1 j_1^2\right) + \frac{1}{2}\rho_1 j_1^2 \left[0.99 - 0.82\left(\frac{j_3}{j_1}\right) + 1.02\left(\frac{j_3}{j_1}\right)^2\right] \tag{7.59}$$

$$j_2 = j_1 - \left(\frac{d_3}{d_1}\right)^2 j_3 \tag{7.60}$$

References

ANSI/ISA-75.01.01-2012 (60534-2-1 MOD) *Industrial – Process Control Valves – Part 2-1* (2012).

Al-Safran, E. and Kelkar, M. (2007). *Predictions of Two-phase Critical Flow Boundary and Mass Flow Rate Across Chokes*. Richardson, Texas, USA: SPE Paper 109243.

Ashford, F. and Pierce, P. (1975). Determining Multiphase Pressure Drops and Flow Capacities in Down-hole Safety Valves. *Journal of Petroleum Technology* 27: 1145–1152.

Azzopardi, B. (1986). Two-phase Flows in Junctions," Chapter 25. In: *Encyclopedia of Fluid Mechanics, Vol. 3, Gas-Liquid Flows*. Houston, USA: Gulf Publishing Company.

Blevins, R. (1992). *Applied Fluid Dynamics Handbook*. Malabar, Florida, USA: Krieger Publishing Company.

Chisholm, D. (1986). Predicting Two-phase Flow Pressure Drop," Chapter 19. In: *Encyclopedia of Fluid Mechanics, Vol. 3, Gas-Liquid Flows*. Houston, USA: Gulf Publishing Company.

Collier, J. and Thome, J. (1994). *Convective Boiling and Condensation*, 3e. Oxford, U.K.: Oxford Science Publications.

Crane Corporation (1988). Flow of Fluids Through Valves, Fittings, And Pipe, Technical Paper No. 410, The Woodlands, Texas, Crane Valves North America.

Delhaye, J. (1981). *Two-Phase Flow and Heat Transfer in the Power and Process Industries*," Chapter 4 (ed. A. Bergles). New York, NY, USA: Hemisphere Publishing Corporation.

Emerson-Fisher (2019). *Control Valve Handbook*, 5e. Iowa, USA: Emerson-Fisher.

Griffith, P. (1978). *Class notes Subject 2.57J/22.36J, "Two-phase Flow and Boling and Heat Transfer"*. Cambridge, Massachusetts: MIT.

Haug, R. (2012). *Multiphase Flow Through Chokes*. NTNU – Norwegian University of Science and Technology, Master Thesis.

Hsu, Y. and Graham, R. (1976). *Transport Processes in Boiling and Two-Phase Systems*. Washington DC, USA: Hemisphere Publishing Corporation.

IEC60534-2-3, *International Electrotechnical Commission, Industrial – Process Control Valves – Part 2-3: Flow Capacity – Test procedures* (2015).

Lahey, R. and Moody, F. (1993). *The Thermal Hydraulics of a Boiling Water Nuclear Reactor*, 2e. The American Nuclear Society.

Levy, S. (1999). *Two-Phase Flow in Complex Systems*. New York, USA: John Wiley & Sons Inc.

Liepmann, H. and Roshko, A. (1957). *Elements of Gasdynamics*. New York, USA: John Wiley & Sons.

Lin, Z. (1986). Two-phase Flow Measurements with Orifices. In: *Encyclopedia of Fluid Mechanics, Vol. 3, Gas-Liquid Flows*, Chapter 29. Houston, USA: Gulf Publishing Company.

Riazi, M. (2005). *Characterization and Properties of Petroleum Fractions*. West Conshohocken, PA, USA: ASTM International.

Salcudean, M. (1986). Effect of Flow Obstructions on Flow Transitions and Pressure Drop in Two-phase Flows. In: *Encyclopedia of Fluid Mechanics, Vol. 3, Gas-Liquid Flows*, Chapter 26. Houston, USA: Gulf Publishing Company.

Shapiro, A. (1953). *The Dynamics and Thermodynamics of Compressible Fluid Flow*, vol. Vol. 1. New York, USA: The Ronald Press Company.

Wallis, G. (1969). *One-Dimensional Two-Phase Flow*. New York, NY, USA: McGraw-Hill Inc.

Whalley, P. (1987). *Boiling Condensation and Gas-Liquid Flow*. Oxford, UK: Oxford University Press.

8

Two-Phase Flow Thermal Modelling

8.1 Introduction

Two types of *operating regimes* and *environments* ascertain the magnitude of the heat losses. Section 8.2 addresses the differences between normal (i.e. quasi-steady-state) and transient operation regimes of the production system. On the other hand, Section 8.3 refers to an offshore versus onshore environment to identify the major drivers of heat losses.

Heat transfer manifests through three basic mechanisms: conduction, convection, and radiation. The impact of such mechanisms on the system heat transfer is briefly reviewed in Section 8.4.

Section 8.5 sets at the heels of the previous section summarizing the best-known convection correlations for internal fluid flow. This section includes forced and natural convective correlations applicable to flowing and shutdown conditions. Section 8.6 is concerned with heat transfer due to external fluid flow over the production system, namely, forced convection due to wind on onshore applications or sea currents in subsea environments.

Heat losses, being a significant risk for system production targets, can be somehow mitigated with *thermal insulation* to reduce the losses to an acceptable level. The latter always leads to the tradeoff between Capex (insulation cost) and production performance. Section 8.7 addresses typical materials and configurations of thermal insulation.

The overall heat transfer coefficient (*OHTC*) combines all the thermal resistances separating the temperature gradient between the production stream and the environment. Section 8.8 addresses the calculation of the *U-value* factor and the *OHTC*.

Section 8.9 touches on the widespread practice of using the Earth's low thermal conductivity as an equivalent thermal insulation system for buried pipelines. Again, this practice is a compromise between Capex (trenching

Multiphase Transport of Hydrocarbons in Pipes, First Edition.
Juan J. Manzano-Ruiz and Jose G. Carballo.

Companion Website: www.wiley.com/go/Manzano-Ruiz/Multiphasetransport

cost) versus production performance (mitigation of phase change by reducing losses).

At a high level, the objective of thermal modeling is to predict the temperature profile along the whole production system, that is, from the reservoir wellbore to the production facility. Section 8.10 covers the additional equations needed to model the system's thermal performance.

Sections 8.11 and 8.12 address two important operating scenarios where thermal modeling determines performance. These scenarios are system cooldown and wellbore heat transfer.

8.2 Normal and Transient Operation

Normal operation resembles a steady-state process but with some small fluctuations due to perturbations affecting the system. For a production system, normal operation takes place 80% or more of the uptime. Transient operations last little compared to normal operations yet heat losses may be important too.

The heat losses for normal operation determine the temperature profile along the production system and are dependent upon the level of thermal insulation (or trenching) the system has. Thermal performance during transient operations is also modified by the effectiveness of the insulation, but in general, the thermal capacitance effect of such insulation is not crucial to prevent appreciable heat losses.

For normal operation, the thermal performance is driven by the *OHTC* and the magnitude of the temperature gradient with respect to the environment. For transient operations, the key factor is the duration of the transient with respect to a thermal energy storage characteristic time. The latter is discussed later in Sections 8.11 and 8.12 via practical applications.

8.3 Offshore and Onshore Pipelines

Subsea flowlines and risers are exposed to a saline environment at constant conditions of temperature and currents' velocity. In a way, for modeling, this environment is simpler due to the steadiness of the prevailing external conditions.

For onshore pipelines, the environment may ever be changing. The system may be exposed to night/day insolation (unless it is buried), varying wind velocity, four seasons, rain, and snow. From the standpoint of simulation, the selection of the worst environmental conditions is necessary to simulate the system, in order to realize the most adverse scenario of internal two-phase flow. For example, if the pipeline's service is for heavy crude and it is not trenched, the oil flow rate can have noticeable swings between day and night, just because of the heat gain due to insolation. For offshore flowlines the thermal gradients are in general steady but of larger magnitude compared to the onshore flowlines.

8.4 Heat Transfer Mechanisms

There are three basic heat transfer mechanisms, namely

- *Conduction.* This is a molecular-level thermal energy transfer driven by a temperature gradient. This mechanism is characterized by the thermal conductivity property k, the temperature difference between the bodies exchanging heat, and the heat transfer area perpendicular to the temperature gradient (Carslaw and Jaeger, 1959). The governing heat conduction equation is designated as Fourier's law (equation 8.1). This mechanism is important for heat transfer through pipe walls and thermal insulation.
- *Convection.* This is a mechanism where appreciable volumes of fluid in motion (at a scale much larger than molecular) transport thermal energy driven by a temperature gradient. The convection mechanism can be either *forced* or *natural* (i.e. free). In forced convection, an external source drives and determines the fluid flow characteristics. The source could be a wind stream, a sea current, a pump, or a gas compressor. Natural convection fluid motion is driven by buoyancy forces generated by temperature differences in the bulk of the fluid. Forced convection is usually the dominant mechanism for heat transfer inside and outside the pipeline. However, natural convection drives internally the fluid motion for shutdown pipelines.
- *Radiation.* Thermal radiation affects onshore pipelines exposed to sunlight, also creating heat exchanges with the surroundings via infrared radiation. The heat transfer rates are driven by the difference in the fourth power of the absolute temperatures of the thermal-exchanging bodies, and proportional to the Boltzmann constant (Incropera et al., 2003). Usually, the radiation mechanism is orders of magnitude lower than the convection mechanism and it is then disregarded.

8.5 Internal Heat Transfer Coefficients

There are many correlations developed to predict forced convection inside the piping. Sections 8.5.1 through 8.5.7 list a few of the best-known correlations for single- and multi-phase flow inside a pipeline when the internal flow pattern may be considered well developed (e.g. far downstream from changes in direction or diameter). The heat transfer convective coefficient is universally distinguished as the *h*-coefficient, or simply *h*.

The trend in the past has been to start with single-phase flow correlations and apply them along the periphery of the internal area depending on the fluid phase in contact with the pipe wall. This is the mechanistic approach (Shoham, 2006) adopted in this book for evaluating the internal heat transfer coefficient. In all cases, the assumption made herein is that the ratio of pipe length to diameter is extremely large ($L/d \gg 1$) justifying the condition of fully developed flow, even in the two-phase flow case.

At the initial stages of a project, there may not be enough information to enable rigorous calculation of the heat transfer coefficient. Therefore, some rule-of-thumb heat transfer coefficients suffice for a first-pass thermal analysis. Table 8.1 shows generic single-phase flow values of *h* coefficients that are useful for quick estimates of heat transfer rates without the drawback of complex correlations.

8.5.1 Turbulent Flow Forced-convection Inside Tubes

The classic and highly referred to Colburn (1933) correlation, equation (8.3), is the best-known equation applicable to developed single-phase turbulent flow. This correlation is expressed in terms of the non-dimensional

Table 8.1 Order-of-magnitude single-phase heat transfer coefficients, Incropera et al. (2003).

Case	h (W/m^2 K)
Forced convection developed gas flow inside tubes	25–250
Forced convection developed oil flow inside tubes	100–20000
Natural convection of gas inside a tube	2–25
Natural convection of oil inside a tube	50–1000

numbers: Nusselt (Nu) – equation (8.1), Reynolds (Re), and Prandtl (Pr) – equation (8.2):

$$Nu \equiv \frac{h_i d}{k_f} \tag{8.1}$$

$$\Pr \equiv \left(\frac{c_p \mu}{k}\right)_f \tag{8.2}$$

$$Nu = 0.023 \mathrm{Re}^{0.8} \Pr^{1/3} \tag{8.3}$$

It is important to highlight that this form of the Colburn correlation is used for cooling flows inside the tubing.

8.5.2 Laminar Flow Forced convection Inside Tubes

Developed laminar flow within tubes takes place when the Reynolds number is below 2000. For this case, the Sieder and Tate (1936) correlation, equation (8.4), is useful to estimate the single-phase flow heat transfer coefficient for very viscous fluids.

$$Nu = 1.86\left(\mathrm{Re}\ \Pr\ \frac{d}{L}\right)^{1/3}\left(\frac{\mu_{bulk}}{\mu_{wall}}\right) \tag{8.4}$$

8.5.3 Two-phase Disperse bubble Flow in Horizontal Pipe

Two-phase flow internal heat transfer occurs in the turbulent regime. For the case of dispersed bubble flow, Shoham (2006) proposes a pseudo-homogeneous no-slip flow model based on the liquid holdup equation (8.5).

$$\lambda_l \equiv \frac{j_l}{j_g + j_l} \tag{8.5}$$

The Nusselt number for dispersed bubble flow adopts the following expression (subscript "H" indicating homogeneous flow):

$$Nu_H = \frac{h_i d}{k_H} = 0.023 \mathrm{Re}_H^{0.8} \Pr_H^{1/3} \tag{8.6}$$

The fluid mixture properties necessary to evaluate the Prandtl number are calculated with

$$c_{p,H} \equiv \frac{c_{p,l}\rho_l\lambda_l + c_{p,g}\rho_g(1-\lambda_l)}{\rho_l\lambda_l + \rho_g(1-\lambda_l)} \tag{8.7}$$

$$k_H \equiv k_l\lambda_l + k_g(1-\lambda_l) \tag{8.8}$$

8.5.4 Two-phase Stratified Flow in Horizontal Pipe

The phases' spatial segregation enables a simple approach for this regime, supported by single-phase correlations applied at their respective flow areas and wet-wall surfaces. Either equation (8.3) or equation (8.4) can be utilized substituting the corresponding diameter by the hydraulic diameter of the phase, defined by equations (8.9) through (8.13),

$$P_{W,l} = d(\pi - \varphi/2) \tag{8.9}$$

$$P_{W,g} = \frac{d\varphi}{2} \tag{8.10}$$

where the subtended angle φ is calculated from the implicit equation (8.11) and the stratified-flow void fraction,

$$\alpha = \frac{\varphi - \sin\varphi}{2\pi} \tag{8.11}$$

and the hydraulic diameters are calculated from equations (8.12) and (8.13):

$$d_{h,l} = \frac{4A_l}{P_{W,l}} = \frac{\pi d(1-\alpha)}{(\pi - \varphi/2)} \tag{8.12}$$

$$d_{h,g} = \frac{4A_g}{P_{W,g}} = \frac{2\pi d\alpha}{\varphi} \tag{8.13}$$

8.5.5 Two-phase Slug flow in Horizontal Pipe

The estimation of the heat transfer coefficient for an intermittent flow pattern is fairly challenging. An extensive experimental work performed by Shoham (2006) for horizontal flow determined that the heat transfer coefficients vary with the mass flow rate of liquid and gas and have the following typical (Table 8.2) values on different zones of a slug unit.

The conclusion from Shoham's (2006) experimental work is that heat transfer coefficients in intermittent slug flow regimes are high enough

Table 8.2 Order of magnitude of heat transfer coefficients in slug flow, Shoham (2006).

Slug location	Coefficient (W/m^2 C)
Top of nose	2900–35,000
Bottom of nose	17,000–34,000
Top of body	2900–20,000
Bottom of body	13,000–25,000
Trailing gas bubble	5000–38,000

to neglect the internal thermal resistance (addressed in Section 8.8) for pipelines that are thermally insulated. At these conditions, the thermal resistance offered by insulating coating is an order of magnitude higher or more than the internal resistance.

8.5.6 Natural Convection in a Stagnant Fluid Within a Horizontal Pipe

During an extended system shutdown, the production stream of oil and/or gas may be trapped while exchanging heat with the surrounding environment. Internally in the pipeline, natural convective currents set in due to buoyancy. The technical literature on natural convection inside horizontal pipes is scarce.

One of the few experimental works on this topic, performed by Martini and Churchill (1960), indicates that the overall internal heat transfer coefficient is insensitive to the natural convection drive force represented by the product of the Prandtl (*Pr*) and Grashof (*Gr*) numbers. Figure 14 of the Martini and Churchill (1960) reference points at a constant Nusselt number of 7.0 for single-phase fluids. The constancy of the Nusselt number indicates a limiting case of heat conduction mechanism for stable processes (long-time approximation).

Detailed numerical simulations with finite differences were performed by Hellums (1960) and by Stewart et al. (1974) for the transient natural convection problem inside cylindrical cavities. Newer work performed by Ludovisi and Garza (2013) utilizing *CFD* (computational fluid dynamics) focused on the four transient stages of natural convection inside cylindrical enclosures. Stage 1 mechanism is heat conduction with a fast drop in Nusselt number. Stage 2 is a developing flow regime that lasts a little. Stage 3 is a steady-state regime characterized by equation (8.14). Stage 4 is a long-term heat conduction regime. For the last stage, Ludovisi and Garza (2013) found that the Nusselt number approaches the value of 5.8 for single-phase fluids when the Rayleigh number (equation 8.15) is negligible, in reasonable agreement with the previous Nusselt number of 7.0 observed by Martini and Churchill (1960).

In case fluid is trapped at the worst thermal location inside the horizontal flowline (i.e. end of the line during a shutdown), the constant Nusselt number for long-term condition could provide a fair estimate for the internal heat transfer coefficient during cooldown scenarios of horizontal flowlines. However, in the general case of non-negligible Rayleigh number, the work of Ludovisi and Garza (2013) recommends equation (8.14),

$$Nu = 1.15\, Ra^{0.22} \quad \text{for } 30{,}000 \le Ra \le 10^{10} \tag{8.14}$$

where the Rayleigh number Ra characterizing the balance between buoyancy and friction is defined by equation (8.15):

$$Ra \equiv \Pr \left[\frac{g\beta(T_{avg} - T_{surf})d_i^3}{\left(\frac{\mu}{\rho}\right)^2} \right] \tag{8.15}$$

8.5.7 Forced Convection in Vertical Two-phase Flow

Two-phase flow in wellbores and risers is characterized by quasi-axisymmetric flow. For a well, the lower sections of the production tubing may be flowing as single-phase fluid, and the heat transfer coefficient can be estimated with equation (8.3). Then, in the well, the fluid may turn two-phase, and a mechanistic approach is recommended starting with the flow pattern determination by sections.

For either dispersed bubble or mist flow patterns the homogeneous flow approach is recommended, utilizing the equations included in Chapter 5. Churn/slug and annular flow patterns can be modeled using the annular flow model and the *annular flow triangular relationship* (Collier and Thome, 1994). For these regimes, there exists a liquid flow film attached to the wall with a core of gas flow toward the centerline. The triangular relationship connects δ, the film thickness, with the velocity profile within the film and the wall shear stress at the interface, using the momentum balance equation. The customary methodology is to estimate the film thickness δ using the triangular relationship and the following approximations:

$$h_i \approx \frac{h_{l,\,film}}{1-\alpha} \tag{8.16}$$

$$1-\alpha \approx \frac{4\delta}{d} \tag{8.17}$$

The film heat transfer coefficient in equation (8.16) can be estimated using equation (8.6) substituting the pipeline's internal diameter by the hydraulic diameter of the film $\approx 2\delta$. Details of this model can be found in Collier and Thome (1994).

8.6 External Heat Transfer Coefficient

External crossflow over pipelines, either submerged in seawater or aboveground pipelines exposed to wind, represents another form of thermal resistance to transfer energy from the production fluid to the environment.

The correlation by Churchill and Bernstein (1977) can be used to estimate the Nusselt number and thereafter the heat transfer coefficient for external forced convection. This correlation is based on the external diameter of the pipeline system, and it is valid for RePr > 0.2.

$$Nu = 0.3 + \frac{0.62\mathrm{Re}^{1/2}\mathrm{Pr}^{\ 1/3}}{\left[1+\left(\frac{0.4}{\mathrm{Pr}}\right)^{2/3}\right]^{1/4}}\left[1+\left(\frac{\mathrm{Re}}{2{,}82{,}000}\right)^{5/8}\right]^{4/5} \tag{8.18}$$

Churchill and Chu (1975) correlation, equation (8.19), can be used for an unburied pipeline in stagnant air or water ambient. This correlation is based on the dimensionless Rayleigh number *Ra* which is product of the dimensionless Prandtl (Pr) and Grashof (*Gr*) numbers. The definition of *Ra* and *Gr* numbers can be found in Incropera et al. (2003).

$$Nu = 0.68 + \frac{0.67Ra^{1/4}}{\left[1+\left(\frac{0.492}{\mathrm{Pr}}\right)^{9/16}\right]^{4/9}} \tag{8.19}$$

For environments in which the fluid may alternate between stagnant and flow conditions, the two previous equations may be applied. However, in order to get a conservative (i.e. larger) estimate of heat losses, equation (8.18) may be preferred regardless of whether natural or forced convection is the dominant mechanism.

8.7 Thermal Insulation

Thermal insulation is the barrier for heat losses for steady-state operation and for limited-time exposure during transients. This section summarizes the most common coating materials utilized in the oil and gas industry for offshore and onshore applications. In time the material will degrade, and the thermal conductivity may slightly increase. For detailed engineering design, it is recommended to seek assistance from the material manufacturer to have access to the most reliable data.

For deepwater subsea applications, the insulation is molded directly around the flowlines, forming the so-called *wet insulation* which does not need an external protective jacket. Such type of insulation has been designed and tested to sustain the high underwater pressures encountered in deepwater applications. Table 8.3 summarizes representative typical values of insulation thermal conductivity.

Table 8.4 shows representative thermal conductivity values for other coating material for onshore and offshore pipelines. Some of these coatings are

Table 8.3 Insulation thermal conductivity.

Material	k (W/m °C)
Aerogel	0.010–0.024
Calcium silicate	0.05
Glass fiber	0.040–0.060
Polypropylene (PP) solid	0.170–0.220
Polypropylene (PP) syntactic	0.150–0.165
Polypropylene (PP) light foam	0.040–0.060
Polyurethane (PU) solid	0.022–0.025
Polyurethane (PU) syntactic	0.150–0.170
PU syntactic deep water	0.08
Polyurethane (PU) light foam	0.02

Table 8.4 Coating thermal conductivity.

Material	k (W/m °C)
Coal tar enamel	0.19–0.22
Concrete (CWC) lightweight	0.10–0.30
Concrete (CWC) medium	0.40–0.70
Concrete (CWC) dense	1.0–1.8
Fusion bonded epoxy (FBE)	0.10–0.30
Neoprene	0.05–0.06
3-layer polyethylene (3LPE)	0.40–0.51
Adhesive PUF	0.22

used to enhance the adhesion of other coating materials. Concrete material coating is typically applied to subsea pipelines to provide negative buoyancy to the system.

Table 8.5 summarizes representative thermal conductivity values of assorted environments for onshore and offshore pipelines.

Another option available to provide the best insulation possible (lowest *U-value*) to the flowlines and risers is the pipe-in-pipe (*PIP*) alternative. This option comprises two concentric pipes and centralizers, with an annulus filled with material (passive insulation) of exceptionally low thermal

Table 8.5 Other material thermal conductivities.

Material	*k* (W/m °C)
Clay (dry to moist)	0.15–1.8
Clay (saturated)	0.6–2.5
Seawater (10 °C)	0.56–0.58
Sand (dry)	0.15–0.25
Sand (moist)	0.25–2.0
Sand (saturated)	2.0–4.0
Soil, clay	1.1
Soil, organic matter	0.15–2.0
Soil, saturated	0.60–4.0
Vacuum	≈0

conductivity (e.g. Aerogel). The PIP alternative is costly in materials and steep costs to reel and lay. Furthermore, very deep offshore applications may have weight limitations.

8.8 Overall Heat Transfer Coefficient (OHTC)

The estimation of the temperature profile inside a pipeline requires considering the path followed by thermal energy, from the fluid temperature to the surrounding environment temperature. The energy conservation principle has to be satisfied to balance the change in enthalpy flow for the production stream with the heat transferred to the external medium. This is a crucial step in any simulation because the fluid properties and the flashing equilibrium are very dependent on temperature and therefore affect the conservation of mass and momentum.

Two operating thermal scenarios may take place: steady-state and transient heat transfer. For a steady-state scenario, the most useful tool is to combine all thermal resistances into a single one referred to as the *OHTC*.

A key component of the *OHTC* is the *U-value* that combines all thermal resistances associated with heat conduction, namely the piping material wall, and the thermal insulation and coatings external to the piping. The other components built into the *OHTC* are the internal and the external heat convection thermal resistances.

Normally, heat radiation has a negligible contribution to the overall thermal resistance except in cases of pipelines exposed to solar or sky radiation. The concept of thermal resistor is addressed below in this section by means of the electrical analogy.

Heat conduction is a diffusional molecular-level mechanism represented by the thermal conductivity property. Heat convection is associated with fluid in motion, forced by either external sources or buoyancy effects (called natural or free convection). In either case, convective heat transfer may have laminar or turbulent fluid motion.

In transient heat transfer modeling, two elements intervene, the thermal resistors or conduction terms, and the capacitances storing thermal energy. The latter include the mass and specific heat of the piping and the coating materials. In general, the analysis of a discretized production field system, using the finite-difference method of solution, demonstrates that the thermal transient terms have a characteristic time orders of magnitude larger than the residence time of the fluid at each domain cell. Thus, very often the thermal capacitances are ignored in the transient simulation by assuming an instantaneous rate of heat transfer as in the steady-state scenarios.

Modelling the *OHTC* or the *U-value* benefits from the concept of thermal resistors (this is analog to Ohm's law in electrical networks). The analog considers heat fluxes as electric current, and temperature gradients equivalent to electric voltage potential. There is a difference in this analogy; Ohm's law assumes constant linear performance of the electrical resistors, whereas thermal resistors are commonly nonlinear and temperature-dependent.

Cylindrical coating layers covering a pipeline have thermal resistances *per unit length* that can be estimated with equation (8.20); each layer is characterized by its thermal conductivity k_m, and diameters D_m (internal) and D_{m+1} (external).

$$R'_m = \frac{Ln\left(\frac{D_{m+1}}{D_m}\right)}{2\pi k_m} \tag{8.20}$$

Figure 8.1 shows an assembly of four coating layers covering a steel pipeline, plus the internal/external convective resistors. The five layers of coating in this sketch can be represented with equation (8.23), and the two convective resistors can be represented with the next two equations:

$$R'_i = \frac{1}{h_i \pi d} \tag{8.21}$$

$$R'_o = \frac{1}{h_o \pi D_6} \tag{8.22}$$

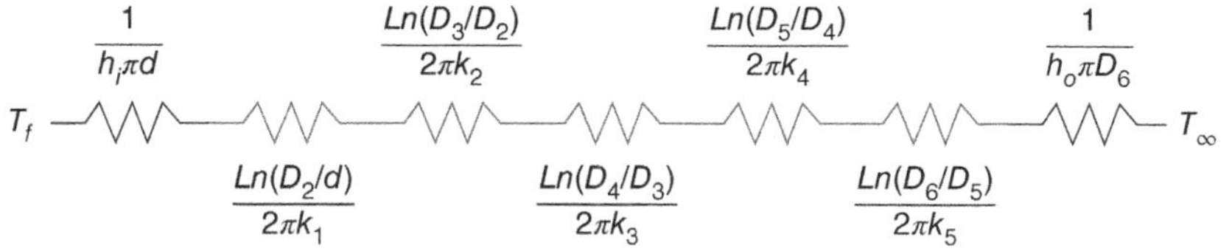

Figure 8.1 Thermal resistance network.

Thus, the *total thermal resistor* for Figure 8.1 results

$$R'_T = \frac{1}{h_i \pi d} + \frac{Ln\left(\frac{D_2}{d}\right)}{2\pi k_1} + \frac{Ln\left(\frac{D_3}{D_2}\right)}{2\pi k_2} + \frac{Ln\left(\frac{D_4}{D_3}\right)}{2\pi k_3} + \frac{Ln\left(\frac{D_5}{D_4}\right)}{2\pi k_4} + \frac{Ln\left(\frac{D_6}{D_5}\right)}{2\pi k_5} + \frac{1}{h_0 \pi D_6} \tag{8.23}$$

The concept of total thermal resistance enables to obtain the heat transfer rate per unit length by

$$\dot{Q} = \frac{\Delta T}{R'_T} = \frac{T_{fluid} - T_{environment}}{R'_T} \tag{8.24}$$

The coatings' *U-value,* referred to the pipeline ID, is defined as

$$U_d = \frac{2/d}{\frac{Ln\left(\frac{D_2}{d}\right)}{k_1} + \frac{Ln\left(\frac{D_3}{D_2}\right)}{k_2} + \frac{Ln\left(\frac{D_4}{D_3}\right)}{k_3} + \frac{Ln\left(\frac{D_5}{D_4}\right)}{k_4}} \tag{8.25}$$

It is convenient to justify the assumption aforementioned that the thermal energy storage in the pipeline wall and the coatings is negligible compared to the thermal energy transferred to the environment. This assumption simplifies the transient simulations by eliminating the energy storage terms. The support for such an assumption rests on comparing the characteristic time for thermal energy storage to the residence time the fluid spends within a discrete element in the pipeline.

For the sake of the argument and to simplify the thermal analysis, let us consider a bare pipeline without any coating. Consider the temperatures of the fluid, the steel wall, and the environment, respectively, as T_f, T_w, and T_o. Thus, an energy balance on the domain element of length ℓ would be given by equation (8.26):

$$\frac{(\rho c)_w \ell \pi \left(D_o^2 - d^2\right)}{4} \frac{dT_w}{dt} = h_i \ell \pi d(T_f - T_w) - h_o \ell \pi D_o(T_w - T_o) \tag{8.26}$$

It is easy to demonstrate that the characteristic time for thermal storage on the pipe wall is given by equation (8.27) where δ is the steel wall thickness:

$$\tau_{storage} = \frac{(\rho c)_w \left(D_o^2 - d^2\right)}{4h_i d} \approx \frac{(\rho c)_w \delta}{h_i} \tag{8.27}$$

On the other hand, the characteristic residence time within the domain element of length ℓ is

$$\tau_{residence} = \frac{\pi d^2 \ell}{4Q_{mix}} = \frac{\ell}{j} \tag{8.28}$$

which leads to the following criterion, equation (8.29), to neglect energy storage in thin-walled piping, by assuming that ℓ is of the order of d. This argument could be validated with typical heat transfer coefficients, steel properties, and mixture velocity.

$$\frac{\tau_{residence}}{\tau_{storage}} \cong \frac{h_i}{j(\rho c)_w} \ll 1 \tag{8.29}$$

For a conventional-coated pipeline by and large, the dominating thermal capacitance lies on the steel, and the criterion of equation (8.29) is of the order of 0.3%, easily extrapolating to such conventional assemblies.

Example 8.1 U-*Value Calculation*
An 8-inch flowline thermal design calls for the assembly listed in Table 8.6, including dimensions and material properties. This type of insulation is

Table 8.6 Example of the calculation of U-value (in bold characters) based on internal and external diameter.

	ID (m)	WT (m)	OD (m)	k (W/mK)	ρ (kg/m^3)	c_p (J/kg K)	R_T (mK/W)
Steel pipe	0.1937	0.0127	0.2191	45.0	7,849	460	0.0004
Coatings							
FBE coating (corrosion)	0.2191	0.00025	0.2196	0.200	1,300	1502	0.0018
Adhesive PUF	0.2196	0.00025	0.2201	0.220	900	2874	0.0017
PP – solid – inner	0.2201	0.00533	0.2308	0.190	900	2820	0.0396
Syntactic PP	0.2308	0.07620	0.3832	0.160	700	1615	0.5043
PP – solid – outer	0.3832	0.00508	0.3934	0.190	900	1611	0.0219
		0.08712					0.5698
U (W/m^2 K)		**2.55**	(based on pipe OD)				
U (W/m^2 K)		**2.88**	(based on pipe ID)				

Table 8.7 Example of the calculation of U-value (in bold character) based on internal diameter of inner pipe for a pipe-in-pipe configuration.

	ID (m)	WT (m)	OD (m)	k (W/m K)	ρ (kg/m^3)	c_p (J/kg K)	R_T (mK/W)
Steel inner pipe	0.1937	0.0127	0.2191	45.0	7849	460	0.0004
Aerogel (annulus)	0.2191	0.03335	0.2858	0.015	10	1900	2.8192
Steel outer pipe	0.2858	0.0191	0.3239	45.0	7849	460	0.0004
FBE (external coating)	0.3239	0.00025	0.3244	0.200	1300	1502	0.0012
							2.8213
U (W/m^2 K)	**0.58**	(based on pipe ID)					

called 5LPP (5 layers of polypropylene). The last column of this table has the thermal resistances for each layer, calculated with equation (8.20). Estimate the coating U-values based on the ID and OD of the pipeline.

The solution proves that U-values based on the outer pipe diameter are lower than the U-values calculated based on the pipe's internal diameter. This is a natural conclusion because the external heat transfer area is larger than the internal heat transfer area.

Example 8.2 ***Pipe-in-Pipe* U-*Value***

Consider a pipe-in-pipe application comprising a 12-inch (286 mm ID, 324 mm OD) casing pipe, concentric and external to an 8-inch (194 mm ID, 219 mm OD) internal piping. The internal annular space between the two pipes is filled with Aerogel. Estimate the U-value based on the ID of the internal pipe and compare the U-value based on the ID with the result from Example 8.1.

The solution to this problem is shown in Table 8.7. The difference in U-values based on the internal diameter, between the flowline of Example 8.1 and the inner pipe of Example 8.2, is striking; the pipe-in-pipe configuration reduces by 80% the U-value with respect to the option of a single pipe insulated with 5LPP.

8.9 Buried Pipelines

It was pointed out before that pipeline trenching is done to utilize the soil medium as an inexpensive thermal insulation material and for safety reasons too. This practice is common for onshore pipelines which also have to cross-rivers and creeks, and clear roads and rail tracks underneath.

The traditional way to calculate the thermal resistance of the surrounding soil is based on two-dimensional (2-D) heat conduction (Incropera et al., 2003). From such 2-D analysis, the *shape-factor* S_{HT} (specified per unit length of pipeline) is defined by equations (8.30) and (8.31) with the help from Figure 8.2:

$$S'_{HT} = \frac{2\pi}{Ln\left(\frac{4\varsigma}{D_o}\right)} \quad \text{for } \frac{\varsigma}{D_o} > 1.5 \tag{8.30}$$

$$S'_{HT} = \frac{2\pi}{\cosh^{-1}\left(\frac{2\varsigma}{D_o}\right)} \quad \text{for } \frac{\varsigma}{D_o} < 1.5 \tag{8.31}$$

The thermal resistance of the soil per unit length for a fully trenched pipeline becomes

$$R_{soil} = \frac{1}{S'_{HT} k_{soil}} \tag{8.32}$$

This thermal resistance, equation (8.32), is added in series with the rest of the network of thermal resistances: fluid-steel-coating and environment to obtain an overall thermal resistance.

The heat losses from a buried pipeline are calculated by dividing the overall temperature gradient by the overall thermal resistance.

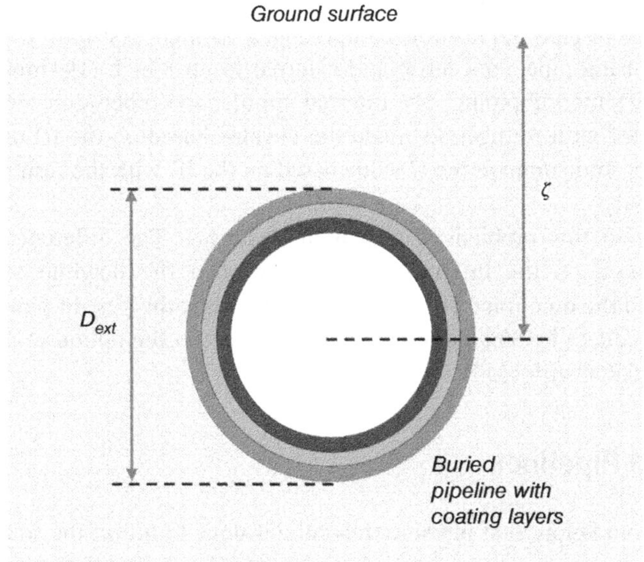

Figure 8.2 Buried pipeline heat transfer domain.

Table 8.8 Example of the calculation of U-value (in bold character) based on internal pipe diameter for a buried flowline.

	ID (m)	WT (m)	OD (m)	k (W/m K)	ρ (kg/m³)	c_p (J/kg K)	R_T (mK/W)
Steel pipe	0.1937	0.0127	0.2191	45.0	7849	460	0.0004
FBE (external coating)	0.2191	0.00025	0.2196	0.200	1300	1502	0.0018
Soil	ζ (m) =	1.11		0.300	S_{HT} =	2.09	1.5947
							1.5970
	U (W/m² K)	**1.03**	(based on pipe ID)				

Example 8.3 ***Buried Flowline***

Consider again the pipeline of Example 8.1, but only covered with the FBE coating and buried one meter to the top of the pipe. Consider that the trenched soil (organic matter) has a thermal conductivity of about 0.3 W/m K. Estimate the U-value based on the ID of the pipe, and compare it against the result from Example 8.1.

The solution to this problem is summarized in Table 8.8. The difference in U-value with respect to the solution to Example 8.1 is −64%, indicating that from the thermal standpoint, it would be much better to bury the flowline than to provide the 5LPP coatings. Of course, the cost of trenching the flowline has to be weighed against the cost of the 5LPP.

8.10 Temperature Profile

The principal objective of the thermal analysis is to determine the local temperature at each cell of the discretized domain. The temperature property is crucial to evaluate the fluid properties that support the conservation principles. Obviously, the result of the thermal analysis is the temperature profile.

For the pipeline design, it is necessary to estimate thermal expansion of the steel material during service start-up, during normal operation, and the ensuing material contraction after shutdown. Thus, a crucial piece of information for the pipeline designer is the set of operating temperature profiles the system may undergo under service and throughout the life cycle. Example 8.4 illustrates the task of finding the temperature profile for a normal operation scenario.

Example 8.4 *Temperature Profile*

Consider the subsea horizontal flowline of Example 8.1 which has a length of 12-km transporting live oil and gas. Use the following data to estimate the average *OHTC* and the temperature profile:

- Flow rate (kg/s): 4.6
- Pressure (bara): 56
- Temperature (°C): 87
- Steel pipeline ID (mm): 193.7
- Coating external OD (mm): 406.3
- Mixture avg. thermal conductivity (W/m K): 0.0876
- Mixture avg. heat capacity (J/kg K): 2,265
- Mixture avg. density (kg/m^3): 485
- Mixture avg. viscosity (s Pa): 0.0041
- Seawater temperature (C): 4
- Seawater current speed (m/s): 0.65
- Seawater thermal conductivity (W/m K): 0.60
- Seawater heat capacity (J/kg K): 3900
- Seawater density (kg/m^3): 1035
- Seawater viscosity (s Pa): 0.0017

The hydrocarbon composition of McCain (1990) is shown in Table 8.9. The calculation of the *OHTC* is shown in Table 8.10.

It is clear that incorporating the internal/external thermal resistances reduces the conductance of the *U*-value found in Example 8.1.

Table 8.9 Fluid composition for Example 8.4.

Component	(% mol.)
C1	36.47
C2	9.67
C3	6.95
i-C4	1.44
n-C4	3.93
i-C5	1.44
n-C5	1.41
n-C6	4.33
C7+	33.29
CO_2	0.91
N_2	0.16

Table 8.10 Example of the calculation of OHTC (in bold characters) based on the internal diameter of the pipe for a long subsea flowline.

	ID (m)	WT (m)	OD (m)	k (W/m K)	ρ (kg/m^3)	c_p (J/kg K)	R_T (mK/W)
Steel pipe	0.1937	0.0127	0.2191	45.0	7849	460	0.0004
Coatings							
FBE coating (corrosion)	0.2191	0.00030	0.2197	0.200	1300	1502	0.0022
Adhesive PUF	0.2197	0.00030	0.2203	0.220	900	2874	0.0020
PP – solid – inner	0.2203	0.00650	0.2333	0.190	900	2820	0.0480
Syntactic PP	0.2333	0.08000	0.3933	0.160	700	1615	0.5194
PP – solid – outer	0.3933	0.00650	0.4063	0.190	900	1611	0.0272
		0.09360					0.5993

Fluid properties		j (m/s)	2.5		
Density (kg/m^3)	485	Re	57,709		
Viscosity (s Pa)	0.0041	Pr	106		
Thermal conductivity (W/mK)	0.0876	Nu	701		
Heat capacity (J/kg K)	2265	h (W/m^2 K)	317	R_T (mK/W)	0.0052
Seawater properties		v (m/s)	0.65		
Density (kg/m^3)	1035	Re	92,337		
Viscosity (s Pa)	0.0017	Pr	11		
Thermal conductivity (W/mK)	0.60	Nu	481		
Heat capacity (J/kg K)	3900	h (W/m^2 K)	1236	R_T (mK/W)	0.0006
OHTC (W/m^{2}C)	**2.66**	(based on pipe ID)			

The conductance reduction due to the internal/external thermal resistances is small of the order of 3%, because those resistances are from liquid-dominated fluids.

It is interesting to note that it is common to perform thermal simulations ignoring the external/internal resistance components, because of the small contribution they add when the fluids are in the liquid phase, as is illustrated in this example. To err on the safe side (i.e. conservative estimate) the simulations may be carried out based on the *U-value* instead of the *OHTC*.

The solution to this problem was obtained with commercial software linked to a compositional model. The temperature profile attained with the software is shown in Figure 8.3 based on the constant *OHTC* of 2.66 W/m^2 K and yielding an arrival temperature of 13.3 °C.

It is very enlightening to solve the same problem by "hand," using equation (8.33), with an average-corrected heat capacity value taken from the same compositional model. The average heat capacity from the commercial software is 1987 J/kg K, a value 12% lower than the original average value provided (2265 W/m K). The resulting estimate by hand is 13.9 °C which is strikingly close to the result from the commercial software.

$$\frac{T_f - T_\infty}{T_i - T_\infty} = \frac{T_f - 4°\text{C}}{87°\text{C} - 4°\text{C}} = \\ = \exp\left[-\frac{OHTC(\pi dL)}{\dot{m}\overline{C}_p}\right] \\ = \exp\left[-\frac{2.66 \times \pi \times 12{,}000 \times 0.1937}{4.6 \times 1987}\right] = 0.1194 \qquad (8.33)$$

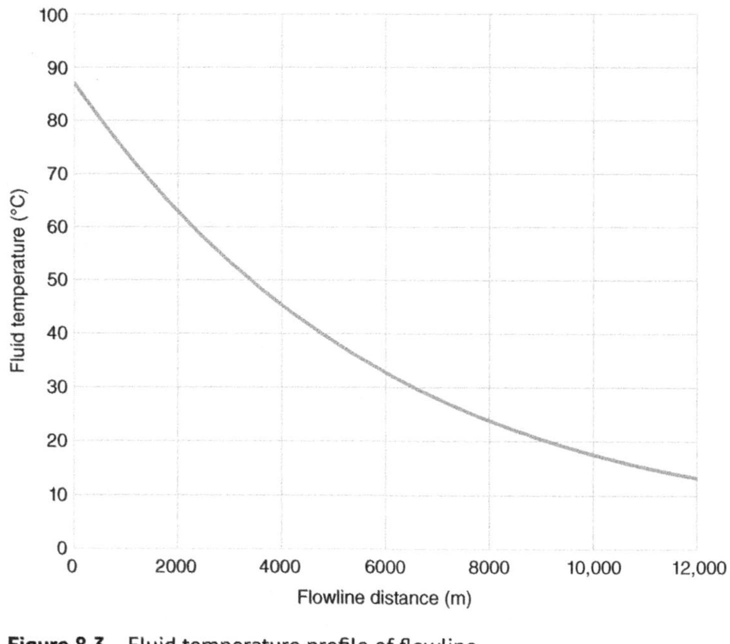

Figure 8.3 Fluid temperature profile of flowline.

A plausible conclusion from this example is that for steady-state thermal analyses the simple model of equation (8.33) is good enough to calculate the flowline temperature profile. However, it is necessary to have a good average estimate for the *OHTC* and for the fluid's heat capacity, both assumed constant along the system. Alternatively, as in this example, the hand calculation serves to cross-check the output from the commercial tool.

8.11 Flowline Cooldown

The transient heat losses occurring during a system shutdown are commonly referred to as the cooldown scenario. Obviously, the system operator is concerned that this situation entails a substantial risk of forming hydrocarbon hydrates that quickly bond, forming a solid crystalline structure plugging the pipeline and impeding the system restart afterwards.

This probable risk highlights the importance of simulating the cooldown time to determine how long it will take to form hydrates. Cooldown time relates to the so-called *no-touch time* that oilfield operators have established in the operating procedures, preventing the consequences of hydrate plugging. The *no-touch time* is the maximum time allowed to remediate any failure in the system that caused the shutdown and to restart the system and setting the live fluid in motion again. Obviously, *no-touch time* includes a safety factor to account for the uncertainties implicit in the heat transfer simulation.

Example 8.5 ***Cooldown***

Estimate the *internal heat transfer coefficient* for the flowline of Example 8.4, at the end of it (base of the riser location), during a shutdown scenario. Consider the following scenarios: (a) the fraction of gas present at the location is negligible; and (b) the liquid holdup ratio at the end of the flowline is approximately 83% (from the steady-state simulation with the commercial code).

Part (b) of this problem is an unlikely scenario because the end of the flowline (i.e. the base of the riser) will eventually receive oil rundown from the riser. Thus, part (a) may be a more realistic scenario. Nevertheless, solving part (b) enables us to compare the effect of two phases present.

- *Part (a).* Assuming a long-term heat conduction scenario following the criteria of Martini and Churchill (1960), we get

$$h_i = \frac{7k_f}{d} = \frac{7 \times 0.215}{0.1937} = 7.77\,\frac{\text{W}}{\text{mK}}$$

- *Part (b)*. Assuming again long-term heat conduction for either phase, following the criteria of Martini and Churchill (1960), and substituting the internal diameter for the hydraulic diameter for each phase, we get

$$d_{h,oil} = \frac{4A_{oil}}{P_{wet,oil}} = \frac{4 \times \pi \times 0.1937^2 \times 0.83}{4 \times 0.4164} = 0.235 \text{ m}$$

$$d_{h,gas} = \frac{4A_{gas}}{P_{wet,gas}} = \frac{4 \times \pi \times 0.1937^2 \times 0.17}{4 \times 0.1921} = 0.104 \text{ m}$$

$$h_{i,oil} = \frac{7k_{oil}}{d_{h,oil}} = \frac{7 \times 0.215}{0.235} = 6.40 \frac{\text{W}}{\text{mK}}$$

$$h_{i,gas} = \frac{7k_{gas}}{d_{h,gas}} = \frac{7 \times 0.0538}{0.104} = 3.62 \frac{\text{W}}{\text{mK}}$$

$$h_{i,total} = \frac{6.40 \times 0.4164 + 3.62 \times 0.1921}{\pi \times 0.1937} = 5.52 \frac{\text{W}}{\text{mK}}$$

It is important to highlight that the solution for part (b) is a gross approximation of the total internal heat transfer coefficient. For a more accurate solution, it is necessary to perform a *CFD* simulation that could provide the required definition of the transport phenomena.

Notwithstanding the simplifications done on part (b), a lower internal heat transfer coefficient occurs when two phases transfer heat in natural convection as compared to the case of the flowline filled with liquid only. Furthermore, to be safe in a risky cooldown scenario, some practitioners prefer to estimate the internal heat transfer coefficient assuming the flowline is filled with gas.

8.12 Document Navigation Heat Transfer in the Wellbore

Wellbore heat transfer is in general a transient phenomenon. The stages of well operation include start-up, stable production flow, and well shut-in. All these stages involve heat transfer to an environment, that is the "well formation", which has an infinite heat capacity. Thus, throughout any of the well operation stages, the heat losses to the formation change the geothermal gradient in the wellbore.

A well completion is a complex geometry comprising distinct size tubulars such as:

- Conductor casing,
- Surface casing,
- Casing liner,

Table 8.11 Representative *U*-values for well completions.

Well interval	*U*-value (W/m² K)
Lower end of tubing	4.8–6.5
Mid-zone of tubing	4.5–6.1
Upper end of tubing	4.0–5.5
Vacuum-insulated tubing (VIT)	0.4–0.46

- Intermediate casing,
- Production casing,
- Production liner, and
- Production tubing.

The gaps between consecutive casings usually are filled with completion fluid which is a brine fluid with additives. The external side of the casings is isolated from the formation by a cement layer to prevent fluid leaks. This is a critical safety element for the well integrity.

It is beyond the scope of this book to discuss the many designs of well completion for an oil or gas well. For further details about completions refer to Bellarby (2009).

The evaluation of the *U*-values applicable to the different intervals of the well completion obviously depends on the specifics of the well design. The assessment of *U*-values per interval entails applying the equations included in Section 8.8. Some typical *U*-values for a wellbore are shown in Table 8.11. In any practical case, the thermal capacitance of the completion layers is ignored for thermal transient simulations of the wells.

Ramey (1962) studied the transient thermal performance of the surrounding wellbore formation and proposed an analytical solution based on the continuous line-source concept (Carslaw and Jaeger, 1959). This solution is expressed in terms of the exponential integral function E_i. This line-source approximation in general is adequate for thermal performance beyond one week after a change in operating condition.

The heat flux per unit length through the completion of the well formation can be represented by equation (8.34), and the total thermal conductance by equation (8.35), Brill and Mukherjee (1999). The function $f(t)$ follows from the line-source approximation. The temperature T_{geoth} in equation (8.33) is the local formation temperature obtained from the geothermal gradient.

$$\dot{Q} = \frac{T_{fluid} - T_{geoth}}{R'_T} = U_d \pi d(T_{fluid} - T_{geoth}) \tag{8.34}$$

$$U_d = \frac{1}{\frac{1}{h_i} + d\left[\frac{Ln\left(\frac{D_2}{d}\right)}{2k_1} + \ldots\ldots + \frac{Ln\left(\frac{D_m}{D_{m-1}}\right)}{2k_{m-1}} + \frac{f(t)}{2k_{formation}}\right]} \tag{8.35}$$

Defining the following parameters

$$t_{diff} \equiv \left(\frac{4t}{D_m^2}\right)\left(\frac{k}{\rho c}\right)_{formation} \tag{8.36}$$

$$\lambda \equiv -\frac{1}{4t_{diff}} \tag{8.37}$$

we could evaluate the exponential integral function E_i with its mathematical definition as

$$f(t) = 0.5E_i(\lambda) = -0.5\int_{-\lambda}^{\infty}\frac{\exp[-t]}{t}dt \tag{8.38}$$

However, this evaluation is cumbersome, and the approximations proposed by Hasan and Kabir (2002) give accurate enough estimates. These equations are

$$f(t) = 1.1281\sqrt{t_{diff}}(1 - 0.3\sqrt{t_{diff}}) \quad \text{for } t_{diff} \leq 1.5$$

$$f(t) = [0.4063 + 0.5Ln(t_{diff})]\left(1 + \frac{0.6}{t_{diff}}\right) \quad \text{for } t_{diff} > 1.5 \tag{8.39}$$

It is important to realize that the line-source model applies to the case of well start-up after an exceedingly long shut-in condition, when the formation temperature corresponds to the geothermal gradient. For pseudo-steady-state scenarios, following well start-up, the line-source model also applies supported by equation (8.37) to evaluate $f(t)$.

A cooldown scenario in a well after shut-in requires more attention. The simulation involves estimating the transient thermal energy flow into the formation caused by the time-dependent temperature of the fluid trapped in the well tubing, and the simultaneous transient penetration of heat into the formation. The analytical solution of the temperature distribution in the semi-infinite formation is much more complex. Hahn and Özişik (2012) reference provides the temperature profile solution in the reservoir domain based on Fourier–Bessel integrals.

Due to considerable flow assurance risk at the top of the tubing, some wells are completed with a section of VIT (vacuum-insulated tubing with ultra-low U-value), to enhance the arrival fluid temperature at the well's tree during normal operation. However, the VIT, due to its low thermal conductivity, prevents the top layers of wellbore from reaching higher temperatures that otherwise would be reached without VIT. Therefore, when a well with

VIT shuts in and starts a cooldown process, the fluid trapped at the top of the tubing will experience a larger temperature gradient that increases heat losses, partially balanced by the larger thermal resistance induced by the VIT section. In conclusion, during cooldown, there are two competing factors that would determine whether the presence of VIT would be beneficial or not.

References

Bellarby, J. (2009). *Well Completion Design*. Amsterdam, Netherlands: Elsevier.

Brill, J. and Mukherjee, H. (1999). *Multiphase Flow in Wells*. Dallas, TX, USA: SPE.

Carslaw, H. and Jaeger, J. (1959). *Conduction of Heat in Solids*. London, UK: Oxford University Press.

Churchill, S. and Chu, H. (1975). Correlating Equations for Laminar and Turbulent Free Convection from a Horizontal Cylinder. *Int. J. Heat Mass Transf.* 18: 1049–1053.

Churchill, S. and Bernstein, M. (1977). A Correlating Equation for Forced Convection from Gases and Liquids to a Circular Cylinder in Crossflow. *J. Heat Transf.* 99 (2): 300–306.

Colburn, A. (1933). A Method of Correlating Forced Convection Heat Transfer Data and Comparison with Liquid Friction. *Trans. Inst. Chem. Eng.* 29: 174.

Collier, J. and Thome, J. (1994). *Convective Boiling and Condensation*, 3e. Oxford, U.K.: Oxford Science Publications.

Hahn, D. and Özişik, N. (2012). *Heat Conduction*, 3e. Hoboken, New Jersey, USA: John Wiley and Sons.

Hasan, A. and Kabir, C. (2002). *Fluid Flow and Heat Transfer in Wellbores*. Dallas, TX, USA: SPE.

Hellums, J. (1960). *Finite Difference Computation of Natural Convection Heat Transfer*," PhD Dissertation. Michigan, USA: The University of Michigan.

Incropera, F., Dewitt, D., Bergman, T., and Lavine, A. (2003). *Principles of Heat and Mass Transfer*, 7e. Hoboken, NJ, USA: John Wiley and Sons.

Ludovisi, D., and Garza, I., "Natural Convection Heat Transfer in Horizontal Cylindrical Cavities: A Computational Fluid Dynamics (CFD) Investigation," Proceedings of the ASME 2013 Power Conference, Power2013-98014, Boston, Massachusetts (2013).

Martini, W. and Churchill, S. (1960). Natural convection inside a horizontal cylinder. *AICHE J.* 6 (2): 251–257.

McCain, W. (1990). *The Properties of Petroleum Fluids*, 2e. Tulsa, Oklahoma, USA: PennWell Books.

Ramey, H. (1962). Wellbore Heat Transmission. *J. Pet. Technol.* 225: 427–435.

Shoham, O. (2006). *Mechanistic Modeling of Gas-Liquid Two-Phase Flow in Pipes*. Dallas, TX, USA: SPE.

Sieder, E. and Tate, G. (1936). Heat Transfer and Pressure Drop of Liquids in Tubes. *Ind. Eng. Chem.* 28 (12): 1429–1436.

Stewart, R., Sabol, A., and Boney, L., "Numerical and Experimental Studies of the Natural Convection Within a Horizontal Cylinder," NASA Technical Note, NASA TN D-7728, Washington D.C., USA. (1974).

9

Advanced Simulations

9.1 Introduction

Section 9.2 defines the implications deriving from time-dependent change upon the performance parameters. The physical significance of a transient two-phase flow is addressed to distinguish between "slow" and "fast" transients.

The normal course of changing the typical field operations, alternating between prolonged periods of "steady-state" and transient states, is covered in Section 9.3. The important realization that the highest operational risks (i.e. *Flow Assurance* risks) arise during some of the transient states points to the value of such simulations throughout the life cycle of production systems.

Section 9.4 revisits complexities of the system of governing equations covered in Chapter 5. The backbone of commercial codes currently employed to simulate two-phase flow are the 1-D models governing the transfer of mass, momentum, and energy, with additional closure models to incorporate radial property variances and multiple interwoven phases. Therefore, the deterministic approach based on first principles for predicting flow behavior constitutes what we call later the "*programming*" methodology, in contrast with the "*learning*" methodology.

The understandable attitude of engineers and technical personnel trusting the output from renowned commercial codes is reviewed and challenged in Section 9.5 concerning the inevitable numerical uncertainty of the programmed models stemming from the data acquisition process used to feed the models.

Section 9.6 provides a vision of where the modeling of transient two-phase flow may head in the short- and mid-term to provide more badly needed insight into this complex field. The emphasis in this section is to briefly

Multiphase Transport of Hydrocarbons in Pipes, First Edition.
Juan J. Manzano-Ruiz and Jose G. Carballo.

Companion Website: www.wiley.com/go/Manzano-Ruiz/Multiphasetransport

highlight the methodologies of *artificial intelligence* (*AI*) and *machine learning* (*ML*). A skin-deep introduction to these evolving fields and some application examples are presented in this section to familiarize the avid reader with the forthcoming developments applied to two-phase flow in pipelines, currently in their infancy. The most important message from this section is the demystification of the "automated" evolution of these remarkable technologies, born from the breathtaking digitalization advances and assumed devoid of human intervention. That is not the reality of such developments because data scientists and engineers are needed to select data sets, pre-process the data set, choose the learning algorithms, validate the models, and test the models on "unseen" data sets. Thus, hardly at least in the short run, we should not expect to count on entirely self-developed artificial intelligence codes without the supervised contribution of data scientists.

Section 9.7 closes this chapter by reviewing the two trending philosophies for simulating two-phase flow in pipelines. This section has been added to alert the reader of a potential forthcoming transition from the traditional deterministic modeling methods toward statistical analysis of field data sets. Brief references are made in this chapter to determine methods that have been covered in previous chapters, and more emphasis is placed on the statistical methodologies of *AI*.

This succinct chapter serves as an opening to the deterministic examples presented in the next chapter, while at the same time providing an overview of the modeling limitations of the state-of-the-art in this subject.

9.2 Nature of Transient Flow

A transient operation implies a change in the operating conditions. Although the normal operation may seem like a steady-state process, reality proves that hydrocarbon production through pipelines carries inherent fluctuations due to external perturbations affecting the system. Thus, a transient operation is herein defined as the transition process from one operating condition to another state, with the objective of changing the flow rate in anticipation of compliance with a production plan, handling an emergency situation, or to follow a maintenance program. Typically this includes start-up operations (e.g. a well, a cluster, or the complete system) and shutdown (again, a well, or the entire system). Obviously, the extreme case of emergency shutdown is included in the category of a transient because of the rapid change of operating state.

The physical significance of a transient two-phase flow is a succession of flow parameters, flow patterns, and changing fluid properties. Hence, it would be desirable to be able to model such transitions as a function of time to predict potential risks within the production system. In this sense, the forecasting of the produced liquids and the gas flow rates, the delivery pressure, and temperature become the task of the simulation work to ensure the system's safety.

Heat losses to the environment will also undergo appreciable change, in particular, when flow rates taper off because larger residence time in the system offers more opportunity to transfer heat. In the ultimate case of a full shutdown, the transient heat transfer becomes a substantial risk for solids deposition and plugging. So again the significance of transient heat losses is a potential severe condition for the technical integrity of the production system.

System blowdown is another important transient operation that entails rapid phase equilibrium changes and declining flow rates. Blowdown operations are commonly executed to preserve the integrity of the system and are triggered by an emergency situation such as during a lost-of-containment event.

Not all the system evacuation operations are considered a blowdown. *System depressurization* is a gradual and limited evacuation process to prevent the formation and plugging risk of hydrocarbon hydrates during a prolonged duration shutdown.

It is essential to define the difference between a regular transient two-phase flow phenomenon and the very fast transients characterized by traveling waves back and forth across the production system. The latter phenomenon, for single-phase flow, is popularly referred to as a *water hammer* phenomenon, involving wave velocities of the order of the speed of sound. The speed of sound is defined as

$$c \equiv \left(\frac{\partial p}{\partial \rho} \right)_s \tag{9.1}$$

The thermodynamics definition of the speed of sound implies waves of infinitesimal magnitude traveling under isentropic conditions in the fluid medium. The measurement of the speed of sound in single-phase flow is straightforward. For two-phase flow, the speed of sound definition and measurement is much more complex and dependent on the flow pattern as well as the phases' properties.

The relevance of a "water hammer" phenomenon is the possibility of "choking" the flow because any perturbation over the system taking place downstream of the choking point cannot travel upstream. The risk of choking a two-phase flow is related to the fact that the speed of sound is lower than the respective value for each of the phases present. For conventional hydrocarbon production systems, the chance of choked flow is low within pipelines, but exceedingly high at narrow transition fittings such as valves, tree chokes, or the exit route during blowdown.

Performance modeling under transient conditions is quite challenging due to the added time dimension intervening in the two-phase flow realm. This additional dimension numerically imposes time iterations upon the domain grid (discretized physical system). Similar to steady-state flow simulation, the physical system domain is discretized into a number of representative cells upon which the conservation principles are applied. One-dimensional numerical integration schemes such as finite differences or finite element are regularly incorporated into commercial codes. One of these standard schemes is the control-volume approach (Patankar, 1980) applied upon staggered grids while selecting numerical-stable implicit integration procedures.

The full three-dimensional physical flow-field simulation for hydrocarbon production is out of the question for typical systems, often spanning distances stretching for many kilometers. The computer storage memory and the shocking computational time requirements do not lend for routine simulations other than for short elements of the system, such as valves, changes in direction, heat exchangers, and separators. Such 3-D codes are usually classed as *CFD* (computational fluid dynamics) incorporating the full Navier–Stokes momentum equations for each phase in either the Eulerian or the Lagrange frame of reference. The difference with transient simulation is therefore the time variation of fluid parameters and properties within the discretized control volumes.

9.3 Transient Flow Applications

Routine two-phase transient operations could be simplistically visualized as the sequence of events shown in Figure 9.1.

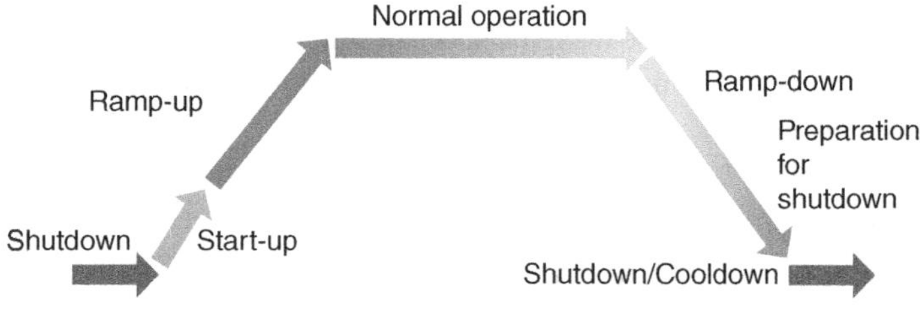

Figure 9.1 Typical transient operations for a production system.

After the shutdown of well clusters the entire system has to be restarted. The restart of a well involves the gradual opening of the tree choke to protect the reservoir from jittery transients conducive to the release of sand fines. A large pressure gradient across the tree choke is a Joule–Thomson risk in gas wells causing large temperature drops and potential hydrate plugging. Thus, the transient simulation of a well start-up requires the accurate modeling of the thermal hydraulics along the production string and across the choke.

The restart simulation of a cluster of wells is a sequence of individual wells started as described above. The overall system restart requires the initial condition in flowlines and risers. Such initial condition is obtained from running the simulation model under steady-state conditions and following up with the ramp-down process and redistribution of phases after shutdown. Moreover, depending on the cooldown process that follows the ramp-down, phase distribution in the system will depend on heat losses affecting it after shutdown.

Once the wells are restarted, the next step is to progressively ramp up production to reach the steady-state production target. Ramp-up should be gradual in order to minimize the risk of liquid surging downstream at the production facility while increasing the production flow rate. In general, the objective of ramp-up simulations is to predict the risk of flooding downstream separators.

Both the wells' start-up process and the system ramp-up transient operation have a time scale of a few hours, depending on the magnitude of the flow rates and the risk of sand production associated with drawdown.

The normal operation period, that is, quasi-steady-state process, will outlast any other transient operation, in pursuit of delivering the production plan. However, system perturbations or preventive maintenance will eventually lead to the ramp-down process followed by the shutdown. Temporary shutdowns to correct a minor problem, lasting a few hours, may not need to flush the live production trapped in the system or the system depressurization as pointed out before. In this case, the operator has a strict "no touch-time" rule in place for when the shutdown lasts no more than 8–10 hours. Beyond such a rule, *system flushing* may be activated with dead oil or hot oil to safely prepare the system for extended shutdown time. The methodology to establish such a precautionary operation is to perform a heat transfer analysis during the cooldown period until the fluids reach the pre-defined high-risk temperature. The flushing operation in itself is another transient two-phase flow application deserving proper simulation.

SURF facilities are constantly exposed to an environment at constant temperature and current velocity. From the modeling standpoint, this is a simpler case compared to onshore flowlines. The latter are frequently subjected to changing environmental conditions of wind, temperature, insolation, rain, and snow, which determine the thermal performance of the system. Partly for this reason, onshore flowlines are regularly buried and have a constant environmental condition.

For safety reasons, production systems occasionally are blowdown to minimize life and environment risks when there is a contingency or a loss of containment situation. A leak, a spill, may cause fire and explosion with severe consequences. Therefore, venting or burning the hydrocarbons in a flare unit is a safeguard to control an unsafe setting. The blowdown is a transient operation while the hydrocarbon inventory in the system is depleted. The initial stage of a blowdown may take place at a constant flow rate if the flow is choked downstream. The system pressure declines in time, and the later stages will experience a declining exit flow rate.

Cooldown is the heat loss phenomenon that takes place after system shutdown. The trapped production fluid would be semi-stagnant, with small internal motion driven by buoyancy. As such, cooldown is not an operation but a consequence of bringing the system to rest. For hydrocarbon systems, the condition of cooldown carries the implicit risk of causing solids deposition and potential plugging. Thus, very often, the operator of the field would attempt to flush the live production inventory with dead oil if an extended shutdown situation is foreseen. In case the last operation is unfeasible, the next option would be to depressurize the system to somehow control hydrate formation risk.

9.4 Transient Modeling Challenges

The fundamental equations of two-phase transient flow have been outlined in Chapter 5. The *TFM* set of equations is the most common one-dimensional tool implemented in most commercial codes, and the pioneering development of OLGA® (Section 5.6.1) in the early 1980s set the course for all other transient codes developed specifically for hydrocarbon multiphase flow in pipelines.

The issue with transient simulations, in general, is the extraordinary complexity of the set of governing equations, demanding exceptional numerical integration schemes applied to staggered grids (Patankar, 1980), while introducing closure equations to knit interface transport phenomena among phases. Parameters such as the area/volume fraction of one phase embedded in another require flow-pattern-specific correlations that are invoked during the execution of mass and momentum balances at each cell of the grid. Thus, the time variable adds another difficulty dimension to the already intricate dynamics of convection, diffusion, and source-term, balancing at every node of the physical domain.

Section 5.6.1 very succinctly addressed the set of equations utilized by the OLGA code in the early 1990s version. At that time, only three phases – oil, gas, and droplets – were considered in dynamical equilibrium. Nowadays, the OLGA code can consider nine concurrent phases pushing the envelope of numerical convolutions even further.

One crucial decision in solving such a daunting set of equations is opting for either the implicit or the explicit scheme of discretizing the physical domain. OLGA code has opted for the implicit scheme because it warrants integration stability for larger cell sizes. On the other hand, the explicit integration scheme, although it is a straightforward time-marching method, leads to unreasonably small cell sizes in order to meet the required conditional stability criterion of *Courant–Friedrichs–Lewy* (*CFL*) (Pletcher et al., 2013).

Regular operations of multiphase flow pipelines do have a "slow" type of transients. In other words, the total time to execute a transient operation is orders of magnitude larger than the residence time of the fluid at any specific location. However, there may be very fast phenomena at which the system dynamics occur at the speed of sound in the medium. One of such event may be an accidental loss of containment of the system, triggering sound and nonlinear waves traveling both ways while bouncing back from boundary system elements. Such phenomenon, previously mentioned in Section 9.2, takes place at the speed of sound in the multiphase medium which usually is slower than the speed of sound in each of the phases involved.

From the mathematical standpoint, "slow" transients, which are the rule in multiphase flow pipelines rather than "fast" *water hammer* events, fall in the category of semi-parabolic systems of equations. *Water hammer* events are characterized by hyperbolic-class systems of equations of the Cauchy problem (Pletcher et al., 2013) and require other integration schemes such as the method of characteristics.

9.5 Current Weaknesses of Simulating Two-Phase Flow in Pipelines

The simulation results from commercial software have a certain degree of uncertainty due to the assumptions, simplifications, and models built into the code, necessary to solve the intricated nature of multiphase flow combined with heat transfer. Such uncertainty in the results is very rarely quoted because of the difficulties in identifying and quantifying the impact of the semi-empirical models utilized to close the set of conservation principles. Commercial software is proprietary code inaccessible to the user interested in assessing the weaknesses of the models used. Furthermore, very rarely do software users have access to the database of multiphase experimental tests that enabled the code developer to generate closure correlations; an exceptional case is the participant members of a JIP (joint industry project). .

A list of sources of uncertainty in the results from deterministic simulation tools follows.

- One-dimensional assumption. Unavoidably, eliminating two flow directions distorts the physics of the flow field, and most codes extend the 1-D conservation principles to cover 2-D and 3-D domains through closure equations (e.g. drift-flux model).
- Averaging processes. Space and time averaging techniques are used to define key parameters such as hold-up and mixture density.
- Flow pattern recognition. For flow maps, this includes the issues related to the experimental work: the universality of the fluids used, the scale of the experimental facility, the extrapolation procedure to field scale, the sharp transitions among flow regimes instead of gradual transitions, the effect of developing length downstream of system components, and the possibility of condensing infinite flow regimes into a single map.
- Production systems real data. Scarce public information regarding accurate and granular field data about production performance to benchmark simulation results. Also, most data are focused on steady-state performance.
- Uncertainty in measured data. Lack of information about the accuracy of the instruments utilized to gather experimental data and a lack of uncertainty propagation analysis for the calculated properties.
- Thermodynamic equilibrium. All codes are based on the assumption of thermodynamic equilibrium for estimating fluids' properties. This assumption is a good approximation for steady-state and slow transients, but it fails at places with strong fluid acceleration or during fast transients.
- Cubic equations of state. The cubic EOS is ubiquitous in multiphase flow simulation for its simplicity and flexibility. However, cubic EOS has accuracy limitations in modeling the presence of more complex phenomena (e.g. heavy-ends characterization, solids precipitation). Moreover, there is debate as to the universal binary interaction coefficients required.

- Intermittent flows modeling. There is appreciable uncertainty for estimating slug frequency, slug length, and holdup for hydrodynamic and terrain-induced slugging phenomena.
- Commercial code uncertainty reporting. Usually, commercial code does not report the confidence level in the results, differentiated by source.
- Numerical errors. The integration of the partial differential equations describing the multiphase flow performance depends on the accuracy of the numerical algorithms (i.e. explicit, implicit, or semi-implicit), the truncation and roundoff error propagation.
- System topology. Every simulation requires a faithful description of the system geometry. The simulation of an existing system could be more reliable than the simulation of a system that is in the design stage.

Nonetheless, despite the list of weaknesses pointed out, the state-of-the-art of simulating multiphase flow has progressed tremendously when compared to the fully empirical black-box models published in the 1940–1950s. The advanced status of modeling tools currently available is based on physical phenomena, field measurements, inference from analogous cases, and expanding the frontier knowledge with expanded databases.

9.6 Future of Two-Phase Flow Simulations

This section presents a personal point of view of incipient developments in the field of multiphase flow in pipelines. Therefore, as a forewarning, this is a biased vision of potentially dominating forthcoming trends.

The traditional way to predict two-phase flow performance in pipelines is by *programming* algorithms derived from *conservation principles* (mass, momentum, and energy), with *fluid properties* obtained through equations of state manipulations, and with the help of *closure equations* and models (e.g. friction factors, holdups among phases) of empirical origin. The simulation case is set up with proper boundary conditions (i.e. system's inlet and outlet properties) and with a specification of the initial state of the fluid before running the programmed algorithms in the commercial code.

The weakness of the programmed simulation approach is the confidence in the validity of the closure models and the predicted fluid properties, as well as the certainty in representing the initial conditions of the overall system. On top of those doubts, there are integration errors creeping in from the application of the numerical scheme utilized. In summary, a simulation result from a code always has an unknown degree of uncertainty despite the remarkable effort put to develop such a tool and the necessary honed skills of the program user.

The state of the art in transient simulators, with the weaknesses, pointed out before, has sparked an interest in alternative prediction methodologies based on the extraction of patterns from large two-phase flow data sets.

The small-scale research laboratories and the semi-industrial scale test facilities have been the established sources of two-phase flow data. However, the digitization phenomenon that has started to thrive in the oil and gas industry in the last three decades has become a particularly important source of field production data. This is the reason why the humongous amount of data nowadays generated creates valuable data sets to "mine" patterns and trends.

Data science (Kelleher and Tierney, 2018) is a field that combines principles, algorithms, and processes in search of concealed patterns from large data sets. The coined term *data mining* is often used interchangeably with data science, although the focus of *data mining* is the extraction and analysis of structured data and knowledge, somehow masked within a vast amount of data. The objective of data mining is to transform the dataset into a coherent structure for further use. *Data mining* involves data management, pre-processing, modeling, metrics, and inference processes. The digital data invading all fields of human activity have driven the impressive size and complexity of data sets. The former manual procedure of gathering experimental data has been replaced by inexpensive and powerful computers.

ML (machine learning) is a novel approach to making use of the extracted patterns and knowledge from a data set. *ML* can be defined (Kelleher and Tierney, 2018) as a field of study that develops algorithms implemented into computer programs in order to identify and extract patterns from data. The field of *machine learning* encompasses the use of computer networks to process pre-selected digital information with algorithms designed/chosen by the data scientist. The vertiginous pace of machine learning evolution has undergone notable and prolific algorithm representations/models such as *artificial neural networks (ANN), cluster analysis, convolutional neural networks (CNN), regression analysis, genetic algorithms, knowledge neural network (KNN), boosting algorithms, decision trees, and vector machines.* Fast computers with large storage capacity apply such specialized algorithms in search of patterns to fulfill the data mining task.

ML is a gradual learning process (Alpaydin, 2021) in pursuit of an optimized search performance of patterns. Therefore, similarly to the knowledge evolution of human beings, the algorithms of the machine learning process require training, either *supervised* or *unsupervised*. *Supervised learning* is a mapping process from the inputs' attributes into the *target attribute*. The supervised process uses the data set as feedback to optimize the extraction model algorithm. This process is iterative, involving many function trials seeking best mapping. As such, supervised learning rests on the skills of the data scientist to choose appropriate models (i.e. preferred algorithms) introducing an element of *learning bias*. The supervised learning process involves at least the following steps: example selection, defining the input representation of the learned function, defining the objective function of the learned function and learning algorithm, running the learning algorithm on the gathered training set, assessing the error of the objective function, and adjusting control parameters accordingly to improve performance of the objective function. Supervised *learning* fits best with numeric types of data sets.

Unsupervised learning is more akin to ordinal data sets (e.g. image processing) rather than numeric data sets. The *unsupervised learning* process has no target attribute, and the selected algorithms look for data regularities, that is, subsets of the database that have similar characteristics. One popular technique employed is *cluster analysis* which iteratively groups the selected subsets and keeps redefining the similarities that identify the groups. The optimization of the algorithm requires similarity metrics during the learning process.

The training of machine algorithms is done with examples involving well-defined inputs (independent variables) and dependent variable(s).

The highest value derived from machine learning is not the capacity to duplicate (i.e. memorize) the data set during the training phase. Memorizing code is nothing more than a "look-up" table algorithm that limits the application exclusively to the training set. The highest value of *ML* lies in the programming capacity to build by itself improved models capable of predicting new performance scenarios beyond the training data set. Such capacity for self-learning and self-improved performance toward new challenges defines the objective of *AI*. It is important to highlight that model improvement is an iterative and gradual process striving to achieve optimum performance, very much as the human brain learns by trial and error to succeed in more complex tasks.

The four more common techniques used in *ML* are *Linear Regression*, *ANN*, *CNN*, and *Decision Trees*. *Linear and Multi-Linear Regressions* focus on the output error defined in terms of the sum of the squared (*SSE*) errors for each instance. The linear relations are weighed with parameters that are changed until the *SSE* reaches a minimum. This process is also referred to as the *least square* method. Obviously, this methodology is useful when the data set shows linear trends.

ANN have become a favorite learning methodology to design extraction algorithms. ANN mimics the human brain's network activity whereby information processing and distribution are realized by communication nodes (neurons) transmitting signals (i.e. inputs) replicating the synapsis of the brain. Nevertheless, ANNs depart from the real brain because the networks are static and symbolic, whereas the biological brain of most living organisms is dynamic (plastic) and analog. In the *ANN* model, the connection "signals" are real numbers, and the output of each neuron is computed by some nonlinear functions (i.e. the *activation functions*) applied to the sum of its inputs. Two of the frequent *activation functions* used in *ANN* modeling are the hyperbolic tangent (tanh) and the logistic functions (Kelleher and Tierney, 2018) because have limited outputs regardless of infinity input. These inputs arriving at each neuron are a *weighted linear function* from the originating neuron which, during the course of the learning process, will be subject to numerous optimizing adjustments in pursuit of minimizing the error objective function applied to the output signal. Thus, the resulting total input on each neuron is a *multiple-input linear-regression* process. Neurons may have thresholds that transmit signals when the aggregate signal overcomes that threshold.

The *ANN* model comprises neurons collected into layers. The first layer is always the *input layer* introducing the attributes from the input signals taken from the data set. The last *output layer* may have one or more neurons, from which the objective function to be optimized is attached to a single of those neurons. In between the input/output layers, there may be several "hidden" layers; depending on the complexity degree, the network is built in. Figure 9.2 shows as an introductory example a simplified schematic of a three-layer *ANN*, to clarify the role of the inputs to each neuron and the applied *activation functions.*

Taking Figure 9.2 as an example that can be generalized to broader networks, inputs from the data set are represented by,

$$I_m \quad \text{data inputs for } m = 1, 2, 3, 4 \tag{9.2}$$

The multi-input linear-regression functions and the variable weights applied to the input signals are

$$\text{multi-input linear-regression function:} \quad Y_{m,n} = \lambda_{m,n} + \omega_{m,n} I_m \tag{9.3}$$

$$\omega_{m,n} \text{ weight coefficient for inputs } I_m \text{ from hidden neurons } n = 5, 6, 7 \tag{9.4}$$

$$\lambda_{m,n} \text{ independent weight coefficient from hidden neurons } n = 5, 6, 7 \tag{9.5}$$

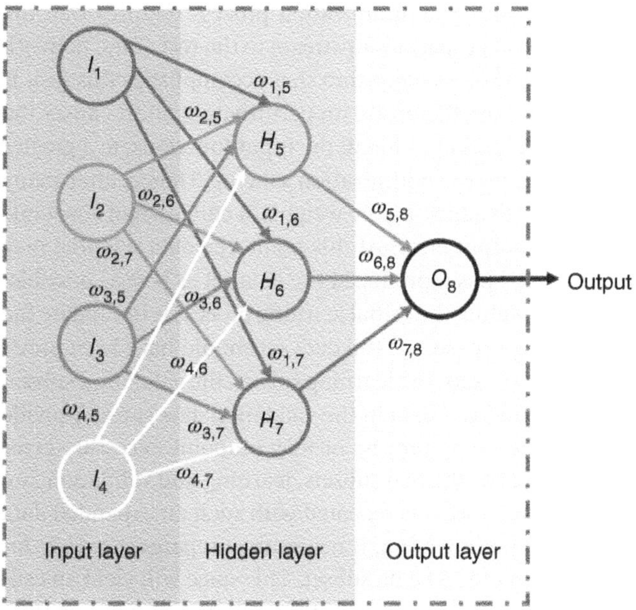

Figure 9.2 Simplified ANN schematic example.

The nonlinear functions resulting from applying *activation functions* to the output signals departing the last hidden layer are represented in equation (9.6). Notice the use of the hyperbolic tangent function to characterize the nonlinear *activation function*.

$$Z_{p,8} = \tanh(\omega_{p,8}H_p) = \frac{\exp(2\omega_{p,8}H_p) - 1}{\exp(2\omega_{p,8}H_p) + 1} \quad \text{for } p = 5, 6, 7 \tag{9.6}$$

The overall output from the example *ANN* in Figure 9.2 has been characterized by the nonlinear *logistic function* as follows:

$$\begin{aligned} Output &= logistic\left(\sum_{p=5}^{p=7} \omega_{p,8}H_8\right) \\ &= \frac{1}{1 + \exp[-(\omega_{5,8}H_5 + \omega_{6,8}H_6 + \omega_{7,8}H_7)]} \end{aligned} \tag{9.7}$$

As pointed out before, the key step of *ANN* is the *learning process* from the data set. This step follows after defining the number of hidden layers and the output neuron(s), and the pre-selection of the algorithm methods (e.g. linear regressions, activation functions) – the so-called *learning bias* effect. This step involves starting from assumed initial weights on the input functions to initiate the iterative optimization that minimizes the error in the output objective function. The optimization process requires that the error-minimization function propagates upstream to the rest of the network neurons. The propagation process algorithm that accomplishes this task is the *backpropagation*. However, the upstream perturbation effect causes the re-adjustment of all the weights $\omega_{m,n}$ for all the input functions, to a greater or lesser extent depending on the optimization scheme. Hence, after many iteration cycles back-to-forth adjusting the weights, a culled set of constants emerges minimizing the output error. At this point, the data scientist may feel confident that the ANN has reproduced the data set as best as possible.

The next step after completing the initial learning process is to validate the models obtained. For this purpose, the full data set should have been previously separated into three subsets: the learning subset, the validation subset, and the test subset. The "unseen" data in the validation data subset provide an authentication of the learning process or may pinpoint necessary error adjustments. In the latter case, the two subsets, learning and validation, are combined and the learning process is repeated with such an expanded data set. Following the second optimization, if necessary, prepares the *ANN* for real prediction testing with the last data subset. This stage will yield an estimate of the accuracy of the machine learning overall process by evaluating the output error against real data. There is criterion freedom to split the total data set into three subsets; some researchers use 50, 30, and 20% ratios when the data set is fully accurate and comprehensive. Other people prefer 40, 25, and 35% for added weight on prediction capability. In any case, there are no rigid rules in this regard. In any event, any *ANN* process will involve *sample bias*.

When the number of hidden layers is large, the learning process takes another twist. This is called *deep learning* and might involve supervised semi-supervised or unsupervised learning. Deep learning adopts several methodologies such as networks, deep belief networks, deep reinforcement learning, recurrent neural networks, and *CNN* that have been successfully utilized in several other fields involving image recognition.

The *decision tree* (*DT*) technique is a powerful way to identify a structure or pattern within a nominal or ordinal data set. The technique works on making path decisions based on key inquiring of the data. The binary character of the process sets a clear-cut way to go through the learning data set.

Despite the very promising developments to date in machine learning artificial intelligence, there are some challenges. Aside from the challenges of *sample bias*, *data pre-processing*, and *learning bias*, one other significant challenge is the concept drift caused by the *stationary assumption*. The latter is the common assumption that large comprehensive data sets do not change over time. Data sets do change, requiring machine learning algorithms' re-training from time to time to keep up with evolution. The list of challenges for machine learning includes additional topics such as nonlinearities, function complexity, bias-variance tradeoff, the sufficiency of training data, data heterogeneity, the dimensionality of input space, and very importantly output noise.

There are some examples of the application of machine learning techniques to the field of two-phase flow. The following examples include a variety of other methodologies not described in this section for which the reader should consult the addressed references.

Khamis et al. (2020) published a paper focused on the optimization of choke-valve size for oil and gas flow, using artificial intelligence instead of the customary deterministic set of available correlations. AI was chosen to assess the optimum choke size to deliver a specified two-phase flow rate. The AI techniques considered in this study were a six-layer *ANN*, *fuzzy logic* (*FL*) with a genetic optimization algorithm, *support vector machines* (*SVM*) and *functional networks* (*FN*). The resulting comparison of the testing correlation coefficients among the methodologies yielded the following predictability confidence coefficients *R*: 99.7% (*SVM*), 84.0% (*ANN*), 81.0% (*FL*), and 81.0% (*FN*).

Arteaga-Arteaga et al. (2021) work on *ML* applications to predict two-phase flow patterns centered on *ML* algorithms and deep learning to address the intricate classification problem. The input variables considered include phases' superficial velocities, viscosities, densities, and superficial tension. Two important additional inputs included were the angle of inclination of the pipe, the flow orientation (horizontal, upflow/downflow), and the internal diameter. Twelve databases (9029 instances) were combined to perform the training and testing of the algorithms. The *ML* models considered included *ANN, CNN*, *Extra Trees* (*ET*), *Random Forest* (*RF*), *Gradient Boosting* (*GB*), *SVM*, *DT*, *KNN*, *QDA*, *GNB, and AB*. The *Extra Trees (ET)* model classified best the data set flow patterns with the highest degree of fidelity, achieving a confidence coefficient *R* of 98.8%.

A third example is the work by Alamoudi et al. (2021). In this paper, the focus was detecting the thickness of depositing scale solids, using the gamma attenuation technique, over the inner wall of pipes conveying two-phase oil and gas. Three flow patterns were investigated in this study, namely stratified, homogeneous, and annular flow. The void fraction for the two-phase flow varied between 10% and 85%. The ML methodology used is the *RBFNN* which is a combination of the *radial basis function* (*RBF*), a type of neural network technique (to process the input data), followed by a neural network algorithm. The successful application of *AI* to this problem yielded a training data confidence coefficient *R* of 98.4% and a testing-data confidence of

98.7%. Such excellent results prove that the *ML* approach properly carried out surpasses any deterministic modeling.

The last example (Ruiz-Diaz et al., 2021) in this section concerns a two-phase liquid–liquid application of oil and water flow. The work used an *adaptive neuro-fuzzy inference* (*ANFIS*) procedure jointly with *ANN* to predict the holdup of a water and oil mixture through vertical pipes. The input parameters used were the pipe's internal diameter, surface velocities, and oil viscosity. The data set for training and testing comprised 722 instances. The best *ANN* model generated had a hidden layer comprising 12 neurons and using the Levenberg–Marquardt training function with the TanSig activation function. The confidence coefficient R was a remarkable 99.9%. Once more the extraordinary results for this AI application confirm the power of this emerging field for oil and gas production systems.

9.7 Simulation Philosophies

At this point, it is convenient to review high-level simulation philosophies for steady-state and transient two-phase flow. We could split the philosophies into *programming* versus *learning*. The *programming philosophy* seeks to create models based on conservation principles complemented by semi-empirical closure equations, to match the number of equations to the number of unknowns. This has been the conventional approach followed during the past 80 years, yielding the superb commercial codes currently available, despite the drawbacks pointed out in Section 9.5. This *programming philosophy* is a deterministic viewpoint.

On the other hand, the *learning philosophy* emphasis is on using up the hidden patterns and structures of reliable data sets comprising numerous instances of accurate field measurement data. The discovery of performance patterns allows to create particularly useful algorithms that with remarkable accuracy are capable of predicting new performance through a proper learning, validation, and testing process. We confirmed the excellent predictability of two-phase flow results with a small sample of examples introduced in Section 9.6.

References

Alamoudi, M., Sattari, M., Balubaid, M. et al. (2021). Application of Gamma Attenuation Technique and Artificial Intelligence to Detect Scale Thickness in Pipelines in Which Two-Phase Flows with Differential Flow Regimes and Void Fractions Exists. *Symmetry* 13 (1198): 1–16.

Alpaydin, E. (2021). *Machine Learning*. Cambridge, Massachusetts, USA: The MIT Press Essential Knowledge Series.

Arteaga-Arteaga, H., Mora-Rubio, A., Florez, F. et al. (2021). Machine Learning Applications to Predict Two-Phase Flow Patterns. *PeerJ Comput. Sci.* 7: e798. DOI 10.7717/peerj-cs.798.

Khamis, M., Elhaj, M., and Abdulraheem, A. (2020). Optimization of Choke Size for Two-Phase Flow Using Artificial Intelligence. *J. Pet. Explor. Prod. Technol.* 10: 487–500.

Kelleher, J. and Tierney, B. (2018). *Data Science*. Cambridge, Massachusetts, USA: The MIT Press Essential Knowledge Series.

Pletcher, R., Tannehill, J., and Anderson, D. (2013). *Computational Fluid Mechanics and Heat Transfer*, vol. 3. Boca Raton, Florida, USA: CRC Press.

Patankar, S. (1980). *Numerical Heat Transfer and Fluid Flow*. New York, New York, USA: Hemisphere Publishing Corporation.

Ruiz-Diaz, M., Hernández-Cely, M., and González-Estrada, O. (2021). Analysis of Liquid-Liquid (Wáter and Oil) Two-Phase Flow in Vertical Pipes, Applying Artificial Intelligence Techniques. *J. Phys. Conf. Ser.* 2046: 1–7.

10

Multiphase Flow Simulations

10.1 Introduction: Simulation Challenges

Multiphase flow refers to the simultaneous flow of three different fluids, typically gas, liquid hydrocarbon, and water, in a single system. In oil and gas systems, a two-phase flow occurs when liquid and gas are produced from a reservoir and transported through pipelines to processing facilities. Numerical simulation of multiphase flow is an essential tool for predicting the behavior of oil and gas systems and for optimizing their design and operation.

The simulation of multiphase flow is a complex task that presents several challenges. Examples include modeling the interaction between the phases, fluid phase change along the system, interaction of the gas and liquid phases with the wall under the different flow regimes, and others. In addition, there are numerical challenges related to the discretization of the governing equations, the selection of appropriate boundary and initial conditions, and the solution of nonlinear equations.

This chapter provides an overview of the challenges involved in numerical multiphase simulation in oil and gas systems. We begin by discussing the fundamental concepts of two-phase flow and the mathematical models used to describe it. We then highlight some of the critical challenges in the numerical simulation of two-phase flow, including modeling, numerical, and computational challenges.

10.1.1 Mathematical Models for Two-Phase Flow Simulation

The mathematical models used to describe two-phase flow depend on the type of flow regime and the nature of the fluids. In general, two-phase flow in horizontal sections can be classified into four regimes: stratified, bubble, slug, and annular flow. In bubble flow, small gas bubbles are dispersed in the

Multiphase Transport of Hydrocarbons in Pipes, First Edition.
Juan J. Manzano-Ruiz and Jose G. Carballo.

Companion Website: www.wiley.com/go/Manzano-Ruiz/Multiphasetransport

liquid phase. In slug flow, the gas phase forms elongated bubbles separated by liquid slugs. In annular flow, the liquid phase forms a continuous ring around the gas phase. In stratified flow, the liquid flows in a layer below the gas phase. In vertical flow, only bubble, slug, and annular flow are possible.

For each of these regimes, different mathematical models are used to describe the behavior of the two phases. In general, the governing equations for two-phase flow include the continuity equation, the momentum equation, and the energy equation. These equations are coupled through the interfacial forces that exist between the two phases.

10.1.2 Challenges in Two-Phase Flow Simulation

Simulation of two-phase flow presents several challenges related to modeling, numerical, and computational aspects. These challenges are discussed in more detail below.

Some of the modeling challenges in multiphase flow simulation include the following:

- *Interfacial Phenomena*: The interaction between the phases is complex and involves several interfacial phenomena, such as surface tension, wetting, and capillary forces. Modeling these phenomena requires accurate measurement and modeling of interfacial properties, which can be difficult to obtain experimentally.
- *Phase Change*: Phase change is a critical aspect of multiphase flow, particularly in oil and gas systems. The phase change occurs when the fluid undergoes a change in state, such as from liquid to gas or vice versa. Accurate modeling of phase change requires knowledge of the thermodynamic properties of the fluids.

The numerical simulation of multiphase flow also presents several challenges related to the discretization of the governing equations, the selection of appropriate boundary conditions, and the solution of nonlinear equations. The numerical methodology used in the simulation has an important effect on the computation time.

10.1.3 Results Validation

Numerical simulation of two-phase flow is an essential tool for predicting the behavior of oil and gas systems and optimizing their design and operation. However, the accuracy of the simulation results is critical for reliable decision-making, and it is important, if possible, to validate these results against experimental data. In this chapter, we discuss the challenges associated with validating simulation results with data gathered in the field.

Validation of numerical simulation results requires experimental data that accurately represent the physical behavior of the system. In oil and gas systems, the data can be obtained from various sources, including field measurements, laboratory experiments, and pilot plant tests. The data can include flow rates, pressure drops, temperature profiles, and fluid properties such as density, viscosity, and interfacial surface tension.

Validation of numerical simulation results is a difficult task. The main challenges in this regard include the following:

- *Variability in experimental data*: The experimental data obtained from the field can be affected by various sources of variability, such as measurement errors, changes in operating conditions, and natural variations in the system (e.g. fluid composition change). These sources of variability can make it difficult to identify and isolate the sources of discrepancies between the simulation results and the experimental data.
- *Sensitivity to model parameters*: Numerical simulations use several model parameters, such as transport fluid properties, interfacial tension, and wall friction. These parameters can have a significant impact on the simulation results and are difficult to measure. Small changes in the model parameters can result in significant differences in the simulation results, making it untenable to validate the simulation results.
- *Intricate system dynamics*: Oil and gas systems can exhibit complex system dynamics (including nonlinear behavior, phase transitions, and multiphase flow regimes) that exacerbate the difficulty to validate the simulation results against experimental data.

Despite the challenges involved in validation, several approaches can be used to validate the numerical simulation results. These approaches include the following:

- *Key Performance Indicators (KPIs)*: Key performance indicators can be used to compare the simulation results with the experimental data. KPIs can include flow rates, pressure drops, temperature profiles, and fluid properties such as density and viscosity. Comparison of KPIs can help to identify discrepancies between the simulation results and the experimental data.
- *Sensitivity Analysis*: Sensitivity analysis can be used to assess the impact of model parameters on the simulation results. Sensitivity analysis involves varying the model parameters over a range of values and observing the effect on the simulation results. This approach can help to identify the criticalness of specific parameters in the simulation results.

- *Model Calibration*: Model calibration involves adjusting the model parameters to improve the agreement between the simulation results and the experimental data. Model calibration requires a thorough understanding of the system dynamics and the accuracy of the experimental data. This approach can help to improve the accuracy of the numerical model and increase confidence in the simulation results, in particular, when field data are available.

10.2 Multiphase Flow Simulation Considerations

Developing an accurate model that represents the physics of the process and delivers results in a reasonable amount of time is the objective of a multiphase flow simulation. This section addresses some of the concerns to develop a multiphase flow simulation model.

10.2.1 Integration Time Step

In numerical methods for solving systems of partial differential equations (PDEs), the time step plays a crucial role in determining the accuracy and stability of the solution. The time step refers to the size of the intervals between two consecutive time iterations, which are used to discretize the equations in the time domain.

The size of the time step must be appropriate for the phenomena studied. For instance, in *water hammer* simulations, the time step scales with the speed of sound of the fluid being studied. In *slow transients*, such as during a well ramp-up, the time step could be larger than that of a water hammer. Choosing an inappropriate time step size could lead to unrealistic results or to instability in the simulation.

The stability of the time step is also crucial. Some commercial simulators attempt to optimize (i.e. increase) the time step based on certain parameters within a determined range. For example, a slow transient simulation with a constant time step of 0.15 seconds will generally produce better results than a simulation whose time step is constantly oscillating between 0.001 and 0.10 seconds.

In general, using a smaller time step can increase the simulation time and the accuracy of the results, all else being equal. However, there are some exceptions to this rule. For instance, a very small time step may be prone to round-off errors that can result in inaccurate results. Conversely, using a larger time step can decrease the simulation time but may also reduce the accuracy of the results. Increasing the time step should be done carefully in many simulators. If the time step is too large and the simulator does not converge, the time step may be automatically reduced, and the simulated time step may be repeated, significantly increasing the overall simulation time.

10.2.2 *CFL* Condition

The *Courant–Friedrichs–Lewy (CFL)* condition, see Pletcher et al. (2013), is a mathematical stability constraint relating the time step and the spatial mesh spacing applicable to PDEs. The condition is named after Courant et al. (1928), who first proposed this condition as a means of ensuring stability in numerical simulations of wave propagation. The CFL condition has become a fundamental requirement for numerical simulations of hyperbolic PDEs in various fields, including fluid dynamics, electromagnetics, acoustics, and structural mechanics.

In multiphase flow simulations, for example, the CFL condition is used to determine the maximum allowable time step size based on the local flow velocities and the grid size. In multiphase flow simulation, the time step should be small enough so that a fluid particle cannot travel more than one cell in a one-time step.

10.2.3 Round-off Error

Round-off errors are a common issue in numerical simulations, especially when dealing with large-scale problems that require high precision. These errors arise due to the finite precision of the numerical representation of real numbers in computers.

In numerical simulations, numbers are represented as floating-point numbers with a finite number of digits. As a result, certain numbers cannot be represented exactly, and small errors may occur in the calculations. These errors can accumulate (i.e. creep in) over time and affect the accuracy of the solution.

Round-off errors can be minimized by using higher precision data types or by implementing numerical algorithms that are less sensitive to rounding errors. In some cases, it may also be possible to scale the problem or change the units to reduce the magnitude of the numbers involved and minimize the effects of round-off errors.

Creating a suitable flowline segmentation and time step is essential for minimizing round-off errors in numerical simulations. The flowline grid segmentation refers to the number and size of the segments used to discretize the domain of the problem. A small segment size can help to capture more details of the solution, but it may also increase the computational time and round-off errors due to the limited precision of the computer's arithmetic operations in extreme cases.

10.2.4 Geometry Representation

Reliable results in flowline simulations require accurate geometry representation. In practice, the real geometry of the flowline may be too complicated, and a simplified model of bathymetry or pipeline routing may be necessary to avoid excessive computational time. This can be particularly important for long flowlines where a high-resolution simulation of the entire bathymetry may be impractical.

An oversimplification of the geometry can result in inaccurate simulation results. For example, the simplified bathymetry may not accurately capture the variations in the flowline geometry, which can lead to significant errors in the pressure drop and liquid distributions, these errors are particularly important in the gravity dominated regions where the flow velocity is low. An example of such geometry is the low points (i.e. valleys) found on quasi-horizontal flowlines, which at low two-phase flow rates tend to accumulate and leave behind liquids at those low points. Therefore, there is a trade-off balance between the level of detail in the simulation and the computational resources required to perform the simulation.

There are several methods to simplify geometry; these methods may involve a combination of predetermined mathematical procedures, engineering judgment, and overall observation of the available data. Regardless of the method used, it is essential to validate the geometry selection by comparing the main simulation results among different degrees of flowline segmentations.

10.2.5 Cell-Size Ratio - the Ratio of "2"

It is generally recommended that the volume of two sequential cells be approximately the same volume to avoid large differences in the resolution between neighboring cells. Large differences in cell size can lead to inaccurate results and may introduce numerical errors in the simulation.

In practice, it may not be possible to have cells of equal size along the entire flowline; in such cases, it is recommended to ensure that the volume ratio of adjoining cells is between 0.5 and 2, a ratio widely used in the oil industry.

10.2.6 Dispersion Effects

In the numerical simulation of sharp front displacement processes (e.g. fast wave phenomena), dispersion effects can occur due to the numerical approximation of the governing equations and will lead to the smearing of the front and uncertainties in the displacement process.

One approach to reduce, but not completely eliminate, the dispersion effect is to decrease the size of the flowline segmentation and the time step; in this way the resolution of the simulation can be increased, allowing for a more accurate representation of the sharp front displacement process.

It is recommended to validate the results of the simulation against experimental or field data, if possible, to adjust the flowline segmentation and time step accordingly.

10.2.7 Constant Composition Versus Compositional Simulation

Simulations can be run as *constant composition* or as *compositional simulations*. Constant composition simulations assume that the total fluid composition is constant in every cell and every time step. While the total composition is constant, the relative gas and liquid amounts do change as a function of the pressure and temperature in every flowline segment. Additionally, thermodynamic equilibrium is assumed in every cell. For instance, a flowline in steady-state production conditions can be simulated as a constant composition system even if the pressure and temperature (and therefore the relative gas–liquid amounts and properties) change in every cell. This constant composition assumption simplifies the simulation process and reduces the computational time required to obtain results.

In contrast, there are operations where the overall composition is changing. In a multiphase flowline depressurization, the composition of the fluids in the flowline changes with every time step. In these cases, a compositional simulation may be required to obtain detailed results of how the composition changes in the flowline with respect to time.

Compositional simulations are more time-consuming than constant composition simulations. In a compositional simulation, a flash calculation is performed in every segment at every time step, and the computational time required to obtain results from a compositional simulation can be significantly larger than that of a constant composition simulation. The number of components used in a compositional simulation will have a large impact on the total simulation time, as well as the option of having a three-phase flash will also greatly affect the simulation time.

Thermodynamic equilibrium is also commonly presupposed for compositional simulations. This assumption can be manually overridden to assume a certain degree of nonequilibrium if data are available.

It is important to note that compositional simulations can potentially be more unstable and therefore prone to a certain degree of inaccuracy even in the most benign cases.

It is the opinion of the authors that a fully compositional simulation is only worth it in very limited and special cases where the predictions of the composition of the fluid are required for nonconstant composition cases. In most cases, however, constant composition simulations are sufficient to predict the overall behavior of fluids in pipelines and wells. In most practical cases, having an approximate result with a clear idea of the margin of error associated is more important than a supposedly "more accurate" compositional simulation.

10.2.8 Initial Conditions

When running a transient multiphase flow simulation, the initial conditions can greatly impact the accuracy and stability of the results. It is important to ensure that the initial conditions are appropriate to avoid instabilities.

The first type of initial conditions is the user-provided initial conditions. This involves manually setting the initial conditions such as the flow rates, pressure, temperature, and holdup at the start of the simulation. While this method is straightforward, it is often not accurate enough, and errors may propagate through the simulation.

The second type of initial condition is when the system is flowing at steady-state conditions. In this case, the simulation starts at a point where the flow rates, pressures, and temperatures have reached steady-state conditions. The simulator steady-state preprocessor can be a great initialization tool for most cases, but in others, serious errors or "not-convergence" of this tool can lead to completely unrealistic results.

The third type of initial condition is restarting from the last point of a previous simulation. This method involves using, in most cases, the last simulated time of a previous case as the initial condition for a new simulation. This method can be useful when running long or complex simulations.

10.2.9 Deepwater Settings

Deepwater settings present unique challenges for oil and gas systems due to the possibly harsh operating conditions, remote location, and involved geological structures. Multiphase flow simulation can be a valuable tool for designing and optimizing multiphase flow systems in deepwater settings.

Some of the specific applications of multiphase flow simulation in deepwater settings include the following:

- *Well performance*: Simulations can evaluate the performance of individual wells in deepwater settings. This includes evaluating well productivity, optimizing wellbore and completion designs, and identifying opportunities for artificial lift.
- *Subsea production systems*: Multiphase flow simulation can be used to design and optimize subsea production systems. This includes selecting flowline diameters, designing subsea manifolds, and assessing the use of subsea processing and separation facilities.
- *Subsea export pipelines*: Simulations enable the design and optimization of subsea export pipeline systems. This includes selecting pipeline diameters, designing pipeline routes, and optimizing pumping and compressor stations.

10.3 Multiphase Flow Applications

In this chapter, we discuss some typical multiphase flow applications to illustrate how engineering companies and oil operators utilize simulation for the design and operation of multiphase flow systems. At a high level, the applications can be ascribed to the two following categories:

- *Design and optimization of oil and gas production systems*: Multiphase flow simulation can be used to optimize the design and operation of oil and gas production systems. This includes designing wellbore configurations, selecting flowline diameters, and designing separation and processing facilities.
- *Prediction of flow assurance issues*: Field production operations can be simulated to predict and mitigate flow assurance issues, such as hydrate formation, wax deposition, and scale buildup. By modeling the fluid flow and heat transfer within the system, the engineer can predict where these issues are likely to occur and suggest preventative measures.

The following subsections describe representative production system applications and include examples to illustrate simulation performance.

10.4 Production Wells Simulation

The objective of multiphase flow simulations for wells is to try to achieve production target rates and minimize the risk of flow assurance issues. Some of the key parameters that are typically modeled include the following:

- *Pressure and temperature profiles*: Pressure and temperature profiles are critical to understanding the flow behavior within the production well, which determines the phase equilibrium and the fluids' properties as a function of the well's depth.
- *Flow rates*: Predicting the flow rates of oil, gas, and water within the production well's string is critical to optimizing production rates and minimizing the risk of flow assurance issues. Oil and gas flow rates are vital to keeping track of performance and cumulative reserves produced.
- *Flow regimes*: The flow regime in a production well can have a significant impact on the flow behavior and the risk of flow assurance issues due to the different flow patterns.
- *Production tubing and completion systems*: Simulation is the proper means to predict the flow behavior within the tubing and ensure that the system size is designed to minimize the risk of flow assurance issues.

Multiphase flow simulation can be used to study various aspects of well start-up, including the following:

- *Slug Flow*: Slug flow is a common flow regime that can occur during well start-up, especially in long horizontal or highly deviated wells. In slug flow, the liquid and gas phases alternate in the wellbore, resulting in large pressure fluctuations and potential damage to the wellbore and surface equipment. The simulation can be used to predict the onset of slug flow and optimize the flow rate (e.g. through tree choke management) and wellbore design to prevent slug flow from occurring.
- *Flow Assurance*: During well start-up, the temperature and pressure conditions can change rapidly, which can lead to flow assurance issues such as hydrate formation, wax deposition, and asphaltene precipitation.
- *Wellbore and Surface Equipment Design*: The simulation can be used to optimize the location and design of choke valves, to control wellbore performance and surface equipment during slug flow or pressure fluctuations.
- *Well clean-up:* The simulation can predict and monitor in real time the behavior of fluids during the clean-up process, including the velocity and pressure of fluids, and the impact of fluid properties and flow rates on the overall clean-up efficiency. By simulating different clean-up scenarios, engineers can determine the most efficient and effective clean-up procedure.

Well shutdown is an essential operation in the life cycle of a production well. It can be either planned or unplanned. A planned shutdown is usually scheduled for routine maintenance, equipment inspection, or to perform interventions such as stimulation or workovers. On the other hand, an unplanned shutdown can happen due to equipment failure, safety concerns, or production issues.

During a well shutdown, it is essential to ensure that the well and surrounding equipment are preserved and protected from blockage. One of the main concerns in subsea wells is the formation of hydrates, which can occur when gas and water mix at low temperatures and high pressure. Hydrates can form in the wellbore and flow lines, obstructing the flow of fluids and causing deferred production.

To prevent hydrate formation, methanol injection is commonly used. Methanol is a chemical that acts as a hydrate inhibitor, preventing the formation of hydrates in the wellbore and flow lines. Methanol can be injected continuously or intermittently, depending on the well's operating conditions and the risk of hydrate formation.

The injection of methanol is usually done through the wellhead and the tree components, which are the valves and fittings that control the flow of fluids from the wellbore to the surface. These components are particularly susceptible to hydrate formation due to their proximity to the wellbore and their exposure to low temperatures (e.g. for deepwater systems) and high pressures.

The well shutdown procedure can be simulated to minimize flow assurance risks, including the injection of methanol (i.e. bullheading operation) to prevent hydrate formation.

10.4.1 Onshore Well Operating in Steady State

In this section, we provide an example of onshore well simulation.

Table 10.1 summarizes the parameters of a simulated deviated well, the production fluids, and the delivery point at a surface separator near the wellhead. Well's elevations are expressed with respect to a datum located at the wellhead elevation.

Figure 10.1 shows the well geometry as a function of the horizontal length and as a function of the total length. In this chapter, the simulated results are presented as a function of the total well length for a better appreciation of the results.

Figure 10.2 shows the steady-state operating conditions of the simulated well when flowing 8,000 blpd, where the lines represent the parameter profiles, and the dashed line represents the well's elevation curve. In all graphs, the well geometry is represented by the dashed line.

During steady-state conditions, the wellhead temperature is approximately 150 °F. The bottom-hole flowing pressure is 2900 psia, and the

Table 10.1 Simulated well characteristics—Onshore well.

Well location	Onshore
Well type	Build and hold ("J" Shape)
Well overall heat transfer coefficient (OHTC)	1.6 btu/ft^2 h °F (based on ID)
Well internal diameter (ID)	6.456 in.
Wellhead elevation	0 ft
Well kick-off point (KOP)	2,000 ft
Well bottom-hole elevation	7,000 ft (true vertical depth—TVD)
Extended reach (horizontally)	5,200 ft
Well tubing internal roughness	0.0018 in.
Static bottom-hole pressure (SBHP)	3,700 psia
Bottom-hole temperature	190 °F
Well brine density	74.9 lb/ft^3
Well brine viscosity	7 cP
Well liquid productivity index (PI)	10 stb/day-psi
Target reservoir drawdown	800 psi
Geothermal gradient	70 °F @ wellhead level; 190 °F @ TVD
Reservoir fluid	Black oil
Fluid gas–oil ratio (GOR)	422 scf/stb
Fluid API	27° API
Fluid water cut (%)	5%
Well choke C_V curve	Linear curve. C_V fully open: 200 gpm/(psi)$^{0.5}$
Separator pressure @ wellhead (steady state)	500 psia
Steady-state well flow rate	Liquid Rate: 8,000 stb/d
	Water Rate: 400 stb/d
	Oil Rate: 7,600 stb/d
	Gas Rate: 3.2 mmscfd

wellhead flowing pressure downstream of the choke is 500 psia. The well drawdown at this condition is 800 psi.

Note that at about 6,000 ft gas flashes out of solution, and the liquid holdup and water holdup decrease as a result.

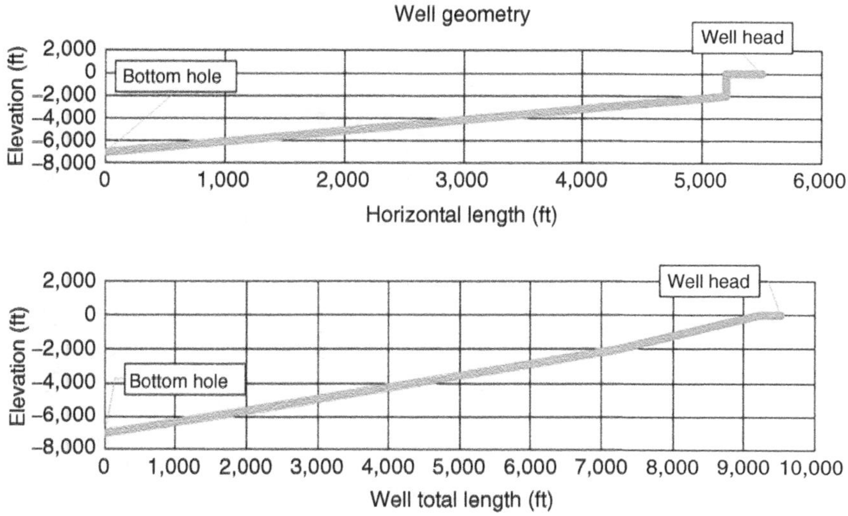

Figure 10.1 Well geometry example.

10.4.2 Onshore Well Initial Start-up Example

After the well described in Table 10.1 is drilled and completed, the well is left filled with completion fluid. In this section, it will be assumed that the completion fluid is a brine (10 ppg density). The well is closed at the wellhead. The normal wellhead flowing pressure is 500 psia. The reservoir static pressure is 3700 psia. Initially, the well is filled with brine, producing a bottom-hole hydrostatic pressure of approximately 3550 psi, and therefore initially the wellhead pressure of the brine-filled well is about 150 psia, which is below the normal operating wellhead pressure of the well (i.e. 500 psia). Under these conditions to start up the well, the wellhead pressure has to be lowered below 150 psia, which means that the backpressure imposed by the gathering system needs to be reduced.

The schedule shown in Table 10.2 was followed for the well start-up simulation.

Figure 10.3 shows the well pressure at three locations: bottom hole, wellhead upstream the choke, and jumper downstream the choke; the choke opening is also shown. The jumper is the connection between the well's tree and either a manifold or a flowline. The solid line shows that for the first 45 minutes, the jumper pressure was lowered to 100 psi. Between 0.75 and 2 hours, the jumper pressure is increased from 100 to 500 psia, its normal operating level. After the second hour, the jumper pressure remains at 500 psia until the end of the simulation.

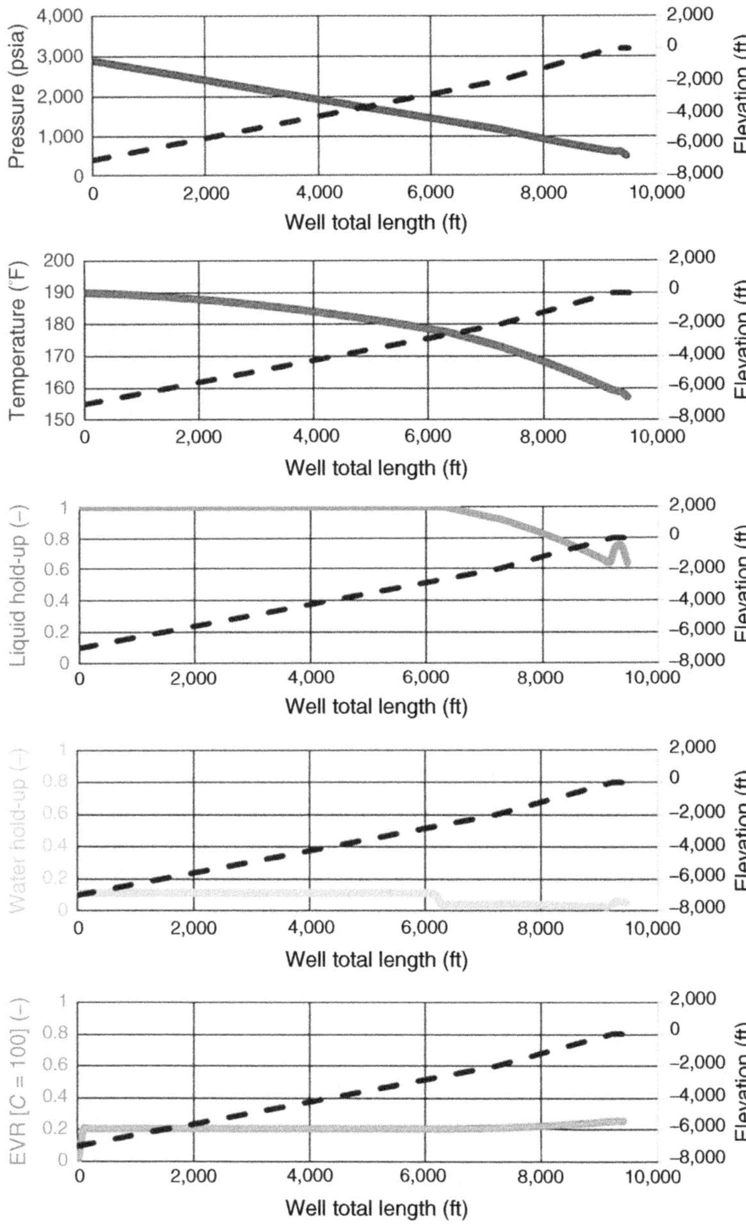

Figure 10.2 Principal parameters profiles for a liquid rate of 8,000 blpd.

Table 10.2 Initial well start-up schedule.

Stage	Initial time (h)	Final time (h)	Liquid flow (blpd)	Well drawdown (psi)
Initial flow	0	8	2,000	200
Flow 2	8	14	4,000	400
Flow 3	14	20	6,000	600
Flow 4	20		8,000	800

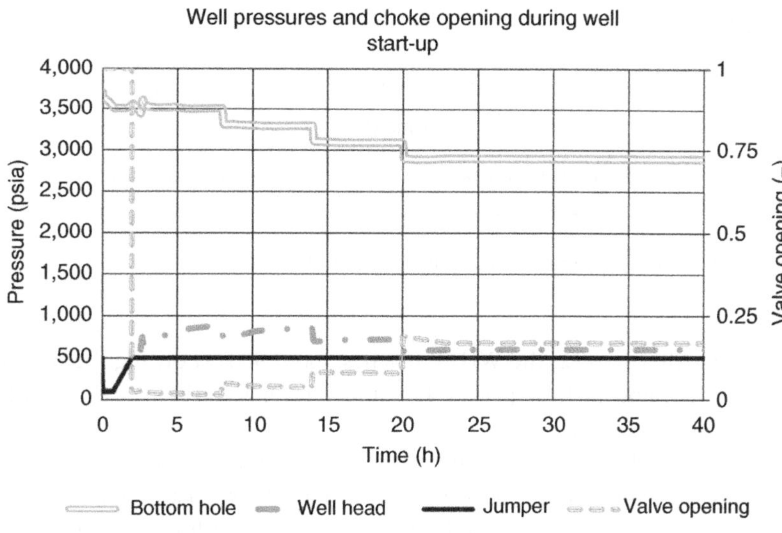

Figure 10.3 Well pressures and choke opening during the well initial start-up.

The well choke remains completely opened (100%) during the first 2 hours of the well start-up. At hour 2, the well choke is quickly lowered to about 2.5% to maintain the reservoir drawdown at 200 psi during the first 8 hours. At hours 8, 14, and 20, the well choke opening is increased respectively to 4%, 8%, and 17% to maintain the scheduled drawdown shown in Table 10.1. The rapid decrease in the well choke opening at hour 2 coincides with the arrival of gas to the surface facilities. This initial behavior is characteristic of many natural-producing wells. The double line shows the bottom-hole flowing pressure, changing to follow the 200 psi drawdown steps' guideline.

The wellhead flowing pressure upstream of the choke (WHFP), for the first 2 hours (while the choke is fully opened), basically matches the jumper

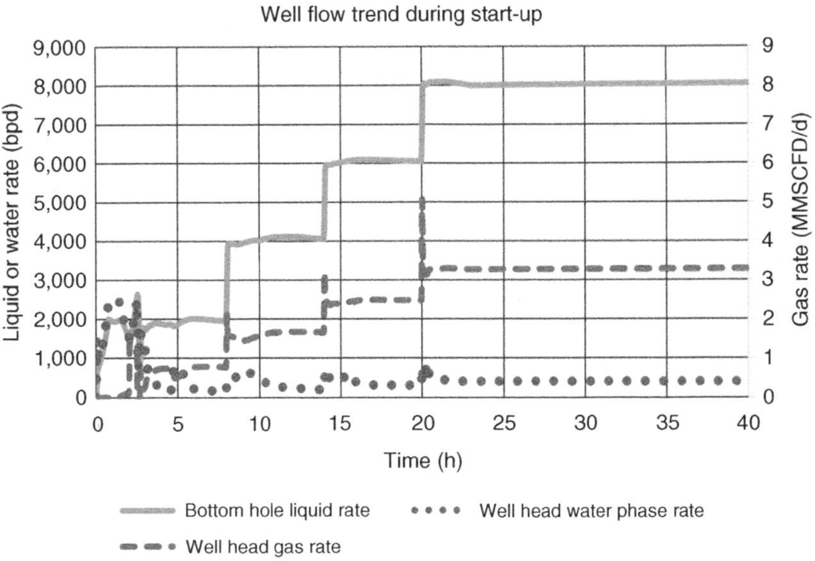

Figure 10.4 Well flow rates during the well initial start-up.

pressure. After hour 2, the WHFP reaches about 900 psi during the initial flow period, and at flow period 4 (liquid rate of 8,000 blpd), the WHFP reaches about 600 psi.

Figure 10.4 shows the flow rates associated with the initial well start-up. The solid line shows the bottom-hole flow rate, where 2,000 blpd flow steps can be observed at 0, 8, 14, and 20 hours. The dotted line represents the water phase flowing at wellhead conditions, showing a distinctive initial water phase burst (brine exiting the well) between 0 and 2 hours. Note that while the initial 2,000 blpd are being produced at the bottom hole, the water flow rate at the wellhead is larger than that, which is due to the removal of the brine and the gas expanding in the well as it travels through the tubing. The dashed line represents the wellhead gas rate production, with gas breakthrough occurring in about 2 hours.

Figure 10.5 shows the well temperature during the initial well ramp-up. In this graph, the double line shows the bottom-hole flowing temperature, in this case, 190 °F. The solid and the dashed lines represent the wellhead temperature upstream and downstream of the choke. Notice that the two lines are very close to each other due to the low-pressure differential across the choke and also due to the low gas flow rate (relative to the liquid) being produced.

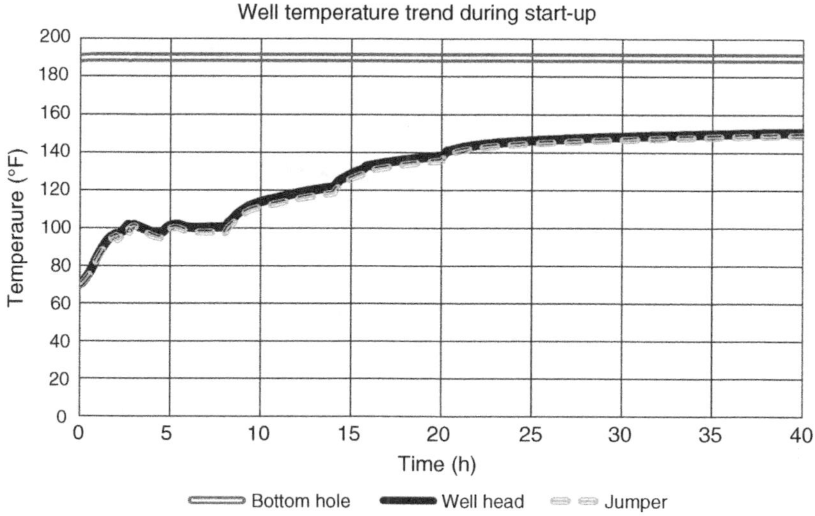

Figure 10.5 Temperatures during the well initial start-up.

Figure 10.6 shows the well holdup profile during the initial start-up at several times. Each graph shows the total liquid holdup (solid line), the water phase liquid holdup (dash double-dot line), and the brine holdup (double line); the well geometry (as a function of the total length) is shown with the dashed lines.

Initially (at 0 hours), the well is filled with brine; note that the brine holdup is 1. At 2 hours, the lower 5,000 ft of the well have about 60% of production fluid and the top of the well has between 10% and 20% of production fluids; note that at 2 hours production fluid reaches the wellhead and there is a gas breakthrough.

At 8 hours, the well has 100% production fluids only below 2,000 ft, while the segment between 2,000 and 9,000 ft has a brine holdup of about 20%.

At 12 hours, brine completely exits the well; the increase of flow rate from 2,000 to 4,000 bpd helps to evacuate the remaining brine. Note that, in this case, a production rate of 2,000 bpd could be considered to be small for a 7-in. well; this is the reason why the brine exits after 12 hours of production.

At 30 hours, the well is in normal operating condition. If the well is maintained at this production rate, the soil surrounding the well will warm up, increasing the wellhead fluid temperature over time.

During the life of the well, it is expected that the reservoir conditions will change. Some of the common changes are decreasing static reservoir pressure, increasing water cuts, increasing GOR, reduction of the productivity

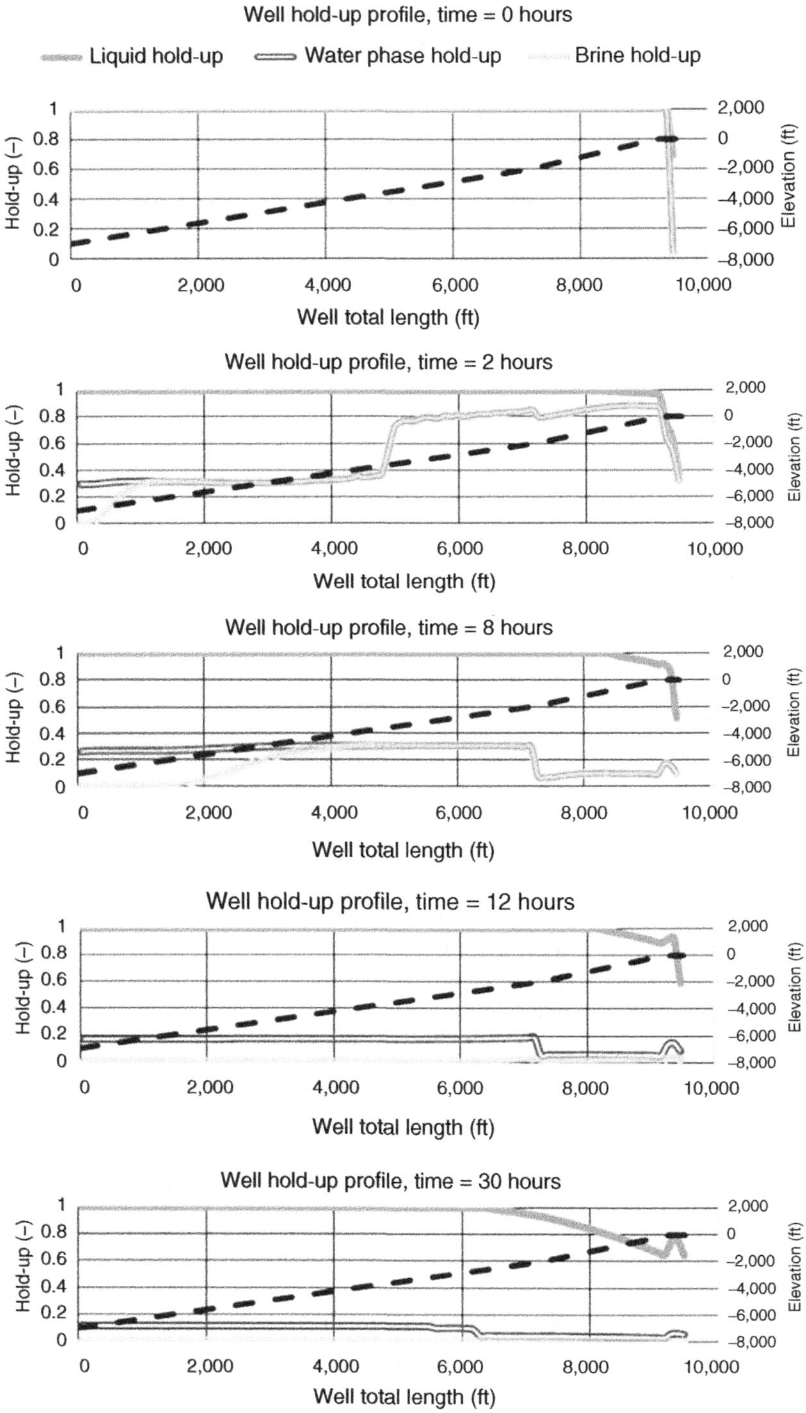

Figure 10.6 Well holdup profile during well initial start-up.

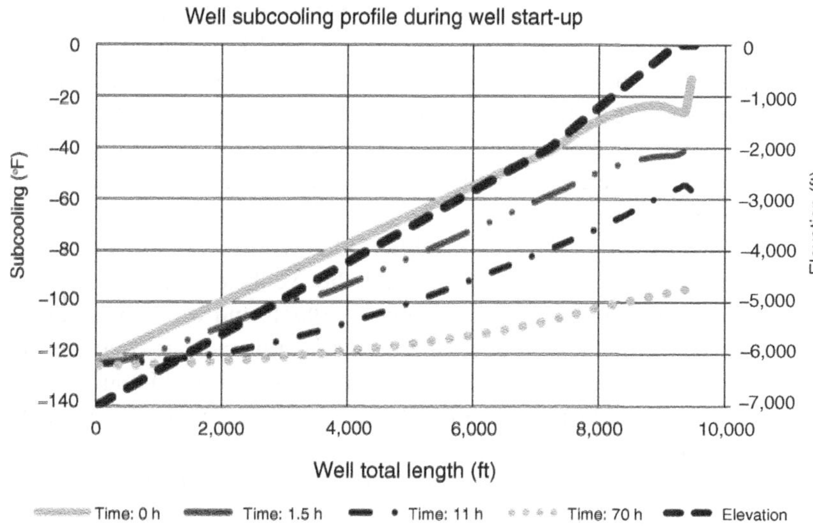

Figure 10.7 Well subcooling profile during well initial start-up.

index, and others. Those cases will alter the normal operating conditions of the well.

A final note, during the well restart, the ambient temperatures, and the well conditions are expected to be outside the hydrate formation region. Subcooling is defined as the difference between the hydrate formation temperature (HFT) at a given pressure and the actual fluid temperature in a given section. Defined this way, a negative subcooling value indicates that the fluid temperature is higher than the HFT at that section; in other words, a negative subcooling value indicates that there is "no risk" of hydrate formation in that section.

Figure 10.7 shows the subcooling profile for the initial start-up operation. The figure shows that the subcooling values are negative for the entire well profile during the entire duration of the well initial start-up.

Having a well initial start-up completely outside the hydrate region is not common for subsea wells. If there are well zones with positive subcooling, careful consideration should be taken to how the well will be started up in order to avoid a hydrate blockage. The details of such an operation are outside the scope of this chapter.

10.4.3 Onshore Well Shutdown Example

The objective of this example is to illustrate the well-fluid segregation during a shutdown. In this section, a simplified onshore well example is studied.

As stated, no hydrate concerns are expected in this well due to the high onshore ambient temperatures. Initially, the well is produced at normal operating conditions at 8,000 bpd. Figure 10.2 shows the normal operating conditions of the simulated well. Notice that the objective of this chapter is to illustrate the typical results of numerical multiphase flow simulations and not to provide guidelines on how to safely operate wells under subsea conditions.

Figure 10.8 shows the well condition after a 120-hour shutdown. The bottom-hole pressure is 3,700 psia and the wellhead pressure is about 1,100 psia. The bottom-hole temperature is 190 °F, and while the wellhead temperature at soil level is approximately 90 °F, this indicates that the well temperature has not reached ambient conditions after a 120-hour shut-in. The ambient wellhead temperature is 70 °F. The slow cooldown in the well is caused by the high thermal mass of the soil surrounding the well. Note that in this simulation it was assumed that approximately 5 ft of soil around the well is warmed prior to the well shut-in. This assumption is valid only if the well has been produced for at least 6 to 12 months prior to the shut-in.

The well liquid holdup remains at 100% except at the very top of the well; this is caused by the low GOR and low API gravity of the simulated fluid.

During a shut-in, the fluid is segregated. The water-to-live-oil boundary can be observed at about 1,000 ft from the bottom of the well. Complete water segregation is expected after 120 hours shut-in.

As it was stated before, the simulated well is not expected to have hydrate issues. In the case of the shut-in of a subsea deepwater well, the hydrate strategy of closing a well needs to be carefully planned and executed, since the upper part of the well, from the SCSSV (surface-controlled subsurface safety valve) to the wellhead, will most likely be inside the hydrate region after an extended shutdown. Additionally, other subsea components, like the wellhead and the well jumper, will likely enter the hydrate region in a matter of hours.

10.5 Offshore Flowlines

In offshore oil and gas production, the flowlines that transport hydrocarbons and other fluids are critical components of the overall production system. Multiphase flow simulation is an essential tool for the design, optimization, and operation of offshore flowlines. This chapter will focus on the modeling

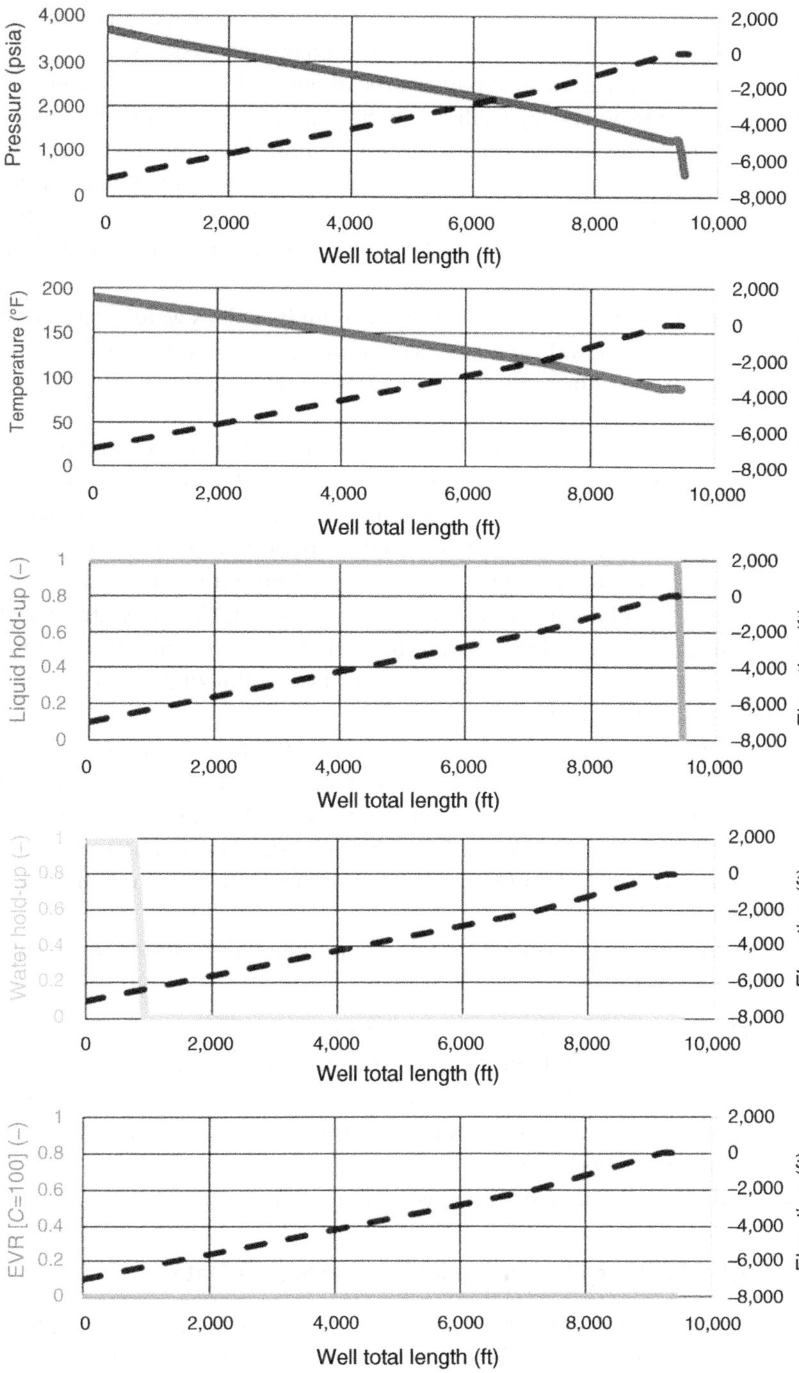

Figure 10.8 Well shutdown conditions after a 120-hour shutdown.

of production and injection flowlines and how engineering companies and oil operators use simulation to design multiphase flow systems.

Production flowlines transport hydrocarbons and other fluids from the wellhead to the processing facilities on the platform or onshore. These flowlines are typically designed to operate at high flow rates, and the fluid properties can vary significantly over time due to changes in production rates and the composition of the produced fluids. As a result, the design and operation of production flowlines require a detailed understanding of the multiphase flow behavior.

Injection flowlines are used to transport fluids such as water or gas from the platform to the wellhead for injection into the reservoir. Injection flowlines are critical for maintaining reservoir pressure and increasing the recovery factor of the field.

Modeling offshore flowlines presents unique challenges due to the operating conditions and the complex geometries of the flowlines. The flowlines are subject to high pressures, temperatures, and corrosive environments, which can affect the fluid properties and the performance of the flowline.

One of the main challenges in modeling offshore flowlines is the accurate prediction of the pressure drop and flow rate. The flowlines are subject to significant pressure losses due to friction, turbulence, and other factors. Accurately predicting these losses is critical for optimizing the design and operation of the flowline.

Another challenge is the modeling of multiphase flow behavior in complex geometries. The flow behavior can be affected by the presence of bends, manifolds, risers, and other components. Modeling these components requires advanced simulation techniques, such as computational fluid dynamics (CFD), to accurately capture the flow behavior and cooldown time (cold spots).

In summary, common applications of multiphase flow simulation in subsea flowlines include

- Optimization of flowline design and operation
- Evaluation of different production and injection strategies
- Prediction of the pressure drop and flow rate under different scenarios
- Assessment of the impact of flow assurance issues, such as hydrate formation and wax deposition
- Planning of maintenance and inspection activities

10.5.1 Offshore Flowline Simulations in Steady State

Steady-state simulations are a common method used to model single-phase and multiphase flow in offshore production and injection flowlines. In a steady-state simulation, the flow variables, such as pressure, temperature, and flow rate, are assumed to remain constant over time. While this assumption may not strictly hold in reality, steady-state simulations are often used for initial design and optimization of flowlines due to their computational efficiency. In a single-phase steady-state simulation, the fluid is either liquid or gas. A straightforward simulation allows us to determine the pressure drop and the temperature change along the flowline or the flow rate.

Multiphase steady-state simulations are more complex than single-phase simulations, as they require the modeling of two or more phases flowing together in the same line. In offshore production and injection systems, the two primary phases are typically oil and gas, and the simulation must consider the effects of their interactions on the flow behavior. Multiphase simulations can also include the effects of water, sand, and other particles in the fluid, which can impact flow behavior and cause erosion or corrosion.

Steady-state simulations can be used to design and optimize offshore production and injection flowlines in normal operation. Steady-state simulations also shed light to troubleshoot issues in existing systems. By simulating different scenarios and parameters, such as changing the flow rate or adjusting the pipe diameter, engineers can evaluate the performance of the system and determine the optimal design for maximum production and efficiency.

10.5.2 Oil Flowline System in a Steady-State Example

Consider the multiphase flowline data summary in Table 10.3 operating in steady state. The task is to simulate the main parameter profiles through the flowline, such as pressure, temperature, holdup, and erosion velocity for a constant liquid flow rate, and to run a liquid rate sensitivity for topsides arrival temperature, system's liquid content, and inlet pressure. The results from this simulation provide ample information to design the flowline and to predict future operation conditions at the topsides.

Figure 10.9 shows the flowline elevation profile as a function of the horizontal length and as a function of the total length. In this section, profile graphs will be plotted as a function of the flowline total length.

Figure 10.10 shows the plot of the flowline's inlet pressure as a function of the liquid rate (i.e. "J curve") as the dash-dot line. The flowline liquid content as a function of the liquid rate is shown as the dotted line. The J curve helps to identify the gravity dominated region (i.e. less than 30,000 blpd) and the friction-dominated region (i.e. flow rates above 30,000 blpd).

In a multiphase system, the pressure drop is a balance between the hydrostatic of the liquid content in the system and the frictional pressure drop. It is recommended to operate the flowline in the frictional-dominated region to

Table 10.3 Simulated subsea oil flowline characteristics.

System	Flowline riser		
Location	Offshore		
Flowline and riser insulation type	Wet Insulation		
Flowline and riser internal diameter (ID)	10.16 in.		
Riser geometry	Assumed vertical		
Flowline horizontal distance	~66,300 ft (20 km)		
Flowline undulation	Yes (PI ~ 80)		
Flowline inlet point elevation	−6,569 ft (−2,000 m)		
Outlet elevation	~ +131 ft (+40 m)		
Flowline and riser internal roughness	0.0018 in.		
Flowline and riser OHTC	1 btu/ft^2 h °F (based on ID)		
Water ambient conditions	Depth (ft)	Temperature (°F)	Velocity (fps)
	0	70	3
	−6,240	40	3
	−6,569	40	3
Air ambient conditions	Air temperature: 70 °F		
	Air velocity: 32 fps		
	Methanol density: 50 lb/ft^3		
	Methanol viscosity: 0.7 cP		
Assumed Methanol properties	Max. Methanol injection rate: 5,000 bpd		
	Dead oil density: 55 lb/ft^3		
Assumed dead oil properties	Dead oil viscosity: 300 cP		
	Max. dead oil circulation rate: 60,000 blpd		
Inlet source type	Fixed mass		
Inlet source GOR	1,000 scf/stb		
Inlet source water cut	25%		
Inlet source temperature	180 F (during steady state)		
FPSO valve diameter	10.15 in.		

avoid flow instabilities in the system. Moreover, due to the numerical uncertainties in the flow simulator, it is usually recommended to avoid operating the pipeline too close to the minimum pressure in the J curve (in this case, 30,000 blpd). In this example, it would be recommended to operate the flowline at a liquid rate greater than 40,000 blpd.

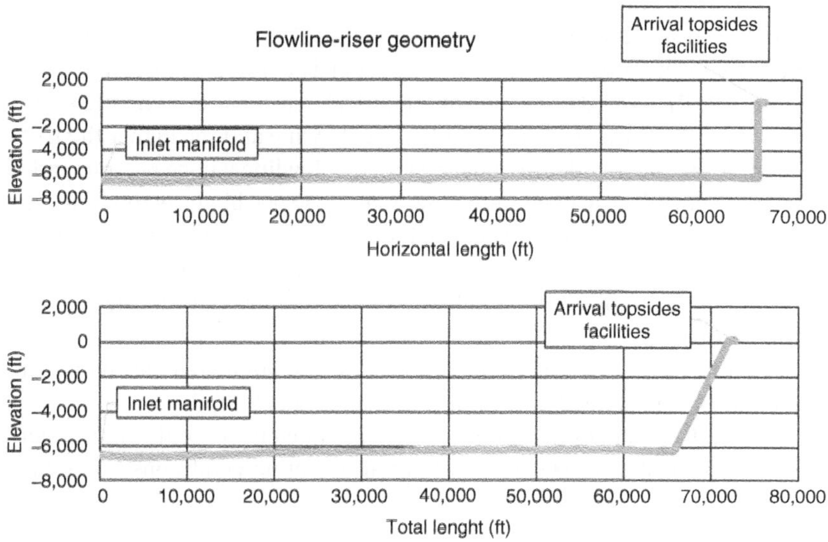

Figure 10.9 Flowline elevation profile as function of horizontal and total lengths.

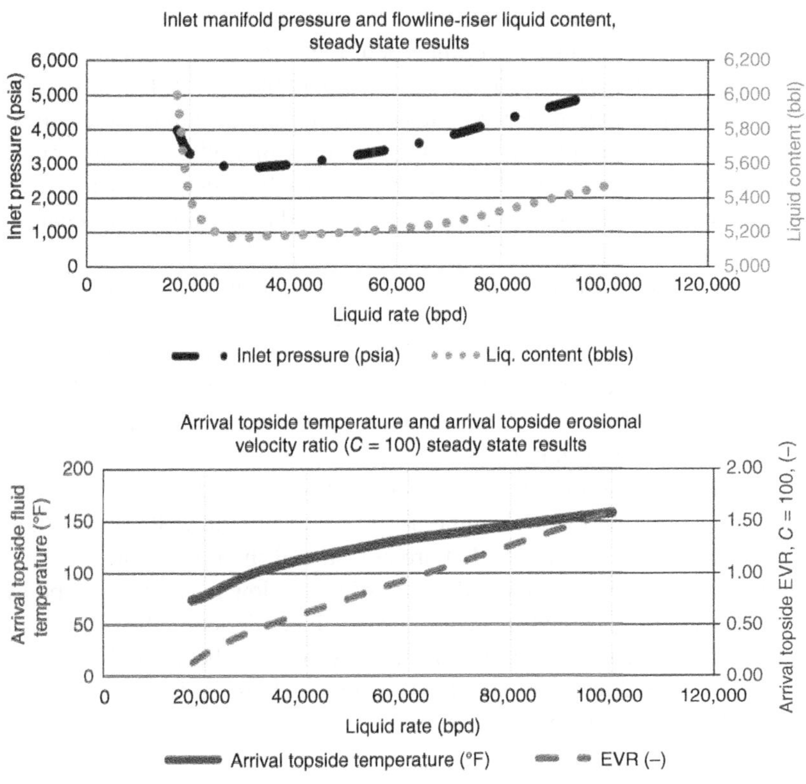

Figure 10.10 Flowline-riser J curve, arrival temperature, and EVR.

The lower graph in Figure 10.10 shows the arrival topside temperature and the *erosional velocity ratio* (EVR). The EVR is the ratio of the mixture velocity to the erosional velocity. The latter is defined by equation (2.14) from API RP 14E (2007), using an empirical constant "*C*." The higher the flow rate, the warmer the fluid within the flowline and the higher the EVR. In this example in order to limit erosion, the flowline will be recommended to be operated at a maximum flow rate of 80,000 blpd; at this rate, the EVR (assuming the conservative value of $C = 100$) is 1.2, and the arrival topside temperature is ~140 °F. The maximum EVR that a flowline may operate is the subject of specialized studies that are outside the scope of this section.

In conclusion, for this example, the flowline simulated can be recommended to be operated in the range of 40,000–80,000 blpd.

Figure 10.11 shows the profiles of five parameters of the flowline system when operated at 80,000 blpd in normal operating conditions (i.e. steady state).

10.5.3 Offshore Flowline Transient Simulations

While steady-state simulations are useful for predicting the behavior of flowlines under normal operating conditions, transient simulations are necessary for analyzing the dynamic response of the flow system to sudden changes in operating conditions, such as start-up and shutdown procedures, power outages, equipment failures, or changes in well production rates. Transient simulations can also be used for analyzing safety scenarios, such as pipeline ruptures or spills, and determining the resulting flow behavior and consequences.

In transient simulations, it is crucial to accurately model the physical phenomena that occur during the transient events. These phenomena can include changes in pressure, temperature, fluid properties, flow rates, and flow regimes. The simulation models must also consider the time-dependent behavior of the system components, such as valves, pumps, compressors, and separators.

One example of transient events in multiphase flow systems is the start-up of a production well or a flowline. During start-up, the well and flowline are usually filled with fluids that are at rest, and the initial flow rates are zero. As the flow begins, the fluid properties and flow rates change rapidly, and the system transitions from a static to a dynamic state. The simulation models must accurately capture the pressure and flow behavior during this transition period to avoid operational issues such as liquid slugging, flow instability, or equipment damage.

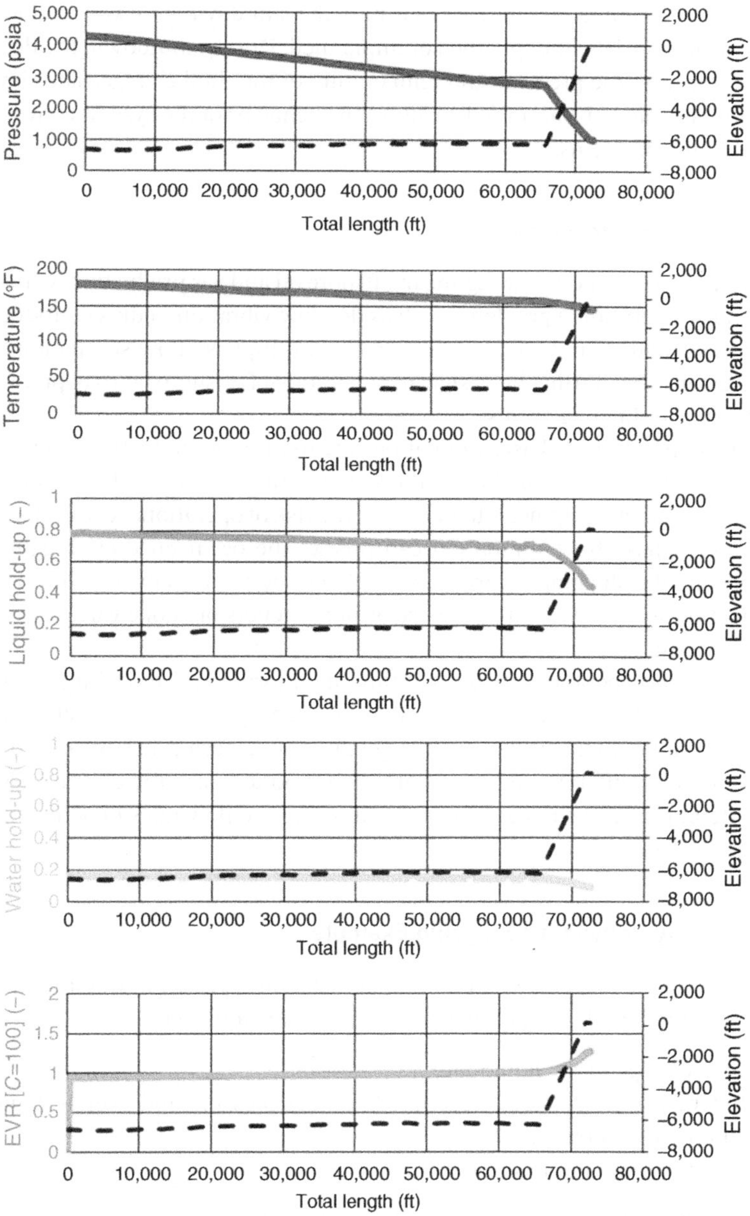

Figure 10.11 Flowline profiles for normal operation (rate 80,000 blpd).

Another example of a transient event is the shutdown of a well or flowline, which can be either planned or unplanned. During shutdown, the flow rates stop, and the pressure and temperature of the fluid change. The simulation models must accurately capture the behavior of the system to avoid operational problems.

10.5.4 Slugging Example

Slugging is a common flow regime in offshore oil and gas production, which can cause operational problems such as flowline vibration, reduced production rates, and even damage to equipment. Multiphase flow simulation is used extensively during the design and operation of offshore fields to predict and manage slugging events.

During the design phase, simulations are used to optimize the design of the production system, including the sizing of the flowline and separators, risers, and other equipment, to minimize the risk of operational disruptions. The multiphase flow simulation can predict the occurrence of slugging, determine the slugging severity, and evaluate the performance of different mitigation strategies, such as the use of slug catchers or choke valves at the arrival point.

During the operation phase, real-time monitoring and modeling of the flowline conditions can help operators to identify and manage slugging events. Simulations can be used to predict the expected behavior of the system under different operating conditions and to assess the effectiveness of different control strategies, such as adjusting the flow rates or activating control valves.

10.5.5 Hydrodynamic Slugging Example

Hydrodynamic slugging is an amplified instability created at the interface between the liquid and gas phases, and it is characterized by

- Rapid change in the liquid rate,
- Moderate or no change in the flowline inlet pressure and overall liquid content of the flowline,
- Moderate or no liquid surge volume.

Modeling hydrodynamic slugging is complex. In general multiphase flow, simulators predict accurately if hydrodynamic slugging conditions exist or not.

Predicting the precise characteristics of hydrodynamic slugging is a more difficult task. Some simulators generate artificial slugs with some arbitrary frequency and let the flow simulator determine if slugging conditions exist (if the slug continues or is dissipated); in this case, the slugging frequency and other characteristics are arbitrary and can easily be manipulated to obtain a different result. In more recent years, the multiphase flow simulators mechanistically predict the slug frequency and claim that the hydrodynamic slug calculations are therefore better.

In reality, it is very difficult to precisely describe the characteristics of hydrodynamic slugging. It is the opinion of the authors that, in most cases, identifying that the flowline will be operating under hydrodynamic slugging is more important than trying to get specific flowing characteristics from the flow simulator.

In flowline design, once a hydrodynamic slugging condition is identified, the flowline and the riser have to be designed for a wide variety of intermittent flow conditions. For hydrodynamic slugging fatigue and proper flowline, anchoring are two of the most important considerations during design.

Figure 10.12 shows an example of hydrodynamic slugging. This graph illustrates a hypothetical hydrodynamic slugging behavior (based on experience, not actually simulated). Figure 10.12 shows a moderate inlet pressure variation, with an average pressure of ~2,900 psi (±15 psi), a rapid fluctuation in the arrival liquid rate from 0 to 60,000 blpd, and an average flow of 30,000 blpd. The liquid content in the flowline stays constant, and the slug frequency in this example is approximately 60 slugs per minute.

In practical terms, hydrodynamic slugging conditions can be present even during the stable production of a flowline system. If the flowline system is operated at low rates (inside the gravity dominated region), it is likely that the oil flowline will be operated under terrain slugging conditions.

10.5.6 Severe Slugging Example

In Section 10.5.2, it was determined that the recommended operating region for the simulated flowline system be from 40,000 to 80,000 blpd. It is expected that operating the flowline at 5,000 blpd could create severe slugging.

Severe slugging is characterized by the following cyclic behavior:

- When the flowline produces at low pressure, the flow rate is also low and the gas phase cannot efficiently remove the liquid, which is left behind as the gas is produced.
- The increased liquid content in the flowline creates an increase in the flowline inlet pressure.

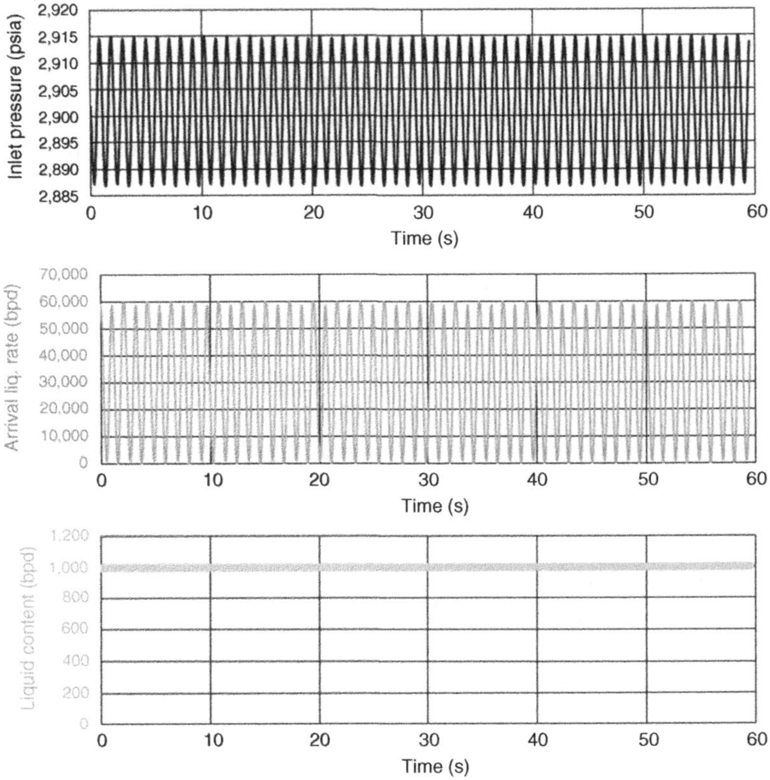

Figure 10.12 Hydrodynamic slugging example.

- When the pressure is sufficiently high, the liquids are evacuated from the flowline and the inlet pressure of the flowline decreases. This process repeats in a cyclical manner.

Figure 10.13 shows the simulated severe slugging conditions of the flowline system from Section 10.5.2 operated at 5,000 blpd. Note the significant pressure fluctuations (between 2,400 and 3,600 psia). The liquid rate is produced in batches, these liquid batches last for less than 1 hour and then the liquid rate stops and is not produced for about 5 hours. The liquid content in the flowline oscillates between 4,200 and 5,300 bbls.

It would be very difficult to operate the flowline system under these conditions. The rapid production of liquids will likely flood the separator. Adding system friction with a topside choke and increasing the separator operating pressure are two of the common methods to mitigate the effects of severe slugging.

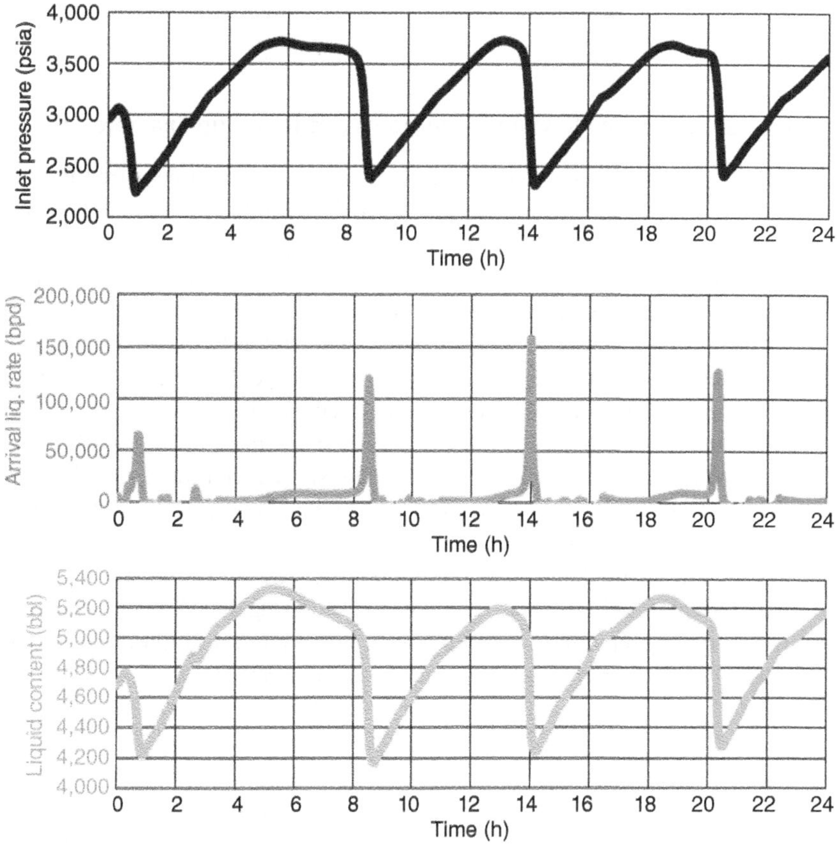

Figure 10.13 Severe slugging example.

10.5.7 Ramp-up and Ramp-down Operations

Ramp-up and ramp-down operations involve the gradual increase or decrease in the flow rate of fluids within the system. These operations are critical in ensuring the safe and efficient operation of the system and in minimizing the risk of damage to the flowline.

Multiphase simulation can be used to optimize the ramp-up and ramp-down process by predicting the behavior of fluids within the system under different flow rates. This includes modeling the behavior of fluids during the gradual increase or decrease in flow rate, as well as simulating the effects of changes in temperature and pressure on the system. By simulating these processes, operators can identify potential issues and develop strategies to prevent overloading, slugging, sudden depressurization, and damage to the flowline.

10.5.8 Offshore Flowline Ramp-down and Ramp-up Example

When a system is in ramp-down, care must be exercised to ensure the flow rate warrants an arrival temperature above the hydrate formation region and the flow stability is preserved to avoid severe slugging.

On the contrary, when the system is in ramp-up, consideration should be made to the liquid slug that will be generated due to the rate increase.

The ramp-up/ramp-down example in this section is based on the flowline described in Section 10.5.2. Figure 10.14 shows the flowline liquid rate (at inlet and topside) and the liquid content during a ramp-down and ramp-up simulation. From 0 to 10 hours, the system operates at 80,000 stb/d. At the 10th hour, the flow rate is instantaneously reduced to 40,000 stb/d. At the 22nd hour, the system is ramped up instantaneously from 40,000 to 80,000 stb/d.

Note that at the 22nd hour, the temporary increase in the arrival liquid rate takes place. This increase generates a liquid slug that should be handled by the topside separator. Assuming an actual separator liquid drain rate of 100,000 blpd, the ramp-up surge volume is about 50 bbl.

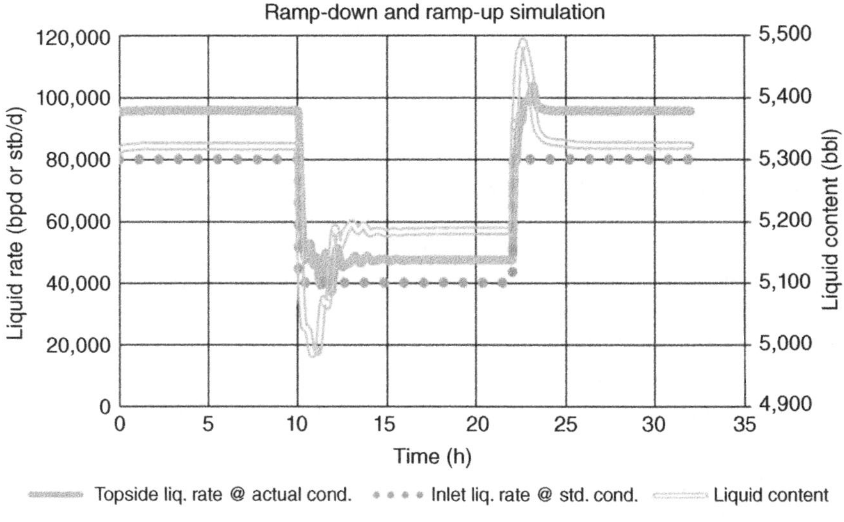

Figure 10.14 Ramp-down to 40,000 blpd and ramp-up to 80,000 blpd.

10.5.9 Cooldown Transient

Cooldown (CD) is the heat transfer process multiphase flowlines undergo after system shutdown. If the shutdown is expected to be momentary and the system will be restarted in a few hours, there is the challenge of managing the temperature differential between the hydrocarbons and the surrounding environment. The temperature differential can cause thermal stresses in the flowlines, leading to deformation, cracking, and failure. Additionally, during cooldown, there is the risk of hydrate formation which can also block and damage the flowlines.

10.5.10 Offshore Flowline Cooldown Example

In this simulation, the flowline system of Section 10.5.2 is initially at a production rate of 80,000 blpd. The flowline is shut (along with the inlet source) at time 0. The flowline is shut-in for 120 hours. This simulation was performed to illustrate a flowline cooldown process while disregarding any hydrate formation consideration, and therefore no methanol inhibition and no dead oil circulation were simulated. The following sections will illustrate flowline inhibition and dead oil circulation as potential safeguard methodologies.

Figure 10.15 shows the flowline profile for the liquid holdup, pressure, and temperature up to the 120th hour. The liquid holdup shows the rapid redistribution of the liquid across the pipeline in the first few hours. The pressure profile shows that after closing the flowline pressure equilibrates and then slowly decreases over time with the temperature decrease. The temperature profile shows that the temperature reduction is quicker in the gas pockets and slower in the liquid pockets of the flowline, due to the higher heat capacity of the liquid with respect to the gas.

After 120 hours, the flowline is expected to be almost at ambient temperature conditions. The flowline hydrocarbon fluid at a low temperature (~40 °F), with free water present and at a pressure above 125 psi, will have all the conditions for hydrate formation. In the following sections, some of the transient operations methods illustrate ways to maintain the flowline system outside the hydrate region.

10.5.11 Depressurization Operation

Depressurization is a crucial process in deep offshore oil and gas systems that involves the controlled reduction of pressure within a flowline or system. This process is particularly important in deep offshore systems due to

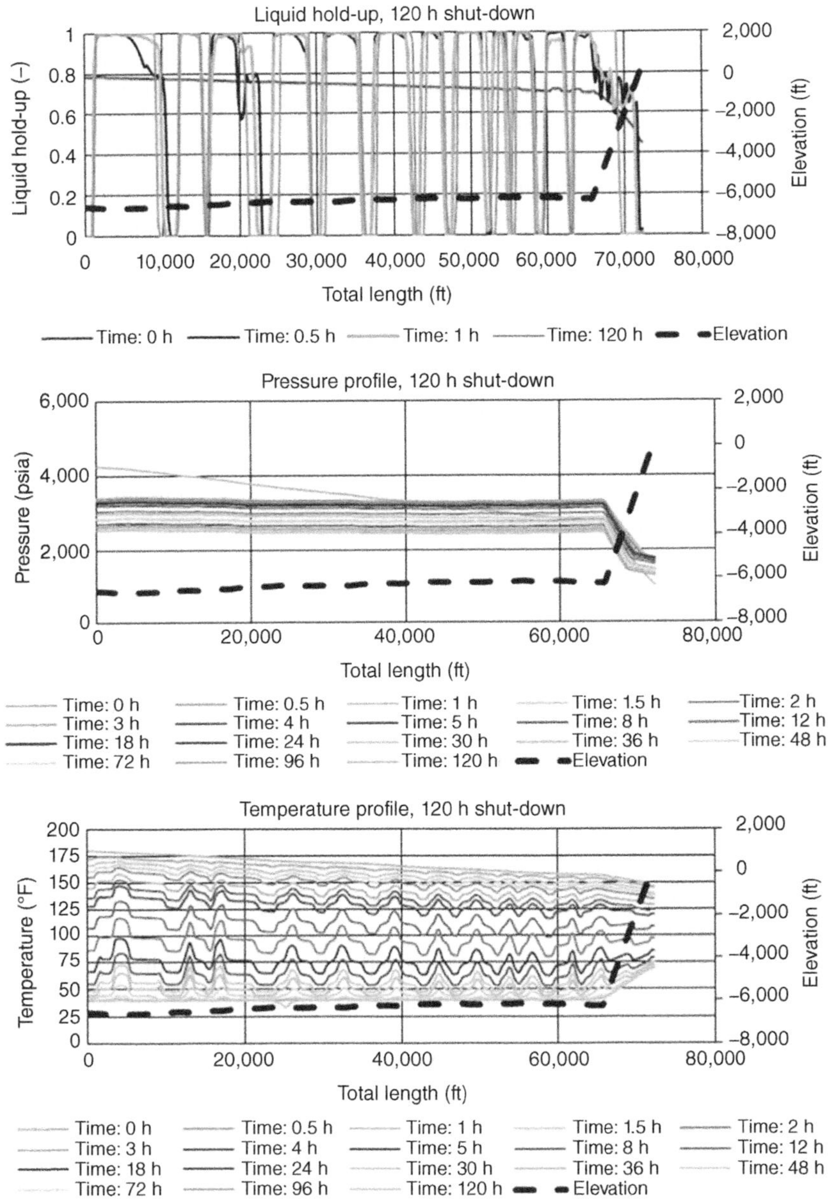

Figure 10.15 Holdup, pressure, and temperature profiles during the cooldown.

the challenging operating conditions, including high pressures and temperatures, harsh environments, and remote locations.

During planned shutdowns, depressurization is carefully planned and executed to ensure the safe shutdown of the system for maintenance and inspection. In deep offshore systems, depressurization is vital as it allows workers to carry out maintenance, inspection, and repair work without risking their safety in challenging operating conditions.

In contrast, unplanned shutdowns or system failures can result in sudden depressurization, which can pose a significant risk to workers and the environment. In deep offshore systems, depressurization must be carried out quickly and safely to prevent damage to the system and minimize the risk of injury or environmental damage.

During operations, multiphase simulation is used to monitor the behavior of fluids within the system and to optimize the depressurization process. This includes monitoring the rate of depressurization, the pressure and temperature within the system, and the behavior of fluids during depressurization.

Slugging can occur during depressurization, particularly during unplanned shutdowns or system failures, and can lead to mechanical damage to the system, as well as a risk of environmental damage due to the release of fluids. Engineers use multiphase simulation to optimize the depressurization process and to design systems that are less prone to slugging. This can include the use of flow control devices, such as choke valves, to regulate the flow of fluids during depressurization and prevent the formation of slugs.

10.5.12 Offshore Flowline Depressurization Example

In this example, depressurization is one of the methods to avoid hydrate conditions in the flowline system. In the following example, the flowline is initially operated at 80,000 blpd, and then the flowline is shut down for 8 hours (i.e. 4 hours of no-touch time plus 4 hours of light-touch time for the unplanned unpacked shutdown). The depressurization simulation starts after 8 hours of the shutdown.

The topside depressurization pressure was assumed to be 50 psia using a 1-in. or a 2-in. topside choke.

Figure 10.16 shows the flowline initial condition profile prior to the depressurization. Note that the flowline settle-out pressure is 3,000 psia.

Figure 10.17 shows the flowline profile after 120 hours of depressurization with a 1-in. choke. Note in this case that the maximum flowline settle-out pressure is about 380 psia.

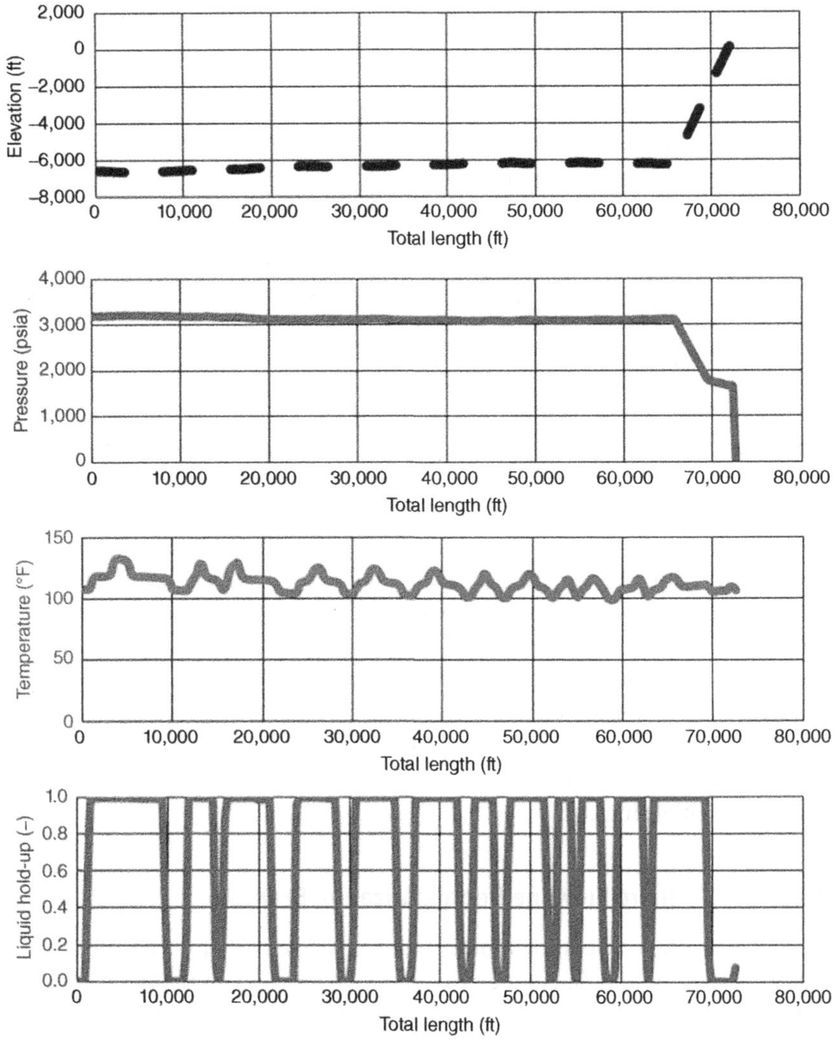

Figure 10.16 Flowline initial condition profile for depressurization (after 8-hour CD).

Figure 10.18 shows the flowline profile after 120 hours of depressurization with a 2-inch choke. Note in this case that the maximum flowline settle-out pressure is about 250 psia.

Figures 10.19 and 10.20 show that the topside liquid and gas flow rates are larger with the 2-inch depressurization simulation when compared to the 1-inch simulation depressurization.

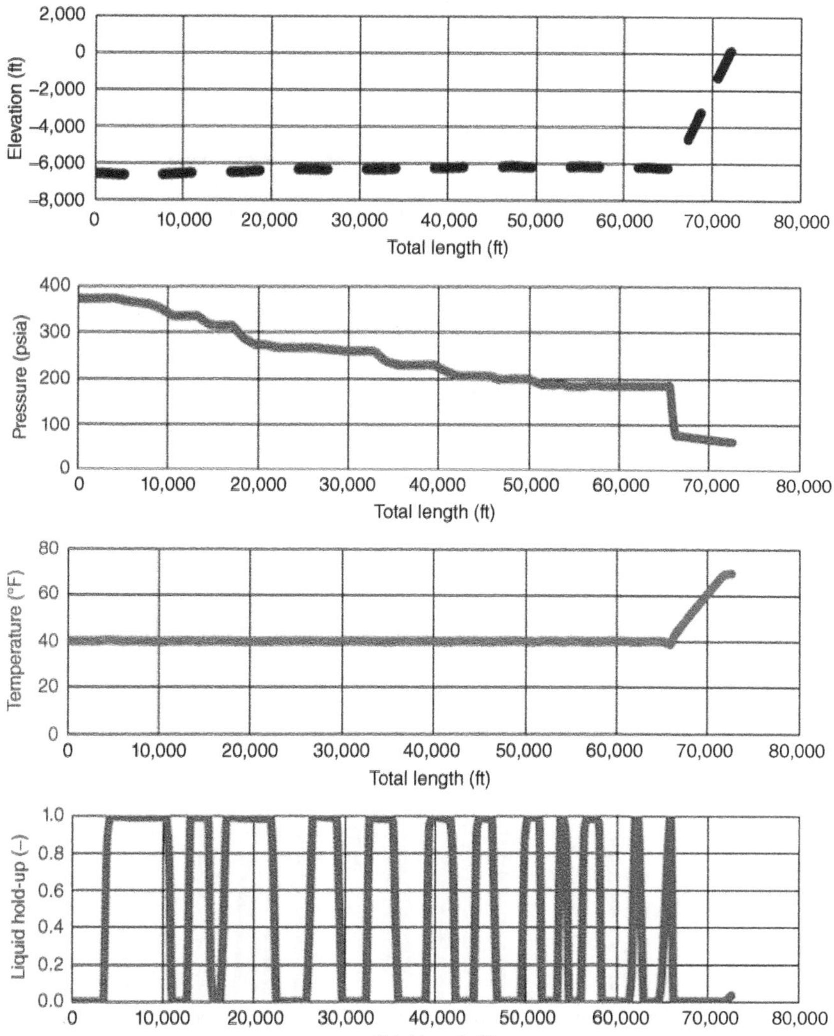

Figure 10.17 Flowline profile after 120 hours of depressurization (1-inch choke).

The ideal objective of depressurization is to reach a settle-out pressure below 125 psi, which is the hydrate formation pressure at 40 °F (seabed ambient temperature). Depressurizing the flowline system is effective in delaying the time when hydrates will form in the flowline, but in this case, the depressurization process is unable to reach the necessary low settle-out pressure to avoid indefinite hydrate formation.

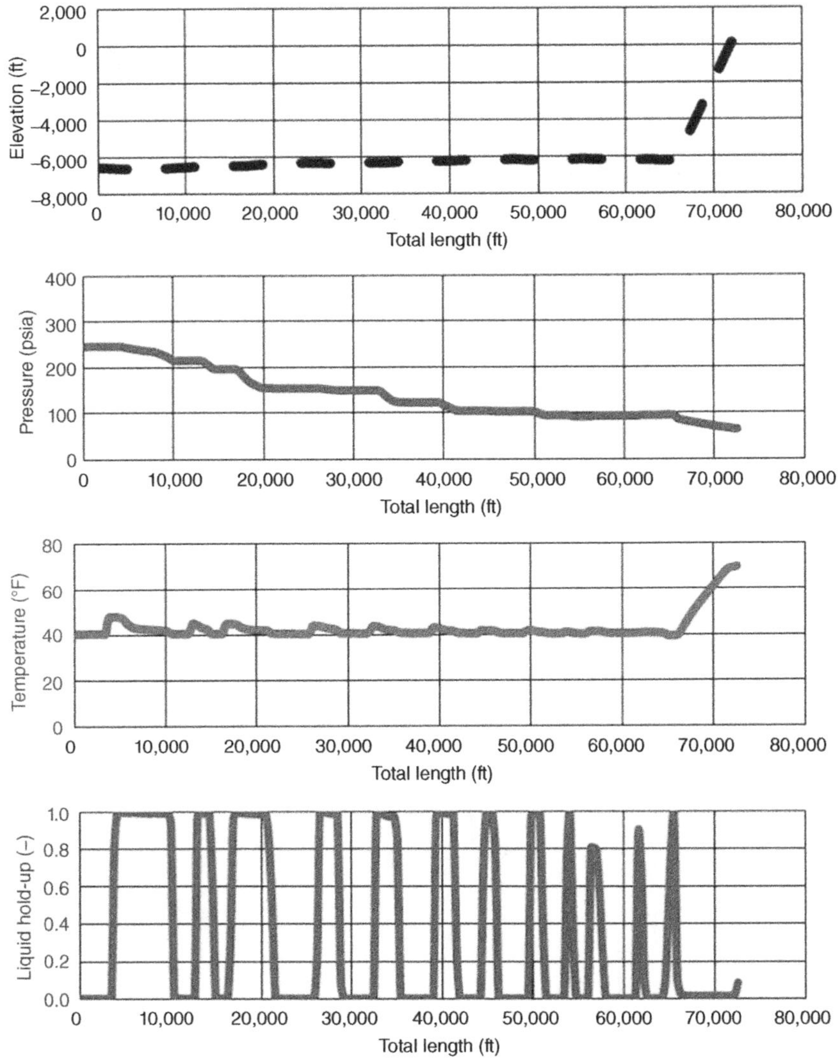

Figure 10.18 Flowline profile after 120 hours of depressurization (2-inch choke).

In general, depressurization is more successful when the fluid is very volatile, when the liquid content in the flowline is low prior to depressurization, and when the water cut is close to 0.

10.5.13 Blowdown (Flowline Rupture) Transient

A rupture blowdown in multiphase pipelines refers to the rapid release of fluids and gases from the pipeline due to a rupture in the pipe. This can occur

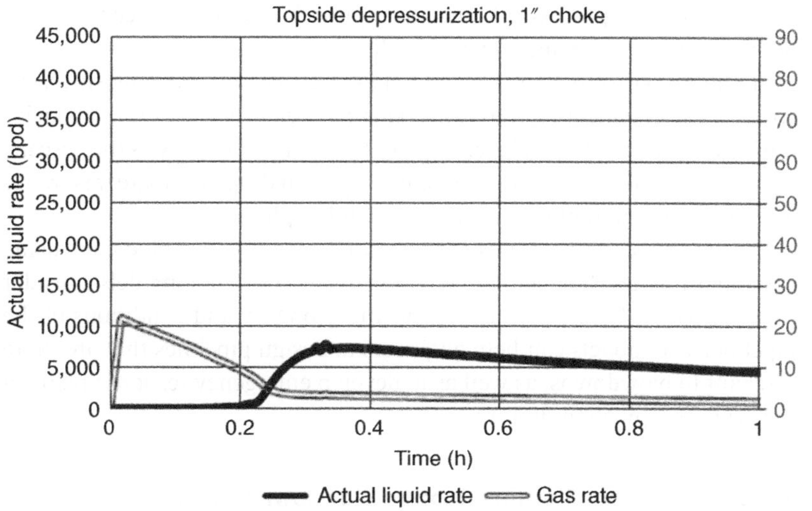

Figure 10.19 Topside gas and liquid rate, 1-inch choke depressurization.

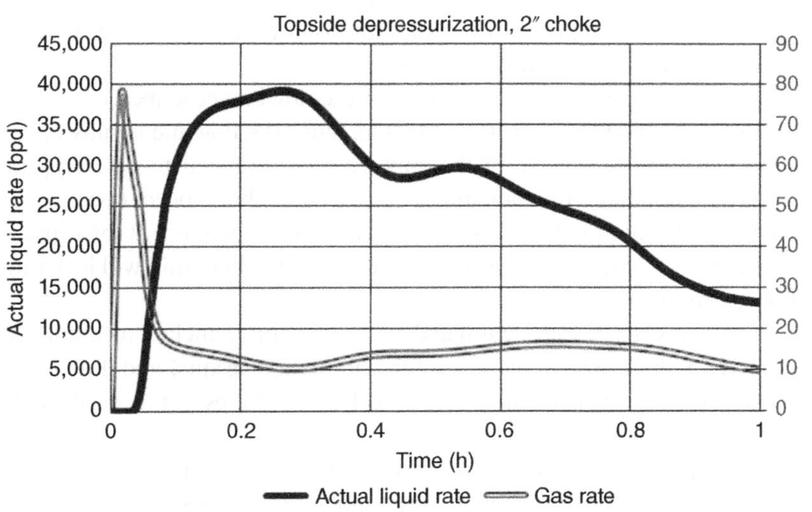

Figure 10.20 Topside gas and liquid rates (2-inch choke).

when the pressure within the pipeline exceeds the strength of the pipe's wall, causing it to burst open. The sudden release of fluids and gases can result in serious safety hazards, including fires, explosions, and environmental damage.

When a pipeline rupture occurs, the pressure within the pipeline drops rapidly, causing the fluids and gases to rapidly expand and escape through the rupture. This process is known as rupture blowdown and can occur in both single-phase and multiphase pipelines.

In a multiphase pipeline, the liquids and gases have different densities, viscosities, and chemical compositions. This can lead to complex flow behavior during blowdown, with different phases behaving differently as they escape from the pipeline. For example, liquids may form droplets or streams, while gases may expand rapidly and create a cloud or plume.

The behavior of the fluids during blowdown can be predicted using computer models that simulate the flow dynamics. These models consider the properties of the fluids, as well as the geometry and conditions of the pipeline. The models can help engineers to design pipelines that are more resistant to blowdowns, as well as to develop emergency response plans in the event of a pipeline rupture.

10.5.14 Offshore Flowline Rupture Blowdown Example

A transient multiphase flow simulation could be helpful in predicting the behavior of the flowline during a pipeline rupture (or leak). Rupture simulations provide important inputs to the HSE team to make fluid dispersion calculations. One of the common questions is how big a leak needs to be in order to be detected by the flowline sensors.

In this example, the simulated leak in the flowline described in Section 10.3 was assumed to occur in the middle. Since the flowline is subsea, the expected pressure outside of the rupture was estimated to be 2605 psia. In this simulation, the flowline is expected to have been operated at 80,000 blpd. According to Figure 10.11, the flowline pressure at the rupture location is ~3500 psia. Under these conditions, flowline fluid will leak into the environment.

Three leak sizes were simulated 0.25, 0.5, and 2 inches in diameter. Table 10.4 shows the flowline rupture simulation results, and Table 10.5 shows the performance differences in leak size with respect to steady-state normal operation.

At this point is necessary to clarify that a flowline system as the one simulated is likely to have pressure and temperature gauges at the inlet and topside outlet, and the flow rate is typically measured only in the topsides.

The bottom line is that a 0.25-inch pipeline rupture (in the middle of the flowline) is likely to cause variations in the inlet pressure of ~0.5% and variations in the flow rate in the order of 1%. Hours may pass before a 1% flow rate difference is even detected, since this small flow variation may be interpreted as a normal variation of the rate (i.e. slugging), slight de-calibration of the measuring devices, a slight loss of well productivity, or other. In general,

Table 10.4 Flowline rupture simulation results.

Leak size (inch)	Leak location (ft)	External pressure (psia)	Manifold pressure (psia)	Arrival temp. (°F)	Topsides gas rate (mmscfd)	Leak gas rate (mmscfd)	Topsides liquid rate (blpd)	Leak liquid rate (blpd)
0	32,800	2,800	4,256	145.6	60	0	95,800	0
0.25			4,235	145.2	59	0.8	94,600	1,200
0.50			4,180	144.2	57	2.9	91,300	4,500
2.00			3,720	135	40	20	63,200	32,600

Table 10.5 Normal operating condition differences during flowline rupture.

Leak size (inch)	Manifold pressure difference (%)	Arrival temperature difference (%)	Topsides gas rate difference (%)	Topsides liquid rate difference (%)
0.25	−0.5	−0.3	−1.3	−1.3
0.50	−1.8	−1.0	−4.8	−4.7
2.00	−12.6	−7.3	−33.3	−34.0

the higher the GOR of the transported fluid, the bigger the leak needs to be in order to be detected promptly.

Note in Table 10.5 that if the rupture has a 2-inch diameter, the pressure and temperature difference is in the order of 10%, and the flow rate difference is in the order of 33%, these differences more easily to be noticed rapidly by the flowline operator.

Leak detection technology in flowlines is a current topic of study in the oil and gas industry. The details of the leak detection system are outside the scope of this chapter.

10.5.15 Shutdown Operation

During start-up and shutdown operations, fluids must be introduced or removed from the flowline gradually to prevent overloading, slugging, sudden depressurization, and damage to the flowline. Multiphase simulation can be used to predict the behavior of fluids within the system under different operating conditions and flow rates, allowing operators to plan and execute these operations more safely and efficiently, this is especially important during shutdown and start-up.

We will briefly address the differences between a planned (Section 10.5.16) and an unplanned shutdown (Section 10.5.17).

10.5.16 Offshore Flowline Planned Shutdown Example

In this example, the flowline system of Table 10.3 is operated to deliver 80,000 blpd with hydrate protection in preparation for a planned shutdown. The strategy is to inhibit the produced water in the flowline using methanol prior to shutdown, minimizing the risk of hydrate formation during the planned shutdown.

For simplicity, it is assumed that an inhibition rate of 0.5 barrels of MeOH per barrel of produced water is sufficient to inhibit the flowline water. The maximum methanol injection rate is assumed to be 5,000 bpd (146 gpm). Under these conditions, the maximum amount of production water that can be treated is 10,000 bpd. The total production stream has a 25% water cut, which limits the maximum liquid flow rate successfully inhibited to 40,000 blpd.

Based on these numbers, the production should be reduced from 80,000 to 40,000 bpd, while the methanol Injection rate is increased from 0 to 5,000 bpd.

The idea is to simulate the time that it will take for the methanol holdup to stabilize and reach the topsides.

Figure 10.21 shows the inhibition results. The figure shows that after 4–5 hours the methanol holdup stabilization is reached. It is customary in the oil field that the flowline should be displaced twice in order to make sure that methanol has fully inhibited the flowline water. Figure 10.21 shows that the assumption of two inhibiting volumes (10 hours injection) is more than a confirmation of the result obtained at the 4th hour.

After the flowline is inhibited, the flowline is assumed to be shut down for 120 hours. Figure 10.22 shows the flowline system profile including pressure and temperature right after inhibition (cooldown time at 0 hours), while Figure 10.23 shows a similar graph after a 120-hour cooldown.

Figure 10.23 shows the liquid segregation expected during a pipeline shut-in. Also, note that the flowline is at ambient conditions (~40 °F).

Notice that the methanol holdup line represents about 33% of the water phase pockets. This methanol holdup is in line with the concurrent flow rates of water and methanol (i.e. 1 barrel of production water plus 0.5 barrels of methanol).

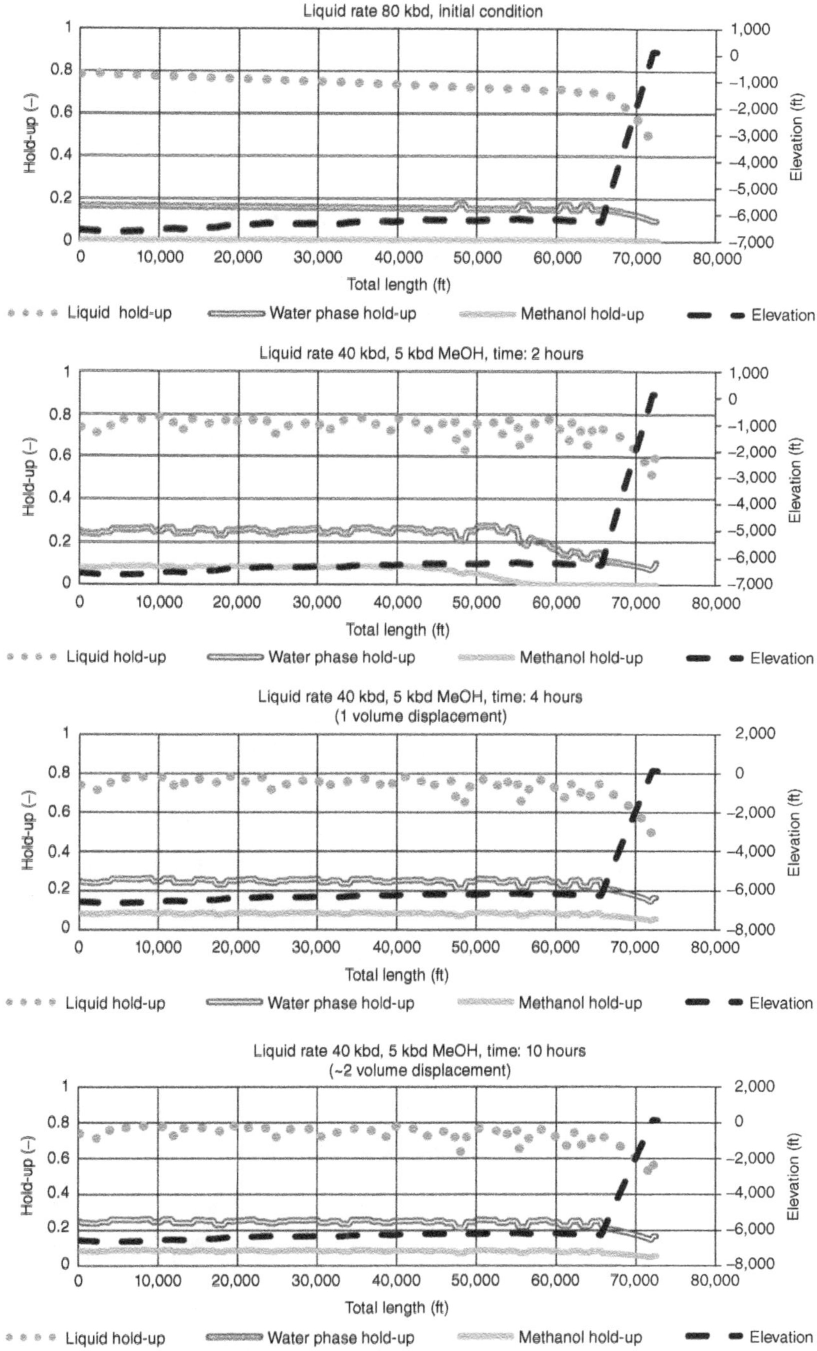

Figure 10.21 Planned shutdown holdup profiles.

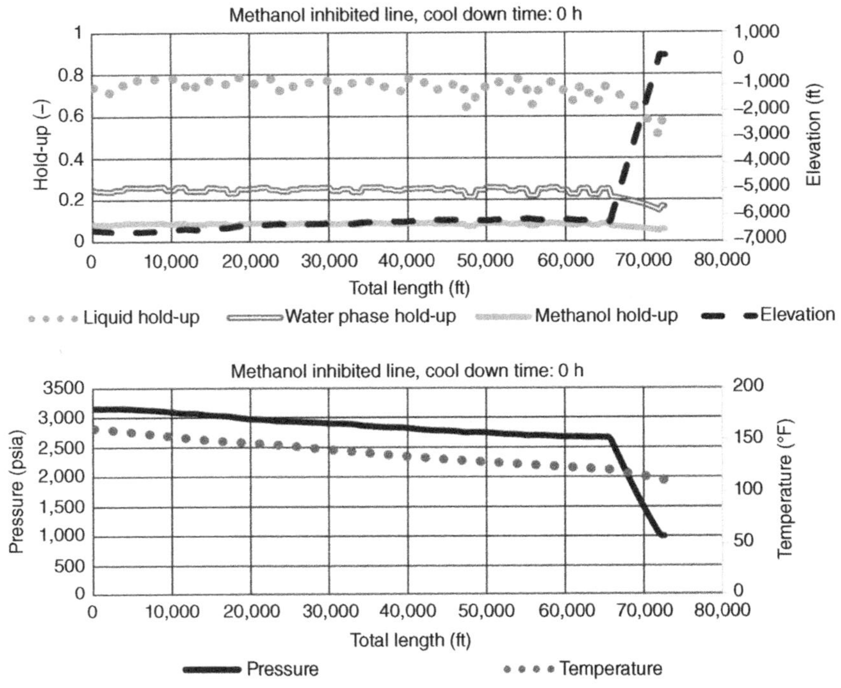

Figure 10.22 Planned shutdown profile at the start of cooldown.

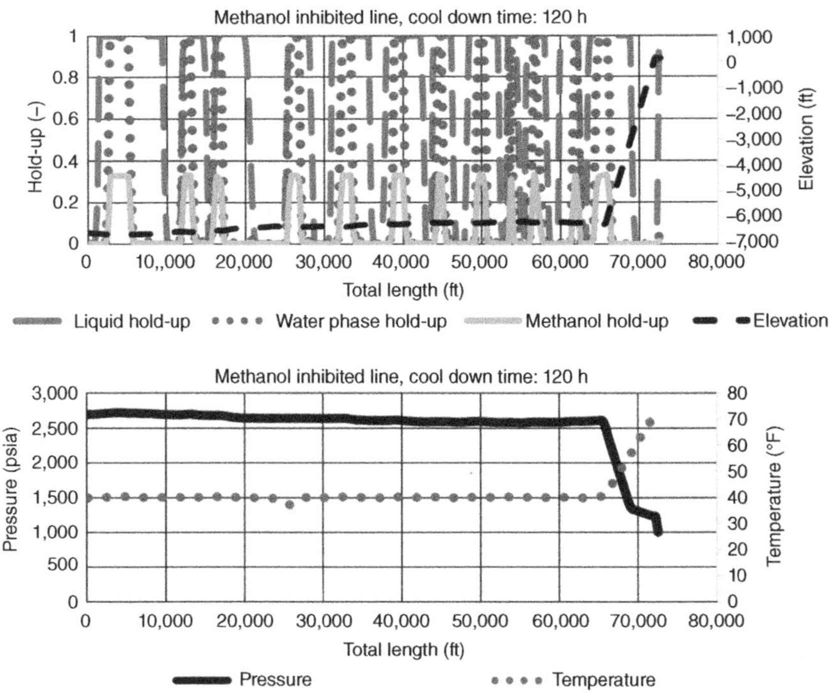

Figure 10.23 Planned shutdown profile after 120 hours of cooldown.

10.5.17 Offshore Flowline Restart After Planned Shutdown Example

After the 120-hour planned shutdown (i.e. Figure 10.23), the system needs to be restarted. The flowline is cold and new production flow needs to be inhibited to prevent hydrate formation until the flowline is warm enough to stop the methanol injection.

Similar assumptions will apply during the restart process; the maximum liquid rate during start-up is 40,000 blpd with a methanol injection of 5,000 bpd in order to fully inhibit the incoming production. In this simplified example, the inlet source was assumed to increase its temperature from 40 to 180 °F in the interval from 0 to 8 hours. It will also be assumed that the flowline system is warm when the topside temperature reaches 100 °F.

Figure 10.24 shows that the flowline arrival topside temperature reaches 100 °F after 14 hours of production. At the 14th hour, the methanol rate stops, and the flowline production rate can be ramped up to 80,000 blpd in 2 hours. This process causes no operational concern because, from the example in Section 10.5.8, the expected surge volume during ramp-up is expected to be less than 50 bbls.

10.5.18 Unplanned Shutdown and Dead Oil Circulation in Flowline Loop

A subsea flowline system typically comprises a loop, that is, a system of two flowlines and two risers, where the produced stream is divided at the subsea

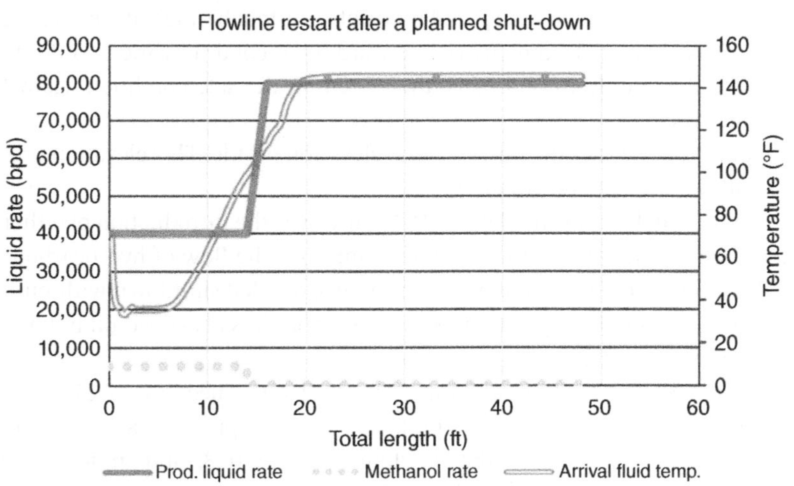

Figure 10.24 Flowline restart after a planned shutdown trend.

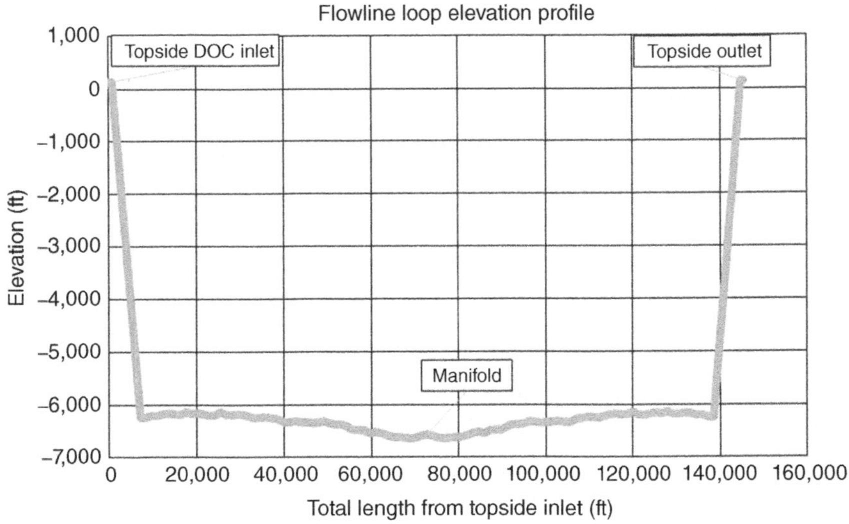

Figure 10.25 Flowline loop elevation profile.

manifold and delivered through the two flowlines toward the topside facilities. A flowline loop configuration is customary in subsea systems to assist in hydrate management and maintenance procedures.

During the unplanned shutdown simulation, the previous example system is producing at 80,000 blpd per flowline (uninhibited), an unplanned shutdown occurs at time 0 hour, and the system remains closed for 8 hours (i.e. 4 hours of no-touch time plus 4 hours of light-touch time), at which time dead oil is circulated from one riser toward the second flowline in order to displace the live fluids and place the system in the safe condition against hydrate risk.

Figure 10.25 shows the flowline loop elevation profile. This elevation profile follows the dead oil circulation path

Hot oil and dead oil circulation (DOC) involves the introduction of either hot or cold dead oil into the flowline to improve the flow of hydrocarbons and to prepare the production system for an extended shutdown with minimum risk of solids' deposition. Notice that hot oil is also dead oil but at a higher temperature.

During the dead oil circulation, a dead oil rate injection condition of 60,000 bpd and 100 °F will be simulated; this rate provides a circulation velocity of about 2 m/s in the flowline of Table 10.3, and under these conditions, 1.5 flowline-riser volume displacement is expected to take place within 9 hours.

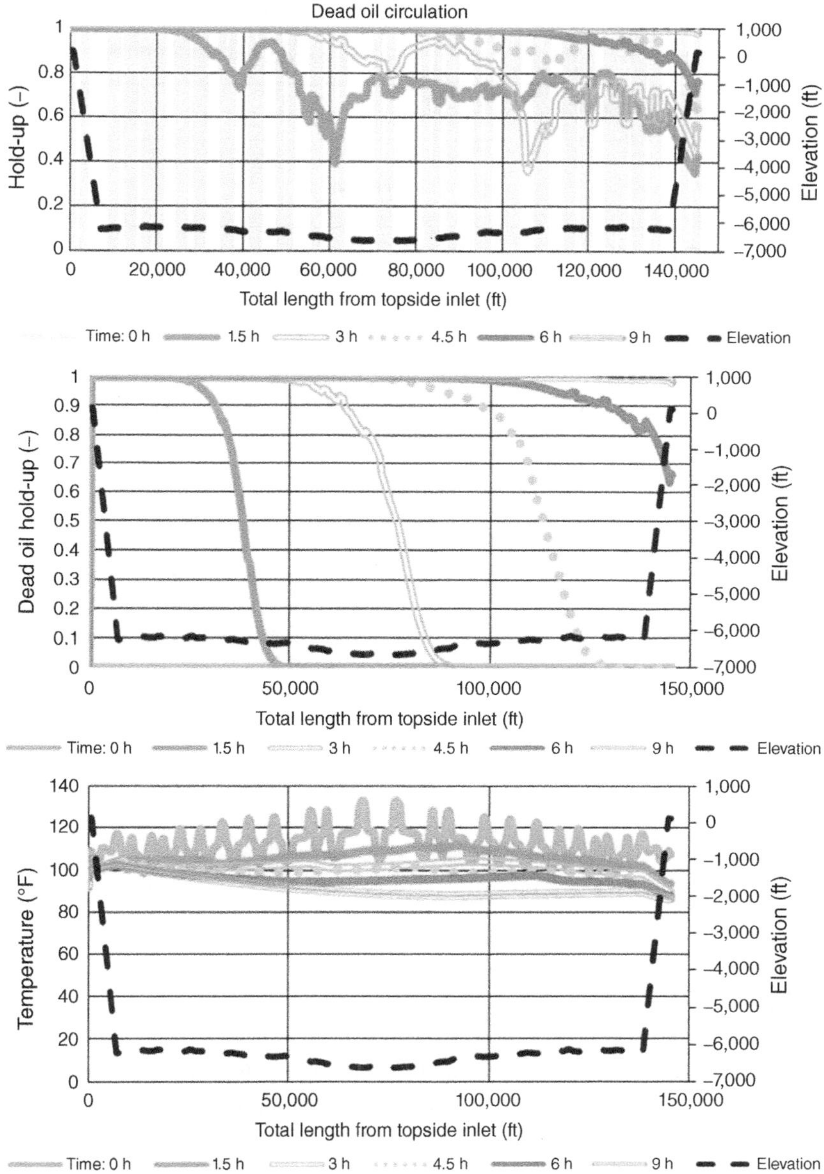

Figure 10.26 Flowline loop dead oil circulation profiles.

Figure 10.26 shows the DOC profile plot. The second graph shows that the dead oil front is somewhat dispersed. The figure also shows that after 9 hours the flowline is filled with dead oil.

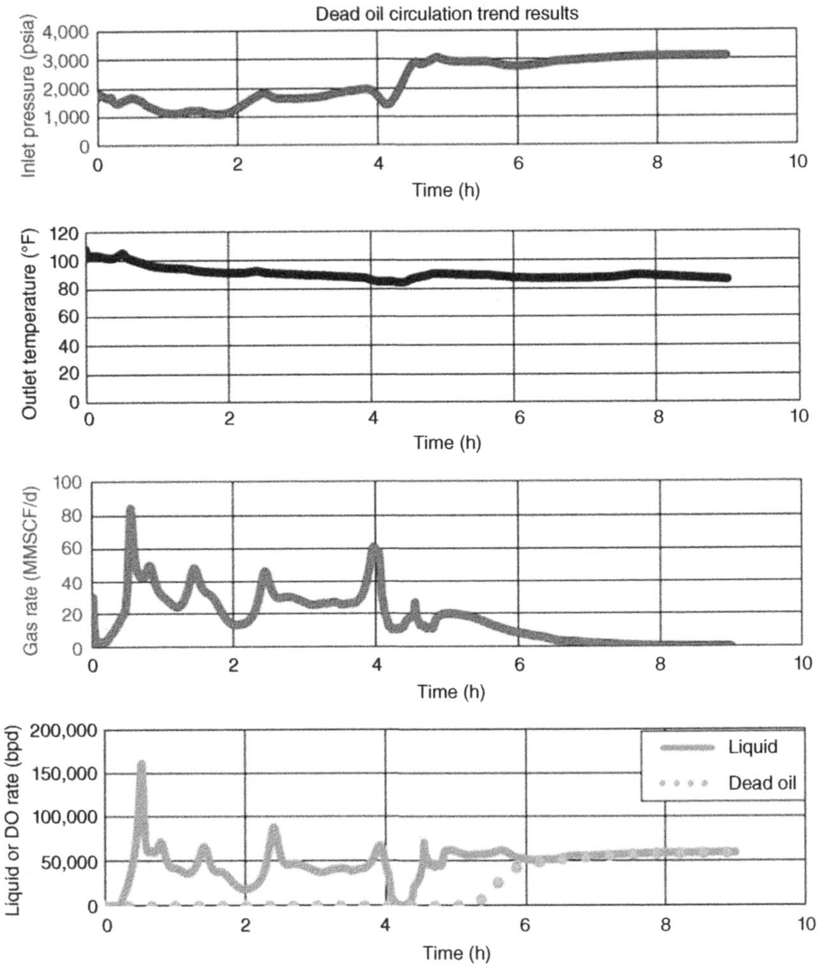

Figure 10.27 Flowline loop dead oil circulation trend.

Figure 10.27 shows the flowline DOC trend plot. Note that the maximum required DOC inlet pressure is ~3,000 psia. The liquid rate plot shows that after 6 hours, only dead oil is received in the outlet node of the flowline loop.

10.5.19 Offshore Flowline Loop with Hot Oil Circulation Example

In this section, it will be assumed that the DOC flowline from the previous section was shut-in for 240 hours; therefore, the system is initially cold. In order to limit the amounts of methanol during a restart, the operators replace the cold dead oil with warm one in a process called hot oil circulation (HOC).

In this example, it will be assumed that the HOC conditions will be 60,000 blpd at 140 °F. The HOC will last for 12 hours (two flowline volumes).

Figure 10.28 shows the HOC profile plot. The first and second graphs show a 100% dead oil-filled flowline system. The figure also shows that after 12 hours the topside outlet temperature is ~80 °F. This simulation shows that the HOC is successful in increasing the flowline temperature.

In general, the HOC increases the temperature sufficiently to help with the hydrate management but not enough for complete wax resolution in case paraffin deposits are present.

Figure 10.29 shows the flowline HOC trend plot; note that the maximum required inlet pressure is ~3,000 psia similar to the one obtained during the DOC. The temperature plot shows that even after 20 hours the arrival topside temperature is still about 80 °F.

10.5.20 Pigging Example

A pipeline inspection gauge or "pig" is used in the oil flowline systems for several purposes such as

- liquid control,
- corrosion control,
- inspection,
- maintenance.

Pigging involves the use of a cleaning tool to remove debris and other materials from the flowline. This operation is critical in ensuring the safe and efficient operation of the system and in preventing blockages in the flowline.

Multiphase simulation can be used to optimize pigging operations by predicting the behavior of fluids within the system under different operating conditions and flow rates. This includes modeling the behavior of fluids during the introduction of the pig into the flowline, as well as simulating the effects of changes in temperature and pressure on the system. By simulating these processes, operators can identify potential issues and develop strategies to prevent overloading, slugging, sudden depressurization, and damage to the flowline. The objective of the pig operation will determine the proper initial conditions, the kick-fluid chosen, and the best pig velocity.

This section includes an example of a pig operation performed on the inhibited flowline loop described in the planned shutdown section. The pig

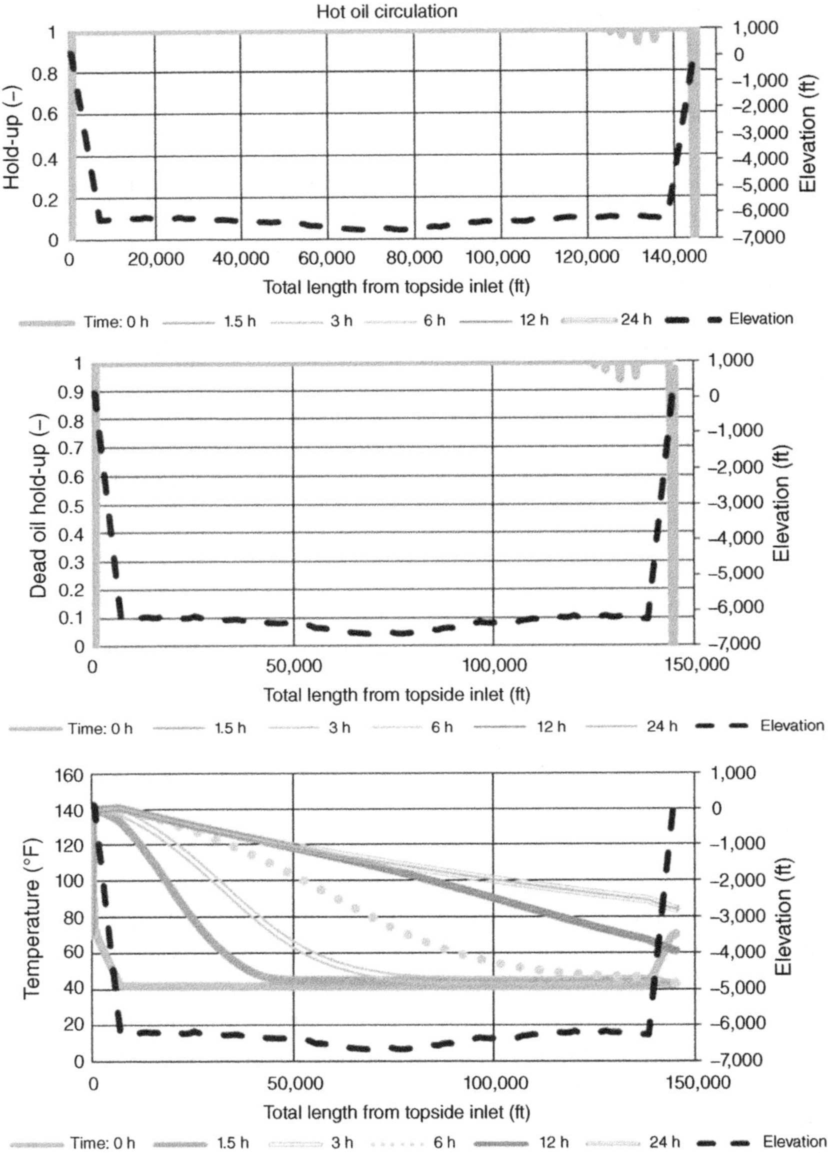

Figure 10.28 Flowline loop hot oil circulation profiles.

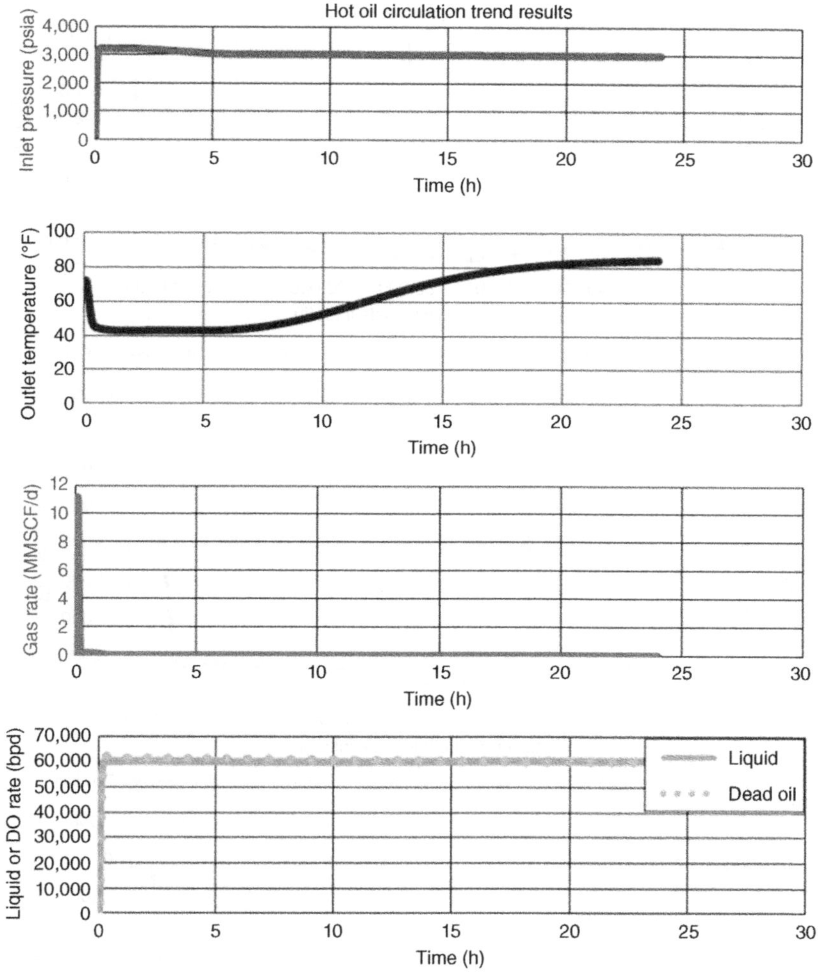

Figure 10.29 Flowline loop hot oil circulation trend.

will be assumed to have a 10.15-in. diameter, and it will be driven with a 60,000 blpd dead oil rate in order to achieve an average pig velocity of about 2 m/s.

Figure 10.30 shows similar behavior to Figure 10.26 (dead oil circulation) but with less dead oil dispersion.

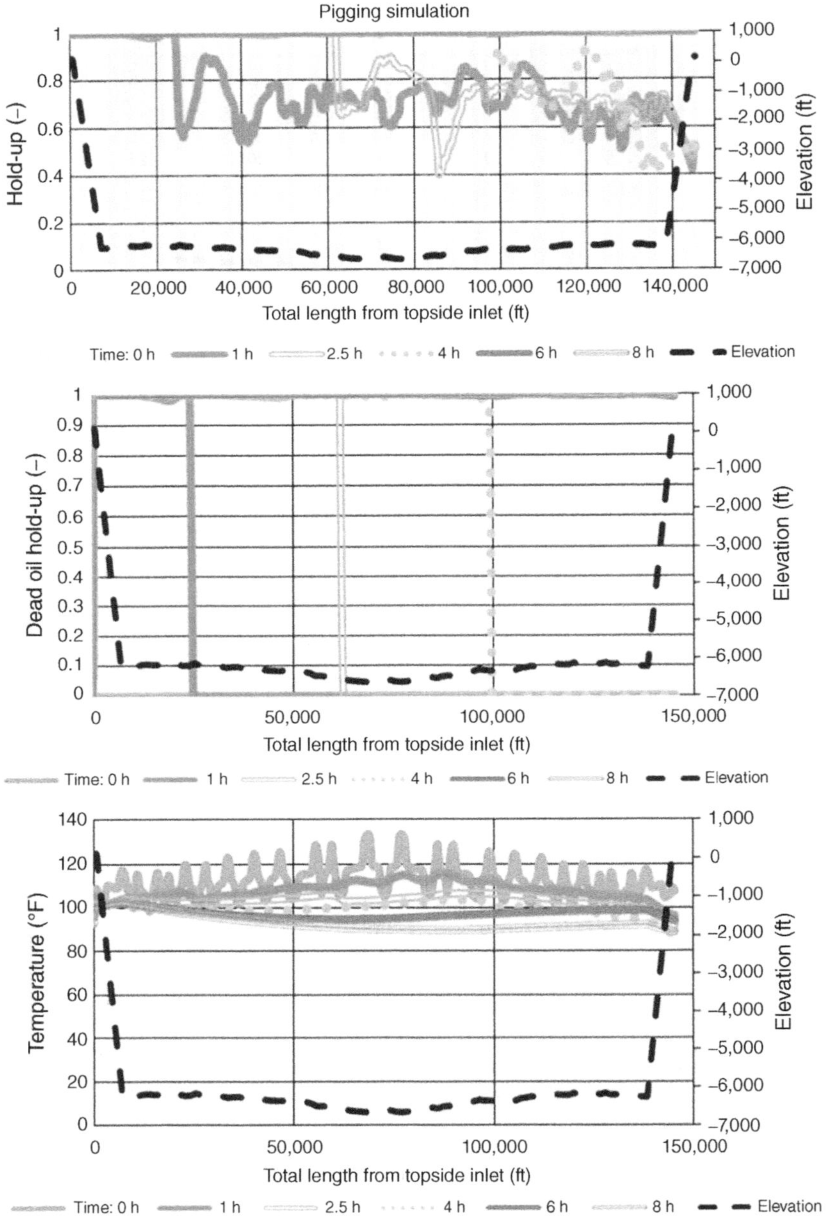

Figure 10.30 Flowline loop pigging simulation profiles.

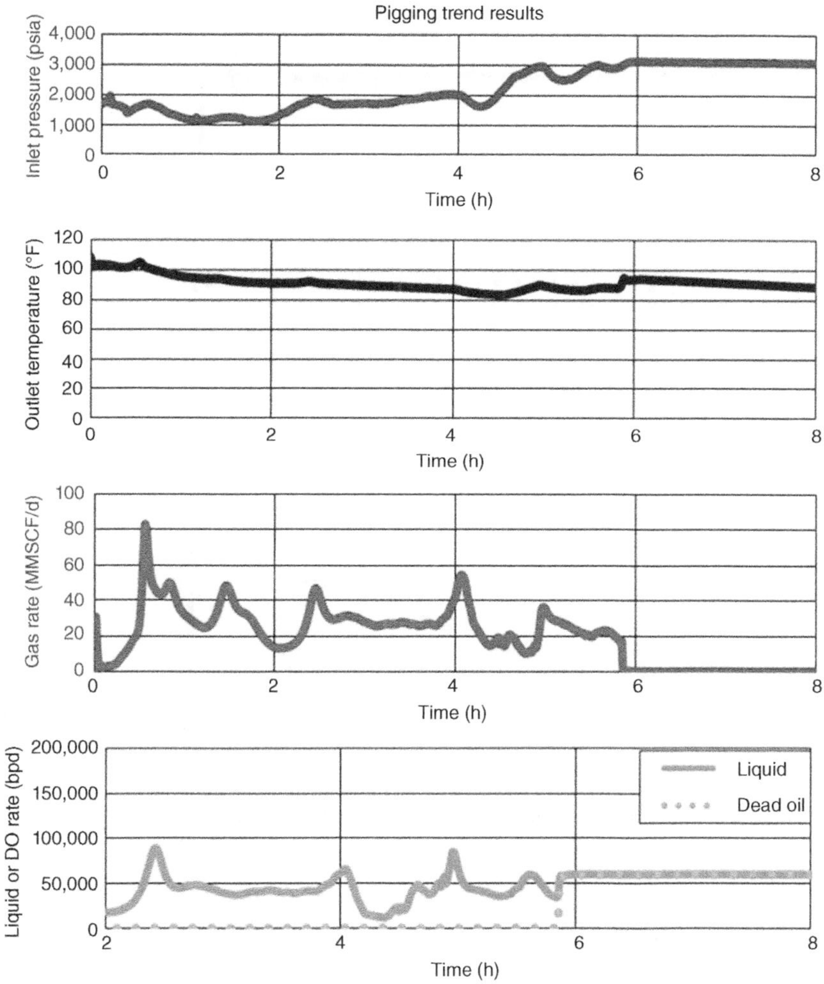

Figure 10.31 Flowline pigging simulation trends (Part 1).

Figure 10.31 shows the flowline pig simulation trend plot; note that the maximum required DOC inlet pressure is ~3,000 psia similar to the one obtained during the DOC.

Figure 10.32 shows the pig velocity, pig average velocity, and traveled distance as function of time. Note that the pig velocity is stable during the entire pigging operation.

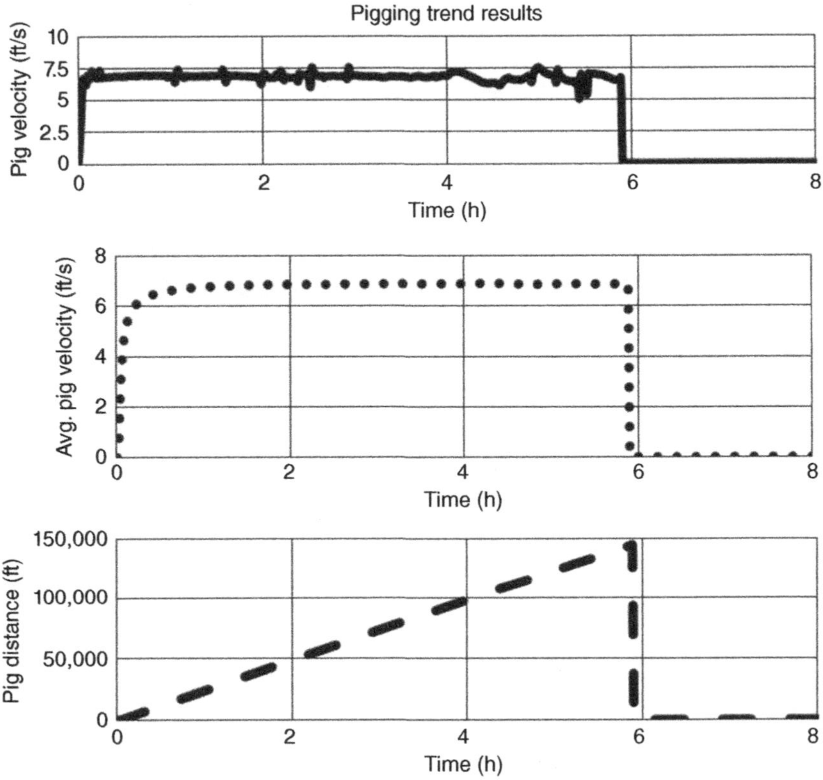

Figure 10.32 Flowline pigging simulation trends (Part 2).

10.6 Summary

This chapter addresses the normal production operation of multiphase hydrocarbon production and also covers the common transient operations that allow it to reach either the normal operation or the system shutdown. Representative examples illustrate the thermal-hydraulic performance of the production system hoping that such examples provide insight for future designs and for the day-to-day operation of such gathering systems.

Needless to say, every project is different and the coverage of examples in this chapter only intends to show typical trends and parameters and to demonstrate the exceptional power of correctly using the simulation tools available to design new systems and to monitor performance in existing ones.

References

API Recommended Practice 14E (2007). *Design and Installation of Offshore Production Platform Piping Systems*, 5e.

Courant, R., Friedrichs, K., and Lewy, H. (1928). Űber die Partiellen Differenzengleichungen der Mathematischen Physik. *Math. Ann.* 100: 32–74.

Pletcher, R., Tannehill, J., and Anderson, D. (2013). *Computational Fluid Mechanics and Heat Transfer*, 3e. Boca Raton, Florida, USA: CRC Press.

References

11

Fluid–Solid Transport

The production of sand (i.e. sand fines) from oil and gas reservoirs is a fortuitous event, and the solids' content is usually of the order of 1% volume or less, creating liquid–solid flow throughout the production system with potentially detrimental consequences. Obviously, higher concentrations are feasible if a cure to the problem is not implemented in time.

Slurry flows are fundamentally an industrial process aimed at transporting bulk solids with water, for chemical processes (e.g. coal transportation in pipelines). Typically, a slurry represents a larger particulate content (more than 10% in volume). Thus, dispersed sand fines through a flowline can be considered a "light" slurry type of dispersed flow, with the suspended and conveyed fines driven by drag forces affecting the pressure drop in the pipe system.

An interesting phenomenon of slurries is the limited range of viable particles velocities as compared to the range of possibilities in the supporting liquid continuum. Slurries could be characterized by the pipe Reynolds number (driving the continuum) and the driving drag forces characterizing the relative motion between the continuum and the entrained solid (Azbel 1981) particles (Reynolds number associated with the particles' size and the slip velocity). Both effects superimpose reaching a balance in slurry flow or end up causing other effects such as solids deposition or even flow blockage.

Gas–solid flows can be found in gas fields prone to produce sand fines and also in several other industrial activities such as pneumatic conveying (Crowe 2006, chapter 4). The gas phase is considered the continuum, and the suspended solids are entrained due to drag forces. The industrial processes (Marchello and Gomez-Plata 1976) are outside the scope of this appendix, and there are quite a few examples of applications such as combustion of crushed coal, pneumatic transport of small pellets, particulate pollution control, sandblasting, and plasma-arc coating, among others.

Multiphase Transport of Hydrocarbons in Pipes, First Edition.
Juan J. Manzano-Ruiz and Jose G. Carballo.

Companion Website: www.wiley.com/go/Manzano-Ruiz/Multiphasetransport

11.1 Characteristics of Solid–Fluid Flow

Single-phase phase flow is quite simple compared to fluid–solid two-phase flow. The behavior and characteristics displayed by solid–liquid flow are driven by particles' parameters such as size, concentration, shape, and density. Also, the fluid's rheological properties have a strong influence on the two-phase flow performance. Major drivers of fluid–solid flow are as follows:

- *Particle–fluid interaction.* The drag force imposed by the fluid flow upon the particles keeps the particles entrained (Azbel 1981) in the liquid medium in a complex balance with other acting forces. This is typically characterized by a drag coefficient (C_D). For an in-depth analysis of the nature of particle–fluid interaction forces, see Wallis (2020).
- *Particle–particle interactions.* Particles clash among themselves, depending on the solid's concentration and spatial distribution, and such collisions determine particles' settling trend or particles' suspension and transportation.
- *Buoyancy force and other forces.* Particles' motion is influenced by gravity and buoyancy, while dragged by viscous force from the liquid medium. Additionally, lift force (e.g. generated by Magnus effects) can be at play contributing to the equilibrium state of acting forces.
- *Particle concentration and spatial distribution.* The concentration and distribution of solid particles produce flow patterns that determine the flow behavior. Dilute suspensions, due to the low particle concentration, are capable of following the flow field. Concentrated slurries have substantial particle–particle interactions, which may lead to bulk agglomeration and potential settling of particles.
- *Flow Regimes.* There are specific flow regimes depending on the prevalence of the previous factors. Particles may be entrained/suspended with the fluid and transported along the flowline. If gravity force dominates upon concentrated mixtures, the flow regime becomes settling on horizontal flowlines. For inclined upward flow, fluidization is another flow regime with the particles levitated/dragged in the bulk of the fluid.
- *Rheology.* Fluid viscosity and yield stress (e.g. mixtures with non-Newtonian behavior) are the two principal rheological properties. Non-Newtonian mixtures exhibit complex deformations under shear–stress depending on the magnitude of the latter.

A rheogram plot provides useful insight regarding the relationship between fluid stresses and the rate of strain due to fluid motion. Figure 11.1 illustrates a generic rheogram for a Newtonian fluid and three other non-Newtonian fluids. A mixture of solid particles and a Newtonian

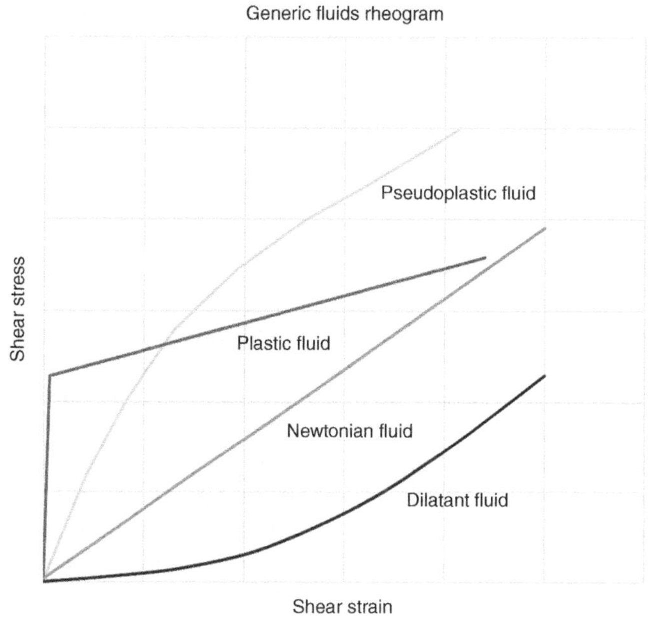

Figure 11.1 Rheogram plot of solid–liquid two-phase fluids.

liquid can exhibit non-Newtonian mixture behavior depending on the concentration and the average size of the particles.

A dilatant-fluid behavior is characterized by increased flow resistance when the strain rate is increased (i.e. shear-thickening when flow rate is increased). Shear-thickening behavior is fairly common in concentrated suspensions of particulates.

A Newtonian fluid shows proportionality between flow resistance and strain rate. Plastic fluid behavior has a limiting shearing stress value before the fluid sets in motion and overcoming such a limit the plastic fluid exhibits proportionality with rate of strain. On the contrary, a pseudoplastic fluid (i.e. shear thinning) is easier to make it flow when the rate of strain is increased.

The non-Newtonian behavior is due to an intricate balance between the particle–particle forces and the particle–fluid forces associated with the fluid's Newtonian viscosity and surface tension. Non-Newtonian rheological properties can be described by simplified models derived from empirical testing and illustrated by the rheogram plot. Accepted models to capture the stress–strain performance of non-Newtonian fluids (Broadkey 1995) include Power Law, Powell–Eyring, Bingham plastic, Casson, and Herschel–Bulkley, among others. Such models lead to defining the apparent viscosity η by equation (11.1) relating shear–stress (τ) and shear strain (γ),

$$\eta \equiv \frac{\tau}{\gamma} \tag{11.1}$$

A conclusion from equation (11.1) and Figure 11.1 is that the apparent viscosity η of a non-Newtonian fluid depends on the shear strain rate (i.e. flow rate), whereas a Newtonian fluid exhibits a constant true fluid viscosity.

11.2 Fundamental Equations

This section addresses the basic *one-dimension*al governing equations (Streeter et.al. 1988) to determine phase velocities and pressure drops. These equations include properties such as the mixture density and the mass diffusivity of the particles into the fluid.

The set of one-dimensional equations included in this section constitutes a rigorous basis for the one-dimensional motion of solid–fluid mixtures in pipelines. However, the set of equations is exceedingly intricate, even simplifying or ignoring the detailed description of suspended particle motion in a turbulent flow field (Azbel 1981), to add any value for a practical field application. Thus, simplified models based on the methods presented in this section provide a tenable approach for the cases encountered in field flowlines. Simplified models are covered in this section and in section 11.3.

11.2.1 Mixture Density

The mass density for solid–fluid mixtures can be defined by equation (11.2), based on the average volumetric concentration c_{solid} ($= 1 - c_{fluid}$) of the particles,

$$\rho_{mix} \equiv \rho_{solid} c_{solid} + \rho_{fluid}(1 - c_{solid}) \tag{11.2}$$

11.2.2 Fluid Phase Mass Conservation

The principle of mass balance for the fluid phase is given by equation (11.3):

$$\frac{\partial}{\partial t}[\rho_{fluid}(1 - c_{solid})] + \nabla \circ [\rho_{fluid}(1 - c_{solid})\vec{V}_{fluid}] = 0 \tag{11.3}$$

11.2.3 Solid-Phase Mass Conservation

For the solid phase, the mass conservation principle is equation (11.4), where D_{SL} is the molecular diffusivity of the solid particles in the fluid medium:

$$\frac{\partial}{\partial t}[\rho_{solid} c_{solid}] + \nabla \circ [\rho_{solid} c_{solid} \vec{V}_{fluid}] = \nabla \circ [D_{SL} \nabla(\rho_{solid} c_{solid})] \tag{11.4}$$

11.2.4 Solid–Fluid Mixture Mass Conservation

Adding equations (11.3) and (11.4) yields equation (11.5) in terms of $\vec{V}_{mix}$ the mixture velocity,

$$\frac{\partial}{\partial t}[\rho_{mix}] + \nabla \circ [\rho_{mix}\vec{V}_{mix}] = 0 \tag{11.5}$$

11.2.5 Mixture Momentum Equation

The momentum equation (11.6) involves the mixture's stress tensor $\overline{\tau}^{*}_{mix}$ as in single-phase flow:

$$\frac{\partial}{\partial t}(\rho_{mix}\vec{V}_{mix}) + \nabla \circ (\rho_{mix}\vec{V}_{mix}\vec{V}_{mix}) = -\rho_{mix} g \nabla z + \left(\nabla \circ \overline{\tau}^{*}_{mix}\right) \tag{11.6}$$

11.2.6 Dilute Suspensions in Turbulent Flow

When the particles' concentration c_{solid} is low, of order 0.1% or less, equation (11.6) can be simplified for dilute suspension neglecting the gravitation and stress terms, yielding equation (11.7). See Brennen (2005) for the criteria to identify the limits of dispersed flow regimes based on a Froude number definition.

$$\frac{\partial}{\partial t}(\rho_{fluid}) + \nabla \circ (\rho_{fluid}\vec{V}_{fluid}) \cong \frac{\partial}{\partial t}(\rho_{mix}) + \nabla \circ (\rho_{mix}\vec{V}_{fluid}) \cong 0 \tag{11.7}$$

11.3 Simplified Horizontal Flow Models

The principal models covered in this section applied to solids transportation through horizontal flowlines are the pseudo-homogeneous and the heterogeneous models. To better understand the logic behind these models, it is important to review the flow regimes observed in real applications.

It is useful to visualize the variety of flow regimes possible for solid–gas and solid–liquid pipeline flow. This visualization provides insight to elaborate models that represent better the physics of the flow. Figure 11.2 illustrates typical flow regimes for horizontal flow of solid–liquid mixtures. Pseudo-homogeneous, or simply "homogeneous," flow may take place when the solid particles are small in size and separated from each other (i.e. dispersed flow). Obviously, the meaningful parameter to identify particles' separation is the distance ratio to the average particle diameter. Pseudo-homogeneous suspensions are characterized by particles' trajectories following the fluid's streamlines.

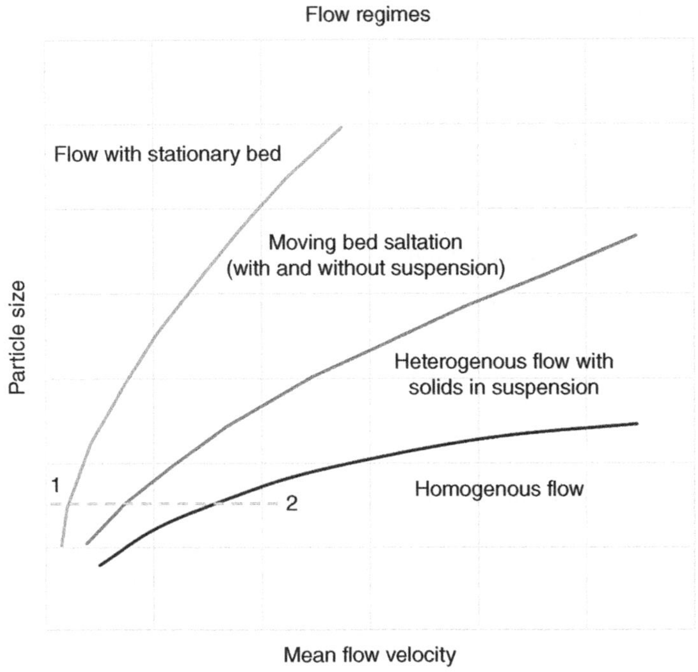

Figure 11.2 Flow regimes for a settling solid–liquid mixture.

Heterogeneous models, on the other hand, embrace a variety of flow regimes ranging from solids in suspension to settled and agglomerated particles at the bottom of the pipe (i.e. stationary bed). These flow regimes though exhibit random flowing patterns and carry larger size particles entrained in the fluid.

From Figure 11.2, we can observe the sensitivity of regime transitions with regard to mixture velocity. For a particular size of particles, we can modify the flow regime from stationary bed, as shown in state 1, to homogeneous flow (state 2) by increasing the mixture average speed. Also, when there is substantial variance in particle sizes, for a given mixture velocity, we could note intermittent transitions among two or more flow regimes.

11.3.1 Homogeneous Flow Model

The homogeneous model is an ideal state where small particles, of the order of 100 μm or less, are entrained in the bulk of the fluid due to drag forces while in balance with buoyance force. Also, typical solids' concentration by volume, c_{solid}, to ensure model validity is of the order of 5% or less. The distribution of particles is uniform, following the streamlines of the carrier fluid.

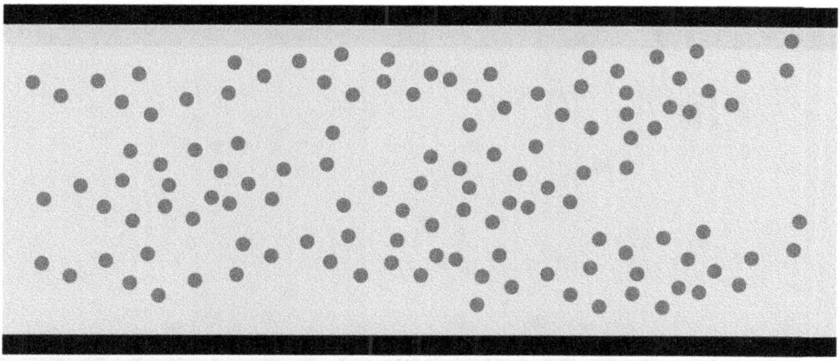

Figure 11.3 Pseudo-homogeneous solid–liquid mixture flow.

A qualitative sketch of pseudo-homogeneous mixture in horizontal flow is depicted in Figure 11.3, showing the dispersion structure of solid particles immersed in the liquid flow.

The modeling of this ideal flow regime follows the steps of homogeneous equilibrium model (HEM) discussed for gas–liquid flow in Chapter 5. The only departing consideration at this point for solid–liquid flow is the use of a mixture viscosity based on Einstein's equation (Govier and Aziz 1972) as per equation (11.8):

$$\mu_{\text{hom},mix} \cong \mu_{liq}(1 + 2.5c_{solid}) \tag{11.8}$$

The density of the homogeneous mixture is also evaluated with equation (11.2). The friction factor to use in the momentum equation depends on the transition Reynolds number, which for a Newtonian mixture is of the order of 2000. Further information on the friction factors for laminar and turbulent flow can be found in Chapter 11 of Nayyar (2000) reference. With such equivalent properties, it follows that calculations for pipeline transport of solid–liquid mixtures can be carried out with the set of equations addressed in Section 5.9. This could include flow rate calculation for a known pressure drop in a given pipeline span, or sizing of the pipeline for given flow rate and pressure drop.

11.3.2 Heterogeneous Flow Model

The heterogeneous flow model considers distinct stages of suspended particles, from fully suspended forming irregular patterns to fully settled/sedimented (stationary bed) with few suspended particles. These generic stages are illustrated by sketches in Figures 11.4–11.6.

The stage between fully suspended and fully sedimented particles is an intermediate stage blending both characteristics. This intermediate stage is referred to as moving bed with saltation (Govier and Aziz 1972).

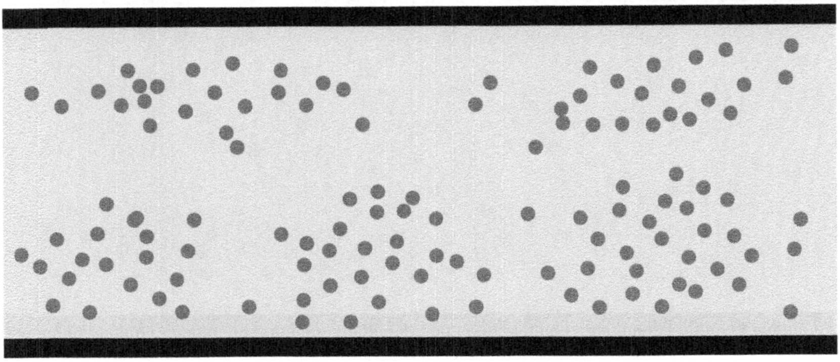

Figure 11.4 Heterogeneous mixture flow with suspended particles.

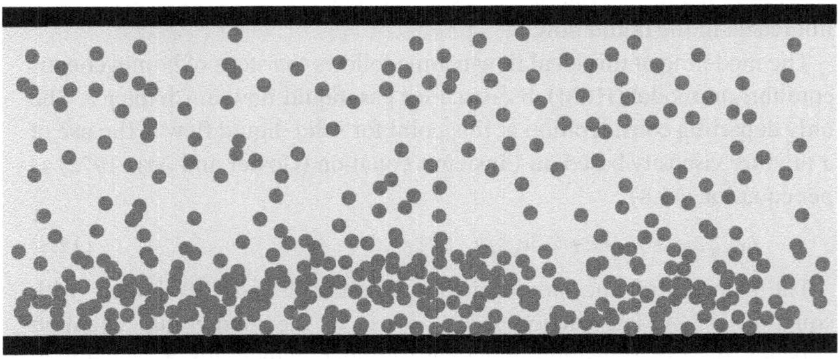

Figure 11.5 Moving bed flow with saltation.

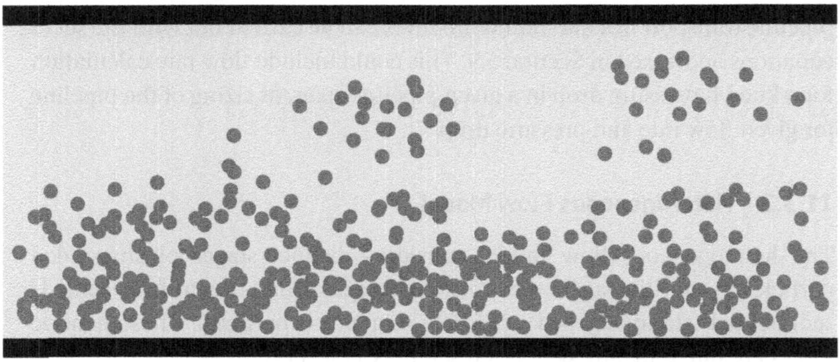

Figure 11.6 Mixture flow with stationary bed.

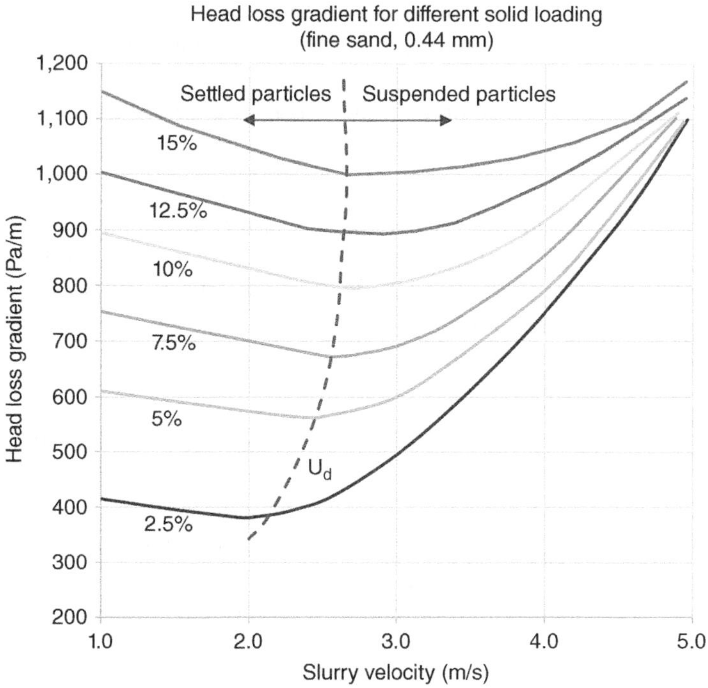

Figure 11.7 Slurry head loss gradient.

The latter refers to the uppermost particles that settle on top of the formed dunes of particles, tumbling over one another causing top bed motion. Larger particles form the stationary bed at the pipe bottom, and there may be fine particles traveling above the bed entrained in the fluid. For an in-depth coverage of the mechanistic models of particulate slurry transport, please consult Chapter 7 of Cheremisinoff (1986) encyclopedia.

The stationary bed stage is dominated by either large-size particles sedimented at the pipe's bottom or by low fluid velocities allowing gravity forces to overcome the drag forces.

Figure 11.7 shows a characteristic plot of head loss or pressure gradient versus slurry velocity for horizontal flow of a mixture of particles with an average size of 440 μm. The particle size may be considered above the limit of small particles, and therefore it is considered in heterogeneous flow. In Figure 11.7, we can observe the following features:

- For mixture concentrations below 5%, there is the possibility of having settled solids.
- The curves of constant solid concentration exhibit a minimum, marked with the dotted line and labeled as U_d (minimum deposit velocity).

- Fluid velocity below corresponding U_d enhances particle settlement.
- Fluid velocity above corresponding U_d enhances head loss but suspends the particles.
- The larger the solids' concentration, the larger the pressure gradient for the same fluid velocity.

A simple pressure-loss model for heterogeneous solid–liquid mixtures in flowlines has been summarized by Blevins (1992), assuming that the fluid velocity exceeds the minimum deposit velocity calculated from equation (11.13). This model can be applied to inclined upward flowlines, varying from horizontal ($\theta = 0°$) to vertical ($\theta = 90°$). The pressure gradient expression for stepwise calculations is given by equation (11.9).

$$\frac{dp}{dx} = \left(\frac{\rho_{liq} V_{mix}^2}{2d}\right)(f_{liq} + f_S \cos\theta) + [\rho_{liq} + c_{solid}(\rho_{solid} - \rho_{liq})]g \sin\theta \tag{11.9}$$

The friction factor f_{liq} in equation (11.9) can be estimated from the liquid flow rate and the equations from Section 5.10. The friction factor f_S in equation (11.9) is obtained from Durand's (1953) correlation referred to by Blevins (1992):

$$f_S = 85 f_{liq} c_{liq} \left[\frac{gd}{V_{mix}^2 \sqrt{C_D}} \left(\frac{\rho_{solid}}{\rho_{liq}} - 1\right)\right]^{1.5} \tag{11.10}$$

The drag coefficient C_D in equation (11.10) can be approximated with Cheng's (2009) correlation for Reynolds number below $2 * 10^5$ by equations (11.11) and (11.12):

$$C_D = \frac{24}{\text{Re}}(1 + 0.27\text{Re})^{0.43} + 0.47[1 - \exp(-0.04\text{Re}^{0.38})] \tag{11.11}$$

$$\text{Re} \equiv \frac{\rho_{liq} d_{solid} V_{liq}}{\mu_{liq}} \tag{11.12}$$

11.4 Minimum Deposit Velocity

This is the fluid velocity limit onset of particles settling at the bottom of a horizontal flowline when the fluid velocity is reduced. An expression to estimate the minimum deposit velocity (Saleh 2002) is equation (11.13):

$$U_d \simeq 1.32 c_{solid}^{0.186} \left(\frac{d_{solid}}{d_{pipe}}\right)^{0.167} \sqrt{2 g d_{pipe} \left(\frac{\rho_{solid}}{\rho_{liq}} - 1\right)} \tag{11.13}$$

Equation (11.13) is empirical, and therefore it serves as a reference value to approximate the minimum velocity required to maintain particles in suspension assuming a heterogeneous flow regime.

11.5 Sand Production

Sand production from reservoirs is due to the weakening of the strength of the rock and associated geomechanical local stresses in the wellbore and/or the perforations (Bellarby 2009). Also, local loads caused by the well hole itself, reduced pore pressure and formation water (Bellarby 2009) are major contributing factors toward the fine production.

Sand production is a concern because it causes erosion in the production system, for example, well tubing, valves, flowlines, and surface equipment downstream, leading in due uncertain time to mechanical integrity risk (i.e. loss of containment, leaks, fire, and explosion).

About 90% of hydrocarbon wells are located in sandstone reservoirs (Bellarby 2009), and approximately one-third of sandstone reservoirs may be prone to release sand. These rough figures explain the incidence of sand fines production in the oil and gas industry. It is not the objective of this chapter to dwell on the topic of rock strength, belonging to the field of geomechanical, but only to point out the source of the commonness of the sand production risk in the hydrocarbon exploitation process.

Often the sand production mechanism follows three distinctive stages (Bellarby 2009), namely

- Transient sand production while the wellbore undergoes plastic deformations, with a characteristic peak production rate,
- A period of quasi-constant sand production rate when the hole is stable,
- Potential sand rate growth if the well drawdown is incremented in pursuit of more hydrocarbon production or during speedy well start-up. Also, late-life well operation at declined reservoir pressure may lead to enhanced sand delivery.

There are two trends for mitigating the sand production issue. One trend is to control sand production at the wellbore, and the second trend is to tolerate some extent of production and cope with the issue (e.g. incremented maintenance frequency and sand disposal). Some operating companies do not accept the first trend because it leads to loss of productivity in the well, potentially causing issues in future well workovers/recompletions, and more costly interventions.

The second trend drives up preventive maintenance costs, system interventions, deferred production during pigging operations, sand collection at separators, and cleaning and disposal costs of sand produced. Thus, it is not a straightforward decision which trend to follow in any other field, but rather it is a matter of judgment and practical experience one-on-one.

11.6 Sand Distribution

Hydrocarbon production has a long history of sand production data. Per ISO 14688 sand fines more often than not fall within the range of 60–200 microns (ISO 2017). Sand grains up to 125 microns are classed as very fine sand, and sand in the range of 125–250 microns is referred to as fine sand. Obviously, every field has its own sand probability distribution, which may shift during production time. An example of a sand grain size cumulative distribution is shown in Figure 11.8. This grain distribution has an average size of 36 microns, and it can be considered very fine sand.

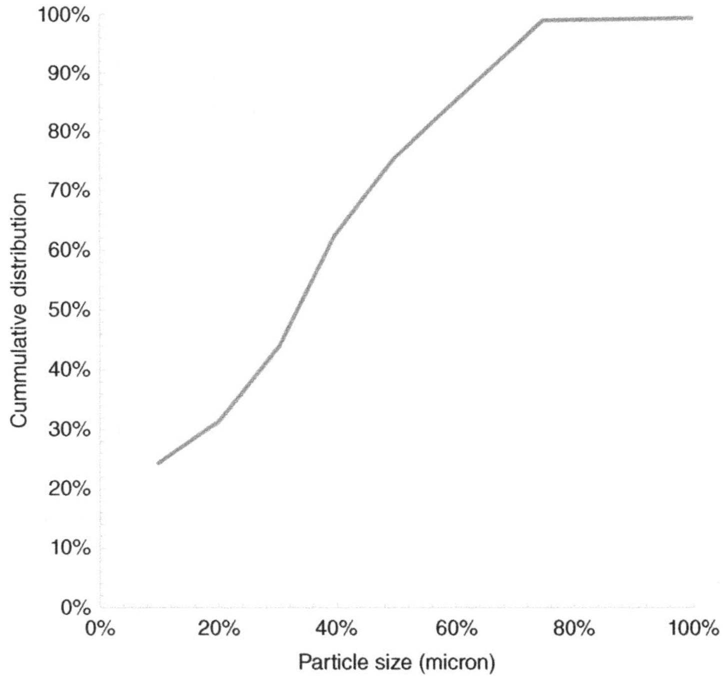

Figure 11.8 Example of sand grain size distribution.

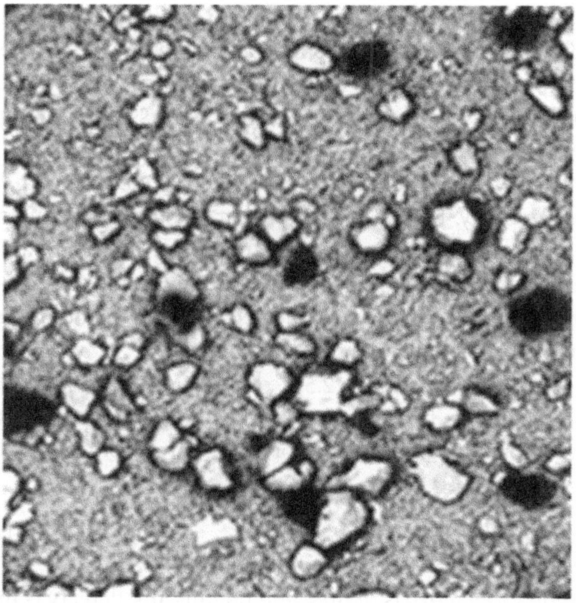

Figure 11.9 Microphotography of sand fine sample.

Sand samples are tested for weight of solids (typically in milligrams) and volume of liquid filtered (usually in liters). Dividing the weight by the volume yields the *TSS* – total suspended solids – expressed in *mg/lt*. In the oil industry, the TSS is converted into the FPS system and expressed in pounds of solids per 1000 barrels by multiplying by the 0.3506 factor.

Sand composition is quite variable depending on the field. However, sand composition usually contains oxygen, aluminum, silicon, potassium, and iron. The silicon and oxygen weight composition dominate (i.e. silicon oxide) over other components. Figure 11.9 shows an example of a microphotography of a sample of sand.

11.7 Pipe Erosion and Erosional Velocity

The presence of sand fines in production flowlines is common in field operations. Sand causes steel material erosion (i.e. metal material removal), in particular, wherever the system exhibits a change in flow direction (e.g. elbows, valve internals). Gas flowlines usually operate at high fluid velocity, and if some sand fines are entrained such applications will carry high integrity risk. Most flowlines in field applications operate in turbulent flow regime, which enhances the damaging effect of erosion due to particle impact driven by the Reynolds' stresses. Pipe erosion sometimes is present in combination with corrosion, causing an intensified process threatening the mechanical integrity of the system.

From a best-practice viewpoint, there is the API RP 14E (2007) standard offering a simple methodology, based on the erosion velocity concept, to verify the risk for flowlines exposed to solid–fluid mixtures. This standard recommends a conservative criterion for the maximum safe operating velocities which some other organizations have indicated could be made more flexible depending on the particulate's concentration.

Erosional velocity is defined in Section 2.5.a. of API RP 14E (2007), to address erosion risk concerns in flowlines, manifolds, headers, and bends transporting solid–fluid flow. Equation (11.14) for erosional velocity per API RP 14E is

$$V_{erosion} = \frac{C}{\sqrt{\rho_{mix}}} \tag{11.14}$$

The velocity in the equation above is in m/s, and the mixture density is in Kg/m^3. The constant "C" in equation (11.14) depends on the type of service (continuous, sand-free, or intermittent). The values for constant C recommended by API RP 14E (2007) are

- $C = 122$ to 152 for continuous solid-free service
- $C = 183$ to 244 for continuous solid-free service with corrosion inhibitor present
- $C = 305$ for intermittent service
- $C < 122$ for continuous service with solids

11.8 Application Examples

Example 11.1

Determine the minimum deposit velocity for sand particles, 220 μm of average size, flowing entrained in a production stream of hydrocarbon condensate having a specific gravity equal to 0.78. The flowline has an internal diameter of 154 mm. Assume the sand density is about 2070 Kg/m^3, and the solids' concentration is 12% volume.

The condensate's density is 780 Kg/m^3. The minimum deposit velocity is calculated from equation (11.13) as 0.67 m/s.

$$U_d = 1.32(0.12)^{0.186}\left(\frac{0.22}{154}\right)^{0.167}\sqrt{2^*9.81^*0.154\left(\frac{2070}{780} - 1\right)} = 0.67\ \frac{m}{s}$$

Example 11.2

Estimate the maximum allowable oil velocity in a flowline of 193.7 mm ID following the erosion criterion stated by API RP 14E (2007). Consider continuous service with corrosion inhibitor and no sand, and continuous service with sand but no corrosion inhibitor. Assume the oil's gravity is a 32.4° API and the concentration of sand fines (density ≈ 2100 Kg/m^3) is less than 5% volume.

The crude oil density is $\rho_{oil} = \left(1000\ \frac{Kg}{m^3}\right)\left(\frac{141.5}{131.5+32.4}\right) = 863\ \frac{Kg}{m^3}$

From equation (11.2), we have

$$\rho_{mix} \equiv \rho_{solid}c_{solid} + \rho_{fluid}(1 - c_{solid}) = 2100^{*}0.05 + 863^{*}(1 - 0.05) = 925\ \frac{Kg}{m^3}$$

Using equation (11.14), and assuming a constant $C = 122$ for continuous service with corrosion inhibitor added, and a constant $C = 80$ for continuous service with solids, we get the following conservative estimates:

- Maximum velocity with corrosion inhibitor $V_{\max} = \frac{122}{\sqrt{925}} = 4.0\ \frac{m}{s}$
- Maximum velocity flow without corrosion inhibitor $V_{\max} = \frac{80}{\sqrt{925}} = 2.6\ \frac{m}{s}$

In general, the effectiveness of a corrosion inhibitor is less than perfect, because it might be unevenly distributed within the bulk of the production stream. For this example, to be on the safe side, it is likely to limit the maximum velocity to 2.6 m/s and the corresponding flow to 0.077 m^3/s (= 41,845 bpd).

References

API Recommended Practice 14E (2007). *Recommended Practice for Design and Installation of Offshore Production Platform Piping Systems*, 5e. Reaffirmed.

Azbel, D. (1981). *Two-Phase Flows in Chemical Engineering*. Cambridge, UK: Cambridge University Press.

Bellarby, J. (2009). *Well Completion Design*. Amsterdam, The Netherlands: Elsevier.

Blevins, R. (1992). *Applied Fluid Dynamics Handbook*. Malabar, Florida, USA: Krieger Publishing Company.

Brennen, C. (2005). *Fundamentals of Multiphase Flow*. Cambridge, UK: Cambridge University Press.

Broadkey, R. (1995). *The Phenomena of Fluid Motions*. New York, USA: Dover Publications Inc.

Cheng, N. (2009). Comparison of formulas for drag coefficient and settling velocity of spherical particles. *Powder Technology* 189: 395–398.

Durand, R. (1953). Basic relationship of the transportation of solids in pipes-Experimental research. In: *Proceedings of IAHR Congress*. Minneapolis, Minnesota, USA: Hydraulics Research.

Cheremisinoff, N. (ed.) (1986). *"Encyclopedia of Fluid Mechanics," Volume 5 "Slurry Flow Technology"*. Houston, Texas, USA: Gulf Publishing Company.

Crowe, C. (ed.) (2006). *Multiphase Flow Handbook*. Boca Raton, Florida, USA: CRC Taylor & Francis Group.

Govier, G. and Aziz, K. (1972). *The Flow of Complex Mixtures in Pipes*. New York, New York, USA: Van Nostrand Reinhold Company.

International Organization for Standardization, (2017), "Geotechnical investigation and testing-Identification and classification of soil-Part 1: Identification and description" ISO 14688-1:2017.

Marchello, J. and Gomez-Plata, A. (1976). *Gas-Solids Handling in the Process Industries*. New York, USA: Marcel Dekker Inc.

Nayyar, M. (ed.) (2000). *Piping Handbook*. New York, New York, USA: McGraw-Hill.

Saleh, J. (ed.) (2002). *Fluid Flow Handbook*. New York, New York, USA: McGraw-Hill.

Streeter, V., Wylie, E., and Bedford, K. (1988). *Fluid Mechanics*, 9e. New York, New York, USA: McGraw-Hill.

Wallis, G. (2020). *One-Dimensional Two-Phase Flow*. Garden City, New York, USA: Dover Publications Inc.

Appendix A

Multiphase Flow Software Tools

This appendix addresses some of the existing codes to simulate multiphase flow through pipelines. The codes described herein are representative but it is not an exhaustive list of tools.

A.1 Thermodynamics and Transport Properties Simulators

Any dynamic simulation code will require a solid estimator of thermodynamic properties such as phase densities, fraction of each phase, compressibility factor, and speed of sound. Likewise, transport properties such as dynamic viscosity, thermal conductivity, and mass diffusivity will be necessary to support the mass, energy, and momentum conservation equations.

The properties' simulators use equation of state (EOS), fundamentally cubic functions capable of doing straightforward flashing calculations to estimate the fraction of either phase. Such simulators are used in two ways, for either generation of look-up tables of properties or as compositional models that estimate point-to-point the necessary parameters at each cell of the grid.

The use of comprehensive look-up tables saves substantial amount of time during the dynamic simulation because the properties are interpolated frequently by a linear method. The property tables are commonly developed using pressure–temperature or pressure–enthalpy independent variables. Thus, a fast interpolation point-to-point provides a good approximation at low cost in terms of computing time. The drawback of this methodology lies when different fluids commingle at tees or manifolds. In this case, the resulting mixture stream does not have a specific pre-calculated table for quick interpolation. Therefore, for such a mixing case, the streams' tables

Multiphase Transport of Hydrocarbons in Pipes, First Edition.
Juan J. Manzano-Ruiz and Jose G. Carballo.

Companion Website: www.wiley.com/go/Manzano-Ruiz/Multiphasetransport

have to be combined at each commingling point, at each time integration interval, to estimate mixed properties weighed with the instantaneous flow rate of the streams.

The compositional modeling of fluid properties may be orders of magnitude more time-consuming than the look-up tables because at each integration step the thermodynamics flashing is carried out. However, this methodology can handle well the mixing case of flow streams and generates more accurate properties' estimates than linear interpolation in tables.

Currently, a short list of popular commercial codes to generate fluid properties are *Pvtsim Nova®, Multiflash®, Aspen Hysys®, Refprop®, and Pvtp®*, because most of them seamlessly interface with the most common dynamic codes. A brief description of each of these software follows. Detailed information has to be sought through the links provided.

A.1.1 Pvtsim Nova® (Calsep, https://www.calsep.com/pvtsim-nova/)

This product from Calsep A/S Copenhagen, Denmark, has been in the market for over 30 years. The code includes several EOSs and offers several packages: Basic PVT, PVT Lab, Reservoir, Flow Assurance, and Wax deposition.

There is also a specific package to interface with Schlumberger's OLGA® and Kongsberg's LedaFlow® dynamic simulators. *Pvtsim* comes as a standard module complemented by specialized modules. The *base module* provides compositional input including molar/weight percentage with molecular weights and densities input for each C_7^+ fraction, or extended gas chromatograph (GC) fluid analyses. Several options of the Peng–Robinson (PR) and Soave–Redlich–Kwong (SRK) cubic EOS are available with volume-translation correction. The PC-SAFT EOS is available for general flash and PVT simulations, and the GERG-2008 EOS is recommended for high-accuracy gas properties' predictions such as gas Z-factors and dew points.

The flash and unit-operation module provides several phase envelope options such as pressure versus temperature, pressure versus enthalpy, pressure versus entropy, volume versus temperature, and enthalpy versus entropy. Phase-equilibrium calculations yield K-constants. Water saturation of the gas mixtures can be performed with this module too. Flash calculations include phase percentage split, fluid composition, molar volume, density, Z-factors, enthalpy, entropy, specific heat, Joule–Thomson coefficient, velocity of sound, viscosity, thermal conductivity, surface tension, and fugacity.

Additional modules, useful for flow assurance risk assessment, include the hydrate module, the wax module, the asphaltene module, and the scale module.

A.1.2 Multiflash® (KBC, https://www.kbc.global/software/advanced-thermodynamics/)

This code offered by KBC-Yokogawa company is an advanced thermodynamics software focused on pipeline simulation work. This product comes integrated with Schlumberger's Pipesim steady-state software for compositional calculations. Multiflash performs PVT modeling to match PVT lab reports, flow assurance risks such as wax, asphaltene, and scale, and predicts physical properties from the reservoir to the refinery. In general, most features available in PVTsim Nova can be found in Multiflash too.

Multiflash is a PVT and physical properties package that allows modeling the phase behavior of complex mixtures and pure substances for upstream flow assurance and production processes simulation in the oil and gas industry. Multiflash can model hydrates, waxes, asphaltenes, and halide scales, as well as track the partitioning of potentially damaging substances such as mercury (Hg) and hydrogen sulfide (H_2S).

Multiflash offers several valuable interfaces such as Excel, Cape-Open, Visual Basic, and API to facilitate the I/O of data and predictions. *Multiflash PVT Lab*® supports an advanced set of tools for EOS tuning that includes fluid characterization from either compositional or black-oil data, model tuning by fitting to data from PVT lab tests, and quality control of the PVT data.

A.1.3 Aspen Hysys® (Aspentech, https://www.aspentech.com/en/products/engineering/aspen-plus)

This code is a broader tool going beyond properties' prediction. This software is a Process Engineering powerhouse to simulate streams, unit-operation equipment, whole plant performance, and even make cost estimations with additional modules. Hysys software evolved as a fine tool from the precursor Hysim developed in the 1980s, one of the first interactive codes for chemical engineers. ASPEN Hysys has a large database of components and a thorough list of EOS at the user's disposal.

Aspen Plus® and Aspen Hysys have a long history in the process engineering market. The two software offer application programming interfacing with the majority of the commercial steady-state and transient codes, making these tools a convenient companion to estimate fluid properties for pipeline multiphase flow simulations. *Aspen Plus and Hysys* go well beyond performing phase-equilibrium calculations and properties' estimation. Each software has a broad spectrum of simulation capabilities for unit operations, to design production facilities, to size process to equipment, to summarize facilities specs, and cost estimate the facilities. These codes have been developed to support not only the upstream, midstream, and downstream hydrocarbon industry but also the chemical industry in general.

Aspen Plus and Hysys are supported by an extensive database of simple substances with all the thermodynamic properties. The database includes default binary interaction parameters that can be changed by the user. The software allows us to define molar/weight mixtures of basic components and to choose from a lengthy list of equations of state. The EOS applies to a broad range of temperature and pressure, including subcritical and supercritical conditions. The software can also model hydrocarbon mixtures that include species such as H_2S, N_2, and CO_2.

Some of the EOS and activity coefficient options are

- Redlich–Kwong
- Soave–Redlich–Kwong (SRK)
- Redlich–Kwong-ASPEN
- Soave–Redlich–Kwong-MHV2
- Soave–Redlich–Kwong-WS
- Peng–Robinson (PR)
- Peng–Robinson-MHV2
- Peng–Robinson-WS
- Lee–Kesler
- Chao–Seader,
- Grayson–Streed,
- UNIQUAC, and
- UNIFAC.

A.1.4 Refprop® (NIST, https://www.nist.gov/srd/refprop)

Refprop is a code developed by the National Institute of Standards and Technology (NIST), USA. This code offers fewer EOS options as compared to the other codes indicated before, a limited number of components to simulate hydrocarbon mixtures, and does not have the capacity to deal with C_7^+ fractions for characterization. Despite these limitations, Refprop is a remarkably accurate simulation tool within the range of application. Most other commercial codes usually offer the option to perform calculations using Refprop subroutines for accurate results.

This code offers the most accurate EOS models currently available: high-accuracy Helmholtz energy equations of state, international standard equations for water, a variety of equations applicable to hydrocarbons, ammonia, carbon dioxide, and others. The Benedict–Webb–Rubin (*BWR*) equation of state is an additional option for the top accuracy predictions near the critical point, and an extended corresponding states model can be used for fluids with limited data. For mixtures of hydrocarbon, the excess Helmholtz energy model gives excellent prediction for fluid properties. The American Gas Association equation *AGA8* (as an alternative to the Helmholtz model) is another choice for natural gas properties.

Viscosity and thermal conductivity are based on fluid-specific correlations (where available), either as a modification of the extended corresponding states model or from the friction-theory model. An extensive list of available properties includes density, internal energy, enthalpy, entropy, specific heats, sound speed, compressibility factor, Joule–Thomson coefficient, gas quality, second and third virial coefficients, second and third acoustic virial coefficients, Helmholtz energy, Gibbs energy, heat of vaporization, fugacity, fugacity coefficient, chemical potential, K-values, thermal conductivity, dynamic viscosity, kinematic viscosity, thermal diffusivity, Prandtl number, surface tension, isothermal compressibility, isentropic coefficient, adiabatic compressibility, and several important derivatives of thermodynamic functions.

A.1.5 Pvtp® (Petroleum Experts, https://www.petex.com/products/ipm-suite/pvtp/)

Pvtp comprises thermodynamically consistent and precise EOS models for fluids, offering lumping and de-lumping algorithms embedded into the program to enable the user to create integrated systems in a straightforward and consistent manner. Pvtp enables the user to perform and validate Pvtp analyses by matching compositions to data available. Special treatment of pseudo-components with quality checks enables a consistent set of parameters while offering a variety of regression techniques.

The Peng–Robinson and SRK EOS models in their original form have well-known limitations when dealing with polar molecules, such as water and CO_2. To overcome these limitations, Pvtp modified the EOS to predict accurately properties of CO_2. Pvtp includes two- and three-phase flash engines to facilitate the iterative calculations.

A.1.6 Dwsim®

This is a new software tool for simulating process plants and estimating fluid properties. Dwsim is an open-source CAPE-OPEN compliant software, and the basic version can be downloaded for free. The thermodynamic models included are GERG-2008, PC-SAFT,

CoolProp, Peng–Robinson, Peng–Robinson (1978), Soave–Redlich–Kwong, Lee–Kesler, Lee–Kesler–Plöcker, UNIFAC, modified UNIFAC (Dortmund), UNIQUAC, NRTL, Chao–Seader, Grayson–Streed, Raoult's law, IAPWS-IF97 steam tables, IAPWS-08 seawater, black oil, and sour water. The software includes the following utilities: phase envelope, hydrate calculations, pure component properties, critical point, spreadsheets, petroleum cold flow properties, pseudo-component creator, bulk C_7^+ component, and distillation curves petroleum characterization, among others.

A.2 Steady-State Multiphase Simulators

The simulation of steady-state performance is necessary not only for regular normal operations but it also is necessary to create the preliminary scenario before simulating a transient process such as ramp-up and rundown, shutdown, and well start-up. Normal operation is by far the longest activity during the field life cycle.

Among the several multiphase flow steady-state operation simulators, we have the following brief list.

A.2.1 Pipesim® (SLB, https://www.software.slb.com/products/pipesim)

The major features of the *Pipesim* software are:

Multiphase flow regime modeling. Mechanistic models based on the prevailing flow pattern can be invoked by the user. For instance, the Tulsa models, the *Olga-S* model, and the *Beggs–Brill* model perform analyses based on the established flow regime at any cell of the system's physical domain. This software can predict the risk of severe slugging in risers, hydrodynamic slugging, and slug size and frequency. These features enable the user to optimize line sizing design and downstream processing facilities.

Variety of gas–liquid models. *Pipesim* includes a diversity of industry-standard multiphase flow correlations as well as advanced three-phase mechanistic models, including *Olga-S* steady state, *LedaFlow*, and The University of Tulsa Fluid Flow Projects (*TUFFP*) *unified model*. These models enable the calculation of flow regimes, liquid holdup, slug characteristics, and pressure loss for all nodes along vertical and horizontal production paths at all inclinations. With this capability, users can design and operate production gathering and distribution systems for different life-cycle scenarios.

Heat transfer losses to the environment. The prediction of heat transfer losses is paramount to predict temperature profile, fluid properties, and solids formation, and to assist the thermal design of the production system. The *Pipesim* simulator performs comprehensive point-to-point energy-balance calculations that involve heat transfer phenomena, such as forced and free convection, molecular heat conduction, Joule–Thomson cooling and heating, and frictional heating. The user can specify the internal and external heat transfer coefficient calculations. Analytical heat transfer models for convection in buried and partially buried pipes are another option to forego complex finite-element numerical methods.

Fluid property estimation. The simulator offers industry-standard black oil correlations and a vast range of EOS for compositional modeling point-to-point. This includes *Peng–Robinson* (standard and corrected), *SRK* (standard and corrected), Cubic Plus Association (*CPA*), *BWRS*, and *GERG-2008*, among others.

Well performance and nodal analysis. The simulator can model vertical and inclined multiphase flow from the reservoir to the wellhead. Artificial lift options such as rod pumps, progressing cavity pumps (PCPs), electric submersible pumps (ESPs), and gas lift are included. Well performance is carried out with nodal analysis to create inflow–outflow plots at any point in the system and perform sensitivity analyses on any system variable.

Slugging. Pipesim offers *severe slugging* and *hydrodynamic slugging* modeling options. These scenarios can occur in systems consisting of a long flowline followed by a riser.

Networks. The complex topology of certain onshore production and injection networks is another upside *Pipesim* for the software. The simulator provides users with an advanced network solver of the set of nonlinear equations, of any size and topology. This capability truly considers the interdependency of wells and surface equipment to determine system deliverability. Furthermore, the built-in GIS network canvas delivers true spatial representation of wells, equipment, and flowlines, to construct networks interactively.

Facilities. One important feature of *Pipesim* is the simulation capability for integrated facilities that define the upstream or downstream boundary conditions for the production system. Equipments such as chokes, check valves, multiphase flow pumps, single-phase pumps, separators, gas compressors, heat exchangers, and chemical injection devices can be added to the production network.

A.2.2 Gap® and Prosper® (Petroleum Experts, https://www.petroleumexperts.com/products/ipm-suite/gap/)

Gap is a multiphase network modeling and optimization tool focused on applications for the oil and gas industry. *Prosper* is a well simulator for multiphase flow production. Key features of these software include the following:

Integration. The software was designed to eliminate artificial boundary conditions in reservoir, well, and surface network models, through the creation of integrated models.

Solver. The fundamental calculations performed by the software balance the principal parameters of pressure, flow, and temperature for the whole production system. This task demands a fast and stable solver built for unraveling fully integrated oilfield networks. Stability is a challenging issue of intricate systems, and optimization is a step further. *Nonlinear global optimization* is facilitated by the software because the user does not have to provide starting points, and intelligence built into the system allows for selecting the appropriate technique depending on the problem at hand.

Well performance. The performance of wells is typically handled by embedding *Prosper* software models in the integrated system, although dynamic well models can be captured through native gap calculations. Wells can therefore be evaluated and optimized over time with respect to the back pressure response of the entire network. Design and performance can be assessed through the life span of each well, considering artificial lift (pumps, gas lift, and plunger lift).

Compositional support Gap can handle different *PVT* descriptions that are used in the reservoir, wells, and surface network. The compositional model can use the lumped composition and can perform de-lumping to a larger composition. Either the *black oil model* or the *compositional model* offers options to enhance performance and computational speed for the large integrated models.

A.2.3 Olga-S® (SLB)

Olga-S is a subset of the popular transient *Olga* software, later described in this appendix. This software uses mechanistic models to simulate each flow regime in pipeline multiphase flow, determining the relevant flow properties and parameters such as pressure drop, holdup, and flow regime. *Olga-S* was developed by SLB in cooperation with the Institute for Energy Technology (*IFE*) and *SINTEF* (Multiphase Flow Laboratory, Trondheim, Norway). *Olga-S* is delivered in binary form as a library that can be dynamically linked to a driver program. Most steady-state pipeline simulators have *Olga-S* incorporated as a flow correlation, in addition to the standard correlations.

A.2.4 Pipephase® (Simsci-Esscor, https://www.sm-iot.net/aveva-pipephase)

This software has been used for planning studies for an entire field. Pipephase Pipeline Network Design covers the complete range of fluids encountered in the petroleum industry, including single-phase, black oil, and compositional mixtures. The field-proven network simulation capability in *Pipephase* enables the user to model large networks of wells, flowlines, and production facilities. The wellbore model has several inflow performance (IPR) libraries, combined with a complete set of surface equipment modules, allowing the user to configure the downstream separation facilities.

The program may also be applied to single-component stream or CO_2 injection networks. *Pipephase* integrates with model reservoirs by developing an interface with reservoir simulators, allowing the simulation of CO_2 sequestration and steam-enhanced oil recovery of extra-heavy oil (Steam-Assisted Gravity Drainage – SAGD).

The software provides thermodynamic property prediction required to simulate complex oil and gas gathering networks, natural gas transmission and distribution pipelines, steam injection networks, CO_2 pipelines, gas lift analysis, heat transfer analysis for heavy oil pipelines, and hydrate risk prediction among other applications

A.3 Transient Multiphase Simulators

Transient codes are the top multiphase fluid flow simulators requiring stable integration schemes such as the implicit methodology. The development of transient codes trails the availability of test data gathered at research institutions. Two examples of these institutions are *SINTEF* in Norway and the University of Tulsa Fluid Flow Projects (*TUFFP*). Currently, two or three software have the lead in transient multiphase simulation of oil and gas flowlines, as described below.

A.3.1 Olga® (SLB, https://www.software.slb.com/products/olga/)

Olga is the oil and gas workhorse simulator developed in the early 1980s, for the analysis of pipeline operations, through the collaboration of *IFE* and *SINTEF*. Along the years, *Olga* has been supplemented with specific modules such as hydrodynamic slugging (*Slugtracking*®), advanced heat transfer (*FEMTherm*®), tracking of individual components (*Comptracking*®), well modeling (*Olga Wells*®), wax (*Olga Wax*®), hydrate risk (*Olga Hydrate Kinetics*®), multiphase pumps, non-Newtonian behavior (*Complex Fluids*®), advanced modeling of the hydraulics (*Olga HD*®), corrosion, and prediction of single-component flows.

Olga is a transient 1-D multiple-phase model, numerically supported with an implicit numerical scheme applied on a staggered grid. Although *Olga* has been developed to deal with slow transients, it also can handle extremely fast transients.

The *Olga Online*® production management system uses live production data to reveal the multiphase flow behavior of the wells, flowlines, and network performance in real time. Accessible through a web browser, *Olga Online* has highly specific capabilities providing results during changing production and field conditions to support operational decision-making and production optimization. This software capability gives insight into critical variables of the operation, such as liquid distribution, pipeline pressures and temperatures, flow rates, and holdup profiles, and it is also a tool for operator training purposes, instrumentation back-up, and data inputs to other disciplines. The solution is delivered with the standard Real-Time Digital Twin Core; other modes are available to provide digital views of the production system under various simulation scenarios.

Two features of *Olga Online* are the *Look Ahead Mode* and the *What If Mode. Look Ahead* can visualize what will happen to your flow over time to anticipate where action would be required. Starting from current operating conditions, the user can run forward-looking dynamic simulations, to forecast future production parameters and potential upsets, and receive advisory suggestions to keep your operation at optimum levels. The *What if Mode* is a tool used offline to change the input variables in order to understand how changes to the operating conditions will impact operating efficiency and optimum performance.

Olga Online also offers the *Leak Detection System*® *(LDS)* option that detects leaks in pipelines and networks using real-time data from the field control system or production historian. The *LDS* is applicable to both single and multiphase flows. This is achieved by comparing field pressure and flow transients with results from the *Olga* simulator coupled with statistical and mathematical algorithms. In addition, advanced pattern recognition logics are used to solidify the solution's accuracy.

Olga's Real-Time Digital Twin Core® is a replica of your wells and production network. It is used for solids formation risk analysis and integrity management purposes, such as erosion and corrosion control, leak detection and, through its offline planning mode, evaluating the impact of planned operational changes to improve operational behavior forecasting.

A.3.2 LedaFlow® (Konsberg, https://ledaflow.com/)

LedaFlow can be used for one-, two-, and three-phase applications and has additional modules: *Engineering 1D*® stand-alone, *Steady-state*®, *Profile 2D*®, *Quasi-3D*®, *Integrated Simulator*®, and *Online*® monitoring tool. The LedaFlow approach includes detailed modeling of water and oil dispersions and gas bubbles in liquid phase. It also allows for modeling of three-phase flows with solid particles such as sand and hydrates.

LedaFlow development has been supported by access to the experimental data from SINTEF in Norway. The other experimental database used is from TILDA® (database of test data acquired at SINTEF's Multiphase Flow Laboratory). The TILDA experiments include up to 12" diameter experiments, up to 90 bar pressure, single-phase flow (gas, oil, or water), two-phase flow (oil/gas, water/gas), three-phase flow (oil/water/gas), and particle (sand) transport in single- and two-phase flow. Furthermore, the development of LedaFlow has benefited from a close collaboration with its industrial partners, who made experimental and field data available to validate the accuracy of LedaFlow code. The latter includes data from gas/condensate and oil-dominated systems, broad range of GOR and WC, pipe diameters ranging from 2″ to 38″, and a large variation in operational pressure.

LedaFlow has a multi-temperature approach for gas, oil, and water as opposed to the conventional mixture temperature approach. This leads to more accurate temperature and heat loss predictions. The software comprises separate momentum equations for gas, oil, and water, whereas other codes may use only two momentum equations. This leads to an improved calculation of the oil–water momentum exchange and slip among phases. The nonequilibrium modeling concept inherent in all LedaFlow modules is superior to existing models assuming equilibrium or near-equilibrium flow.

A unique formulation has been developed which enables simulation and visualization of waves, slugs, droplets, and bubbles in a multiphase pipeline flow. This formulation has been implemented in the *Quasi-3D (Q3D)* and cross-sectional (2-D) profile modules. The Q3D model resolves the dynamics of the flow and distribution of phases in multiple dimensions based on a consistent CFD formulation. The profile model (2-D) gives information on the velocity profile in the cross-section of the pipe.

LedaFlow fully integrates all its models: steady-state, 1-D transient, profile model (2-D), and quasi-3D model (*Q3D*). Thus, the steady-state model is used for the initialization of the 1-D transient calculations. The 1-D model will give a fast transient solution for long pipes. The profile model (2-D) can use the values from a 1-D simulation as input for its calculation of the phase velocities and volume fractions across the pipe diameter. The quasi-3D model will be used to get detailed information on the critical pipe segments of the production system (looking glass).

LedaFlow can be used as an operation monitoring tool to ensure optimal performance during several operational stages such as commissioning, well start-up and shutdown optimization, chemical use minimization, optimal well allocation, monitor well and flowline performance, estimate flowline holdup, detect leaks, look ahead, and calculate cooldown time among other objectives.

A.3.3 Gap Transient (Petroleum Experts, https://www.petex.com/products/ipm-suite/)

Gap Transient® is another enabling tool for field operation management, optimization, forecasting, and development. This tool can model terrain-induced or hydrodynamic slugging, liquid loading, wax formation, and hydrate formation with easy switching between steady-state and transient modeling.

Gap Transient harnesses the functionality from *Pvtp* and *Prosper* to model well and flowline integration. Transient models require accurate boundary conditions to capture the physics and the changes that are happening in the field. Gap transient models transient multiphase flow as well as transient reservoir performance. The integration of transient reservoir performance response allows more accurate prediction of early-time behavior of wells that dominate in the timescales of transient modeling. Well start-up and the impact of changes due to optimization of the well controls are drastically affected. Unconventional reservoir production, in tight-oil/shale oil and gas plays as well as more complex reservoir performance in heavy oil, or dual-porosity reservoirs all exhibit reservoir behaviors that can create complex boundary conditions for transient multiphase flow modeling.

Index

Multiphase Transport of Hydrocarbons in Pipes, First Edition.
Juan J. Manzano-Ruiz and Jose G. Carballo.

Companion Website: www.wiley.com/go/Manzano-Ruiz/Multiphasetransport

d

e

Printed and bound by CPI Group (UK) Ltd, Croydon, CR0 4YY

17/09/2026

14964252-0004